Waste Material Recycling in the Circular Economy - Challenges and Developments

Edited by Dimitris S. Achilias

Published in London, United Kingdom

IntechOpen

Supporting open minds since 2005

Waste Material Recycling in the Circular Economy - Challenges and Developments
http://dx.doi.org/10.5772/intechopen.95163
Edited by Dimitris S. Achilias

Contributors
Rashed Adnan Islam, Jacki Laiz, Syed Zameer, Ayyana Chakravartula, David Chen, Anthony Yu, Daria Frączak, Dimitris Georgantzis Garcia, Sven Kevin van Langen, Macdenis Onyekachi Egbuhuzor, Peter Ogbobe, Chima Umunnakwe, Palanichamy Kalyani, Nazim Anvarsha, Chinnamayan Sudharsana, Thakku Rangachari Banuprabha, Albert Ten Busschen, Samir Ibrahim I. Abu-Eishah, Manal Raheem, Fatma Aljasmi, Fatima Alameri, Amna Alblooshi, Inesar Faisal Alnahdi, Cevza Candan, Sedef Uncu Aki, Banu Nergis, Neslihan Sebla Önder, John A. A Glaser, Endalkachew Sahle-Demessie, Te'ri L Richardson, Rita Kol, Steven De Meester, Kim Ragaert, Martijn Roosen, Sibel Ügdüler, Dimitris S. Achilias, Kevin M. Van Geem, Peter A. Kumble, Dmitri Goljandin, Priit Kulu, Fabio Borgia, Maria Anna Charitopoulou, Eleni Alexopoulou, Panagiotis Alexiou

First published in London, United Kingdom, 2022 by IntechOpen
IntechOpen is the global imprint of INTECHOPEN LIMITED, registered in England and Wales, registration number: 11086078, 5 Princes Gate Court, London, SW7 2QJ, United Kingdom
Printed in Croatia

British Library Cataloguing-in-Publication Data
A catalogue record for this book is available from the British Library

Additional hard and PDF copies can be obtained from orders@intechopen.com

Waste Material Recycling in the Circular Economy - Challenges and Developments
Edited by Dimitris S. Achilias
p. cm.
Print ISBN 978-1-83969-680-0
Online ISBN 978-1-83969-681-7
eBook (PDF) ISBN 978-1-83969-682-4

For EU product safety concerns:
IN TECH d.o.o., Prolaz Marije Krucifikse Kozulić 3, 51000 Rijeka, Croatia, info@intechopen.com or visit our website at intechopen.com.

Meet the editor

Dr. Dimitris S. Achilias is Professor of Polymer Chemistry and Technology, Department of Chemistry, Aristotle University of Thessaloniki (AUTh), Greece. He completed his BSc and Ph.D. in the Chemical Engineering Department, AUTH, in 1985 and 1991, respectively. His research on polymer thermo-chemical recycling, synthesis of polymer nanocomposites, and polymerization/degradation kinetics has resulted in the publication of more than 170 research papers. Dr. Achilias has edited two books on recycling and is a member of the editorial board of the journal *Polymers*. He has participated in several research projects and holds three patents.

Contents

Preface

The problem of waste recycling remains more urgent than ever. The societal value of plastics is demonstrated in its many applications, from manufactured durability to single-use applications. As the production and consumption of materials are constantly increasing, so too are the quantities that end up in waste streams. Disposing of waste to landfills is undesirable due to legislation pressures and the poor biodegradability of commonly used materials. Therefore, recycling is the best solution. Nowadays, more specialized recycling methods are required to manage a wide variety of wastes.

The challenge in today's recycling projects, in line with the principles of the circular economy, is not only to reuse the materials but also to produce secondary value-added products, reducing the consumption of natural resources and the amount of energy needed, while lowering CO2 emissions in the environment. In this context, this book suggests solutions to specific contemporary issues such as the removal of contaminants from plastic packaging, delamination of multilayer structures, advanced techniques for the mechanical and chemical recycling of polymer-based materials, the fate of single-use plastics, separation of microplastics, and more. It targets professionals and recycling companies, as well as researchers, academics, and graduate students in the fields of waste management and polymer recycling in addition to chemical engineering, mechanical engineering, chemistry, and physics.

This book includes fourteen chapters written by forty-four authors from Europe (Belgium, Czech Republic, Estonia, Greece, Poland, The Netherlands, Turkey, United Kingdom), Asia (India, United Arab Emirates), America, and Africa (Nigeria). The chapters are organized loosely into three sections. The first section consists of nine chapters dealing with polymer recycling. The second section includes three chapters on the recycling and valorization of waste materials. The third section includes two chapters that examine the role of society in recycling and the circular economy.

Section 1: "Recycling of Polymeric Materials"

Chapter 1: "Are Reliable and Emerging Technologies Available for Plastic Recycling in a Circular Economy?"

This chapter presents the main environmental and sustainability issues concerning plastic disposal and recycling. It discusses thermal, chemical, and biological depolymerization processes, including pyrolytic technologies. It concludes that evaluation of new processes requires performance assessment for understanding how plastic recycling technologies contribute to the environment and the sustainable reuse of plastic materials.

Chapter 2: "Recent Advances in Pre-Treatment of Plastic Packaging Waste"

This chapter examines the pre-treatment processes used to remove contaminants from plastic packaging waste and the advances that are being developed and/or optimized to achieve closed-loop recycling. Some of the techniques presented include chemical washing to remove inks, extraction to remove undesired plastic additives, delamination, and dissolution-precipitation.

Chapter 3: "Current Topics in Plastic Recycling"

This chapter presents an overview of current trends in plastics recycling focusing on pyrolysis. Emphasis is given to three case studies where special polymer recycling techniques are required: polymeric blends, multilayer plastic packaging, and brominated flame retarded plastics originating in waste electric and electronic equipment (WEEE).

Chapter 4: "Chemical Recycling of Polyolefins (PE, PP): Modern Technologies and Products"

This chapter presents several process paths for the chemical recycling of polyolefins (polyethylene and polypropylene), such as pyrolysis for the recovery of intermediates in petrochemical plants, fuels, or gasification from which a mixture of H2, CO, and CO2 is produced, which can be further transformed into chemicals and fuels or used directly to produce energy. The chapter also provides examples of existing technologies and their level of technology readiness and perspectives for scale-up.

Chapter 5: "Polyethylenes: A Vital Recyclable Polymer"

This chapter discusses the production, properties, processing, and recycling of polyethylene. It discusses the processing of recycled polyethylene as well as the limitations of polyethylene recycling. Finally, the chapter examines hazardous situations that may arise from recycling polyethylene.

Chapter 6: "An Evaluation of Recycled Polymeric Materials Usage in Denim with Lifecycle Assesment Methodology"

This chapter analyzes the advantages and challenges of different chemical and biological recycling technologies in denim production. Life Cycle Assessment (LCA) analysis is used to evaluate the environmental impact of recycled polymeric materials usage in denim fabrics. It concludes with the opportunities of using waste in denim production as a raw material to design circular systems.

Chapter 7: "Retreatment of Polymer Wastes by Disintegrator Milling"

This chapter presents mechanical recycling by *disintegrator milling* of several composite polymers. The groups of polymer materials studied include pure brittle and soft polymers (PMMA, HDPE and IER), blends of plastics (ABS+PMMA, PC+ABS), reinforced plastics (PMMA+GFP), elastomers (rubber and tires), and printed circuit boards (PCB).

Chapter 8: "Effect of Environmental Aging on Tensile Properties of Post-Consumer Recycled (PCR) Polycarbonates"

This chapter compares tensile properties of different grades of post-consumer recycled polycarbonate (PC) plastics with conventional or virgin PC before and after different aging conditions. It seems that with the presence of temperature and humidity aging, tensile strength starts to decrease over time, but recycled PC shows similar aging behavior compared to virgin PC.

Chapter 9: "Performance Analysis and Modeling of Microplastic Separation through Hydro Cyclones"

This chapter highlights the capabilities of a hydro cyclone at separating microplastics from water using mathematical and computational fluid dynamics (CFD) modeling. The results show that hydro cyclone microplastic separation can achieve 98% efficiency.

Section 2: "Valorization of Waste Materials"

Chapter 10: "Industrial Re-Use of Composites"

This chapter examines a method involving machining obsolete thermoset composite products into strips or flakes for re-use as reinforcing elements, which, when combined with fresh resin and fiber, enable the production of new components. This method has been proven in manufacturing retaining walls as well as guide beams for canals and bridge decking, all using strips or flakes from end-of-life composite products.

Chapter 11: "Phytomass-Derived Multifunctional Activated Carbon as a "Wonder-Material": A Paradigm Shift of Filth-to-Wealth"

This chapter explores the idea of recycling the renewables in nature phytomass for producing activated carbon that could find several applications, such as in catalytic supports, the removal of pollutants, gas storage, and so on. It also presents research insights into the identification of unexplored phytomass or wastes that could lead to carbon with novel properties.

Chapter 12: "A Zero-Waste Process for the Treatment of Spent Potliner (SPL) Waste"

This chapter presents a deep analysis of an environmentally friendly process to recover all valuable minerals contained in spent potliner such as graphite carbon and aluminum fluoride (AlF3) and production of sodium sulfate (Na2SO4) and gypsum (CaSO4) when H2SO4 is used as the leaching agent.

Section 3: "Role of the Society in Recycling and Circular Economy"

Chapter 13: "Urban Mining of e-Waste and the Role of Consumers"

This chapter focuses on the circular economy, urban mining, and their intersection with consumer behavior by providing a review of existing and emergent EU regulations for enhancing the collection rate of household WEEE. Then, the chapter critically analyzes the literature on the intersection between consumer behavior and closed-loop supply chains for EEE, identified through a systematic keyword search to ensure replicability.

Chapter 14: "Compost, Social Sustainability, and Circular Economy in Guatemala"

This chapter examines several issues in Guatemala City, specifically the dynamics for initiating a new start-up composting business operating under the principles of circular economy, employment opportunities for at-risk youth, the potential market for compost both in an urban environment and for farmers, and the requirements involved in making compost in the challenging high-altitude climatic conditions of a developing world economy.

I want to express my sincere thanks to all the contributors who provided their expertise and enthusiasm to this project and the publisher IntechOpen for making this work possible. I would also like to thank my family for their patience and love throughout the editing of this book. Finally, I would like to dedicate this book to my father Savvas Achilias for his continuous support throughout the years.

Dimitris S. Achilias
Professor,
Department of Chemistry,
Aristotle University of Thessaloniki,
Thessaloniki, Greece

Section 1

Recycling of Polymeric Materials

Chapter 1

Are Reliable and Emerging Technologies Available for Plastic Recycling in a Circular Economy?

John A. Glaser, Endalkachew Sahle-Demessie and Te'ri L. Richardson

Abstract

A spectrum of plastics has been produced in the last 70 years, and plastic production has increased faster than any other manufactured material. Current recycling of all plastic materials is pegged at 10% or less. The social value that plastics enjoys is reflected in its myriad uses for engineered durability to single-use applications. Disposable or single-use plastic items have become a significant problem. Plastic debris has become ubiquitous to the landscape and aquatic resources, leading to human health, ecological concerns, and sustainability issues. Past disposal practices relied on waste plastic flows to certain countries for disposal, but these have been summarily curtailed, needing alternatives as productive and environmentally conscious recycling technology. Waste plastics can be repurposed using purification, decomposition, or conversion processes that are based on established and emerging mechanical and chemical technologies. Plastic recycling technologies, such as thermal, chemical, and biological depolymerization processes, including pyrolytic technologies using plastics-to-fuel strategies, are under development ranging from bench-scale demonstrations to full-scale implementation. The ideal of closed supply chain constraints offers optimal solutions to plastic recycling. Evaluation of new processes requires performance assessment to understand better how plastics recycling technologies contribute to the environment and the sustainable reuse of plastic materials.

Keywords: plastic recycling, circular economy, chemical and mechanical recycling, emerging technologies

1. Introduction

Plastics are chemical success stories of the past 100 years that have touched every aspect of modern life. However, their success has a huge downside. Today more than 40 kg of plastic waste per person is produced each year globally. In 2018, the total amount of municipal solid waste (MSW) was 292.4 million tons, which is 2.2 kg/per capita per day (**Figure 1a**). This is a 40% increase from the 1990 waste generation. Although 23% of this waste is recycled, there is considerable variability in the type of waste [1]. Solid waste management conducted by local municipalities is a global challenge requiring aggressive attention. Poorly managed waste flows from terrestrial and air-born sources contaminate oceans and other water bodies.

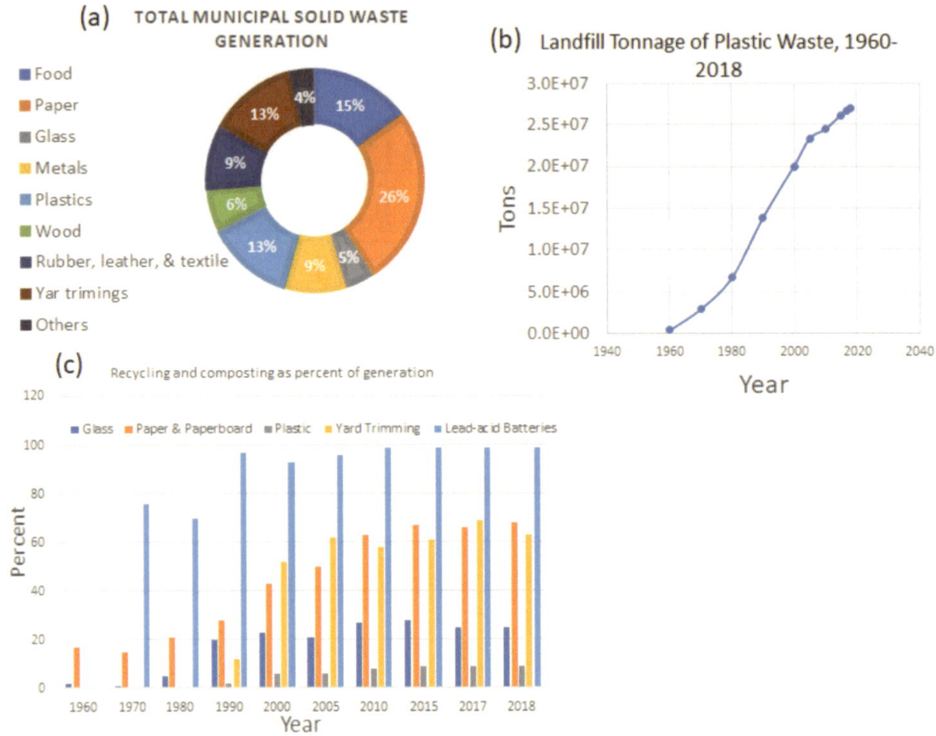

Figure 1.
(a) Total municipal solid waste generation by material in 2018 (total 262 million tons) (b) generation trend of plastic waste in the United States between 1960 and 2018, showing (c) recycling and composting rates as a percentage of generation (Source: US EPA, 2018).

Flooding from clogged drains, air pollution emissions from burning, and contaminated urban areas provide significant input to this global dispersal of plastic waste. The complexity of solid waste management is a combined effect of social, environmental, and economic factors contributing to a multi-dimensional challenge demanding solutions. Inadequate solid waste management practices contribute to environmental problems in treating ecosystems, impacts of human health, and undesirable socioeconomics.

2. Current global plastics problem

Plastic-related waste has increased by two orders of magnitude in the past half-century [2]. Paper and paperboard recycling is currently 66.5%, whereas plastic recycling was estimated at 4.4% in 2018. Some 5 million tons of plastic particles derived from the 300 million tons of annual global virgin plastic production are transported by rivers and deposited in the oceans. Mechanical degradation of macro size plastics leads to the formation of a large volume of microplastics that when combined with manufactured microbeads from personal care products lead to harmful conditions in aquatic ecosystems. The raw material for these virgin plastic resins is derived from petroleum. Its short use-life and high durability in the environment translates into an environmental lifetime exceeding hundreds of years underscoring plastic disposal/reuse as a significant environmental concern [3–6]. Although almost all commodity plastics carry the recycling symbol (**Table 1**), plastic recycling in the United States is crude, energy-intensive with recycled materials characterized by lower qualities than virgin plastic.

Resin Type	Structure	Use	Percent manufactured (million ton)	Percent recycle
1 PETE	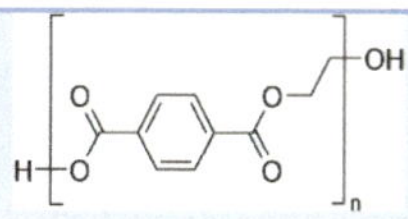	 Soft drink bottle, miner water, Fruit juice container, cooking oil, Single use	4.5	19.5 Recyclable
2 HDPE	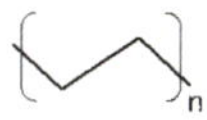	Milk jugs, cleaning agents, shampoo bottles, washing bottle	5.5	10.3 Recyclable
3 PVC	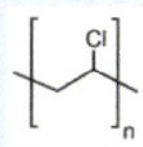	 Pipe, Food trays and packaging carpet backing, flooring	0.9	0.0 Converted to garden hose, floor mat
4 LDPE			7.5	5.3
5 PP	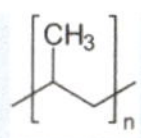	Packing material, chairs, textiles, rope, kitchen and laboratory equipment	2.2	0.5
6 PS	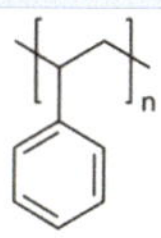	 Single use application Hot cups, shipping packaging cushion, cartons, trays		0.9
7 OTHER	Other plastics, acrylic, polycarbonate, polylactic acid, nylon, fiber glass	5-gal wager bottle, CD, baby feeding bottle, bulk food containers		

Source for recycling rates ref. [1].

Table 1.
Types of plastic resin and resin identification codes (RIC).

A large portion of plastic products is used to make short-lived products such as packaging discarded after a single use [7]. European reports indicate that 70% of marine plastic litter is caused by single-use plastics [8]. Recycling is one of the sustainable ways to reduce the impacts of plastics and product reuse and energy recovery as fuel.

3. Plastic economy

Traditional plastic production is based on fossil feedstocks, where over 90% of currently manufactured plastics is derived from virgin fossil feedstocks with a huge carbon impact. Plastic production has been growing at an average of 40% in the past 50 years, surging from 15 M tons in 1964 to over 320 M tons in 2018, as plastic use increased in every application. Plastic packaging is one of the main applications of plastic use, representing 26% of the total volume of plastic use. Plastic packaging has many economic benefits ranging from protecting products during shipping, preserving and reducing food waste, and reducing fuel consumption for transport due to lightweight. However, 95% of plastic packaging and 90% of total plastics are not recycled, because of which an estimated over 400 billion is lost to the economy annually [9].

Although universal recycling symbols were introduced 45 years ago, only a small fraction of plastics are recycled. Successful recycling requires reliable and efficient post-consumer collections for similar products, consistent consumer demand, and economical remanufacturing. These factors are interdependent, since the economy of scale requires effective collocation of used materials that are not too contaminated, located close to recycling sites, and require a community that is committed to sorting properly based on Resin Identification Codes (RIC; **Table 1**). However, there are too many chemical variations of plastic products distributed among too few categories. Packaging represents 26% of the total volume of plastics used and out of the single-use plastic, mainly packaging materials, which have short first-use cycle one third escape collection system [9]. Of the different resins listed, polyethylene terephthalate (PET) and high-density polyethylene (HDPE), codes 1 and 2, are the more easily recycled plastics. The cost of plastic waste on oceans, clogging urban infrastructure, in addition to greenhouse gas emission during manufacturing, was estimated to exceed USD 40 billion [10].

Sustainable solid waste management requires improved collection infrastructure that includes the separation at the source, policies that encourage reuse/recycling, and supporting recycling [11]. The cost of recycling depends on the market, volume, required sorting, and decontamination of the feedstock. The flow of recycled plastic materials makes it uneconomical to justify separating low-volume streams. Many promising technologies emerging in the market offer opportunities to recover the value of some low-volume recycle streams. Optical and robotic sorting techniques along with digital watermarks are expected to increase the efficiency of sorting. Further investigative research is required to determine how to retrofit existing facilities to include new sorting technologies optimally.

4. Current plastic linear value chain economy

Plastics have been used in many applications, and their uses follow a linear economy, take-make-waste model. The fraction of global recycled plastics is low; 30% in Europe, 25% in China, and 9% in the United States. The current plastic economy is best understood as a "linear" model of value creation where the product lifecycle begins with extraction and ends with landfill disposal or release to the environment. About 4% of the world's petroleum product is used as plastic feedstock and a similar amount is used for processing products and transporting them. The raw materials are extracted and transformed into products. Generally, at the end of their use phase, the items are environmentally mismanaged in the disposal. Value is created in this economic system by producing and selling as many products derived from virgin polymeric materials as possible. The linear economic model has been in practice since the industrial revolution, has achieved economic growth, and has improved the standard of living. The model does not consider the natural depletion of resources or the disposition of the products after use, which often is landfilling, incineration, or other means of release to the environment, nor the value of the disposed of material for recycling or reuse (**Figure 2**). The linear plastic economy has proven unsustainable for its resource consumption and environmental and human health impacts.

Figure 2.
Schematic representation of linear economy of plastic.

The negative influences of the linear economic model of plastics, including the depletion of nonrenewable resources, climate change, and severe ecological impact, leave a significant footprint [1, 9]. The high reliance on an extracted virgin resource subject to high price fluctuations, the need for an increase in trade, and the geopolitical interconnectedness for raw material, and end-of-life disposal are major challenges in a linear economy. As the population and welfare of emerging economies have grown over the past few decades, the number of middle-class consumers has also been increasing. This, in turn, resulted in an increased demand for newer products and a shortened product lifespan that accelerated the disposal of more plastic waste. The move on to a sustainable economy can be achieved through the concept of the circular economic (CE) model that incorporates strategies for continuously reuse of material and resources [11]. Whereas recycling techniques like energy recovery through incineration generate air pollution, CE of plastics economy involves waste recovery using a range of recycling techniques to minimize plastic waste disposal by converting it to valuable products.

5. Plastics and the circular economy

The global recognition of a fundamental shift in the design, use, and reuse of plastics is implicit to his new economic order. A CE advances a closed-loop system where materials are kept in the product lifecycle and not disposed of to reduce raw material usage and energy demand. The concept of a circular economy was proposed over 30 years ago as the impacts of the traditional open-ended economy with no recycling and the environment was being treated as a waste sink became obvious. Previously, the policies of many governments have been focused on encouraging pollution reduction, promoting closed-loop waste management by increasing recycling and dematerialization. For example, the Pollution Prevention Act of 1990 in the United States encouraged reducing pollution at the source and the three R-strategies: reduce, reuse, and recycle. Similarly, the 1996 German law Closed Substance Cycle and Waste Management and the 2002 Japanese law for Establishing A Recycling-based society [12–14].

A shift to a CE could create a disruptive change that is restorative and regenerative by intention, including green by design that generates no waste, keeps products and materials in use, and regenerates natural systems (**Figure 3**). The plastic industry CE deals with environmental impacts, resource scarcity, disruption in traditional end-of-life management options, and simultaneous economic benefits. To create a successful CE of plastics, various approaches should be followed,

Figure 3.
Schematic of the circular economy of plastic.

including eliminating problematic plastics, innovating to ensure that unused plastics are reusable and recyclable, and circulating the plastic in the economy. The complete lifecycle of plastics should be analyzed with reference to their ability to circulate back for reuse to the economic system to achieve the maximum value.

One of the cornerstones of CE is the creation of an effective after-use plastics economy requiring more material value and higher resource productivity. This requires designing products for recyclability, including designing nonrecyclable packages to become recyclable, which can be achieved by using a mono-material design modifying protective coatings, introducing components that will increase compatibility with processing, developing film-orientation technology, and simplifying the product design for ease of recycling. Many of the most common plastics are recycled at meager rates in the United States. When recycled, they are often recycled into lower-value products due to the degradation of polymer properties resulting from mechanical recycling. To support increased recycling, new plastic recycling protocols should ensure that the monomer stream generated after depolymerization is free from contaminants and colors, a common problem with the chemical recycling of plastics.

6. Reliable and emerging technologies for plastic recycling

Existing recycling offers a spectrum of applications to single polymer compositions and to mixed plastic compositions (**Figure 4**).

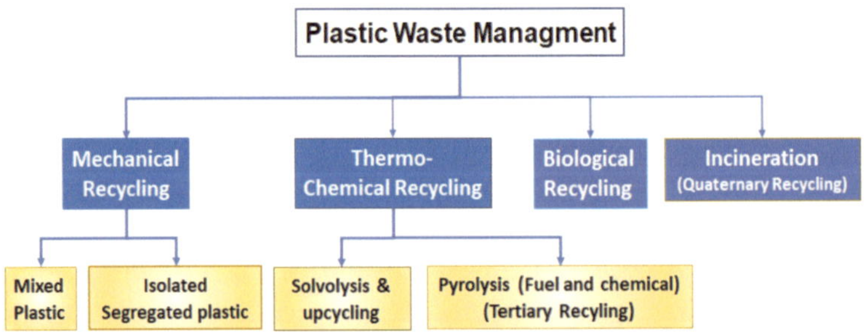

Figure 4.
Major Plastic waste recycling technologies (primary or closed-loop, and secondary recycling – mechanical recycling, tertiary recyling – chemical recycling, AND recycling energy recovery and valorization, quaternary – incineration).

Significant different processing strategies are required to accommodate the complexity of waste feedstock components. Changes to existing and emerging recycling technology are anticipated to convert feedstock to desired products optimally. Conventional and emerging mechanical and chemical technologies ranging from bench-scale demonstrations to full-scale implementation are becoming investment targets across the globe. Notable examples involve thermal, chemical, and biological depolymerization processes (**Table 2**).

Reactor type	Advantage	Disadvantage	References
Traditional methods			
Batch reactor	High conversion efficiency Low quality product	• Lower quality products (e.g., formation of PAHs) • Higher energy consumption	[1, 2]
Fixed bed reactor	Simplicity in design	• Difficulty with high viscous low thermal conductivity • Small catalyst surface area • Used in tandem with a noncatalytic stage	[3, 4]
Fluidized bed reactor	• Yield of liquid products is more than 90 wt % • Low level formation of gas coke	• Complex operation • Difficulty in separating catalyst particles from the from exhaust gas • Reactor erosion	[5, 6]
Conical spouted bed reactor	• Allows process flexibility • Can treat large particles with density disparity		[7, 8]
Free-fall under vacuum	Produces important liquid chemicals such BETX, and naphthalene		[9, 10]
Alternative methods			
Catalytic pyrolysis	• Lowers pyrolysis temperature • Reduces the heat energy requirement • Favors scale-up to larger scale • Produces commercial type final products		[11, 12]
Solvent-assisted pyrolysis	• High heat and mass transfer rate • Reduce operating temperature • Higher liquid yields		[13, 14]
Microwave-assisted pyrolysis	• Extremely rapid heating • Higher production rate • Low production cost • Energy saving	Economics of process scale-up Large-scale heterogenous Hotspots on heat adsorber due to low MW adsorption and conductivity of plastic	
Supercritical fluid	• Efficient heating and fast dissolution of plastic homogeneous enhanced depolymerization	Challenge in scale-up	
Plasma reactor	• High monomer recovery due to efficient heating	Low technology readiness	[15]

Table 2.
Plastic recycling process.

6.1 Plastic recycling strata

The recycling process for plastics has been subdivided into four components: primary recycling, secondary recycling, tertiary recycling, and quaternary recycling [16–18]. Recovery of production scraps as a pure polymer stream or the extrusion of preconsumer polymer has been identified as the operational component of primary recycling. Secondary recycling involves sorting of waste polymer feedstock, waste polymer size reduction, cleaning the waste polymer flow, drying and extrusion to form desired recycled polymer products [19–24]. Multiple cycles of recycling at primary and secondary stages can be used to recover many polymers while maintaining expected performance characteristics.

Conversion of the polymer feedstock to its constituent monomers can be selected when the first two recycling stages do not offer the desired recycling flexibility, economic opportunity, or product quality [25]. This tertiary stage recycling can complement the earlier, more traditional recycling technology strata (**Figure 4**). A selective approach to converting the feedstock polymer to its monomeric components can retain significant feedstock value with the ability to convert the recycle stream to a spectrum of polymer applications. This depolymerized feedstock can be utilized in a spectrum of pathways leading to chemical production syntheses other than reforming recycled polyethylene terephthalate (rPET) as an upcycling endeavor. Chemical recycling utilizes reactions with the chemical structure of the polymer to convert the polymeric mass to monomers and related chemicals [15, 26–30]. The chemical recycling process permits the removal of material consigned to downcycling recycling technology such as colorants, process additives, and other contaminants. This removal provides a material that can be used as a feedstock for opportunities of upcycling plastic waste to access value-added chemicals, thereby enhancing economic performance (**Figure 5**).

The nonselective technologies found useful at the last recycling stage are generally thermolytic such as pyrolysis and hydrocracking technologies that convert the polymer feedstock into monomer and pyrolytic oils [31, 32]. At a quaternary stage, incineration is employed to recover the energy content of the waste polymer feedstock [33–35]. The relegation of polymer feedstock to quaternary recycling can be attributed to feedstock complexity, processing economics, waste unsuitable to other forms of recycling, market forces, and waste polymer treatment demands and capabilities [24, 36, 37]. This stage of recycling recovers little value of the feedstock and may contribute to greenhouse gas emissions having other unintended environmental consequences. Selective chemical recycling of polymers has become more

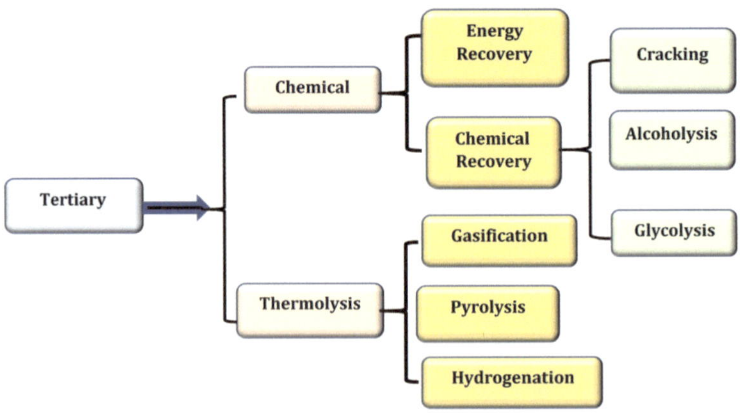

Figure 5.
Tertiary plastic recycling strata.

attractive to engage significant process funding recently, since chemical depolymerization is designed to form the original monomers of the recycled polymer in successive recycling to approach continuous recyclability [38–40].

6.2 Chemical recycling example—Polyethylene terephthalate (PET)

PET is the most common thermoplastic polymer of the polyester family of organic polymers [41, 42]. It is a naturally transparent and semi-crystalline plastic, which is used for packaging, for manufacturing stretch-blown bottles, sheet, and thermoforms, and for producing fibers for textile products [43]. The important characteristics of PET include resistance to water, high strength to weight ratio, and wide availability as an economic and recyclable plastic [44].

As condensation polymers, polyesters are constructed from the reaction of multifunctional carboxylic acids and polyols in which water or alcohol is eliminated to form a growing polymeric chain. The condensation polymer construction of PET provides a unique chemistry leading to its formation through the elimination of water or alcohol depending on the composition of monomers. PET as a condensation polymer has unique chemistry controlling its assembly or disassembly. It is important to understand the synthesis of PET to enable the depolymerization option to recycle PET optimally.

Exhibiting significant mechanical properties, PET's rigid polymer chains and high melting point contribute resistance to fatigue and plastic deformation as desirable features. Resistance to solvents, chemicals, and hydrolysis at normal application temperatures identify PET as a highly desirable component to a host of applications. PET's manufacture is affected with the unattractive disadvantage of a relatively slow rate of crystallization leading to increased processing time of a slow cooling cycle necessitating nucleating agents. As a crystallizable polymer, PET exhibits properties of different desirable states of physical order ranging from amorphous (transparent) to crystallized that are transparent and opaque [45].

6.2.1 PET synthetic technology

PET synthesis is configured in two steps [46]. An esterification of terephthalic acid (TPA) or dimethyl terephthalate (DMT) reaction with ethylene glycol (EG) can be used to form a prepolymer mixture product composed of bis(2-hydroxyethyl) terephthalate (BHET) and short-chain oligomers (**Figure 6**).

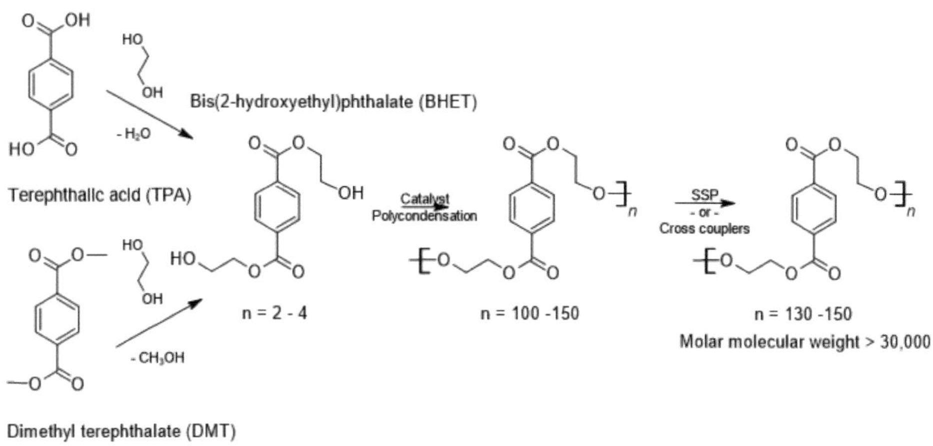

Figure 6.
Polyethylene terephthalate (PET) synthesis.

The condensation reaction requires the removal of water or methyl alcohol to force the reaction to completion. The growing polymer formed by equilibrium esterification is susceptible to depolymerization by hydrolysis or methanolysis, so the removal of any solvolytic agent is important to build molar mass.

A second step is transesterification in melt phase leading to the polycondensation of short chains to form the molar weight range and intrinsic viscosity required for an intended application [47]. The prepolymer formation step can be conducted with DMT or TPA as one monomer with EG or another polyol as a monomeric partner. Even at the prepolymer step, the oligomeric mixture is difficult to purify and this emphasizes the need for high purity starting materials. DMT can be purified by distillation or crystallization from a melt. Heating TPA beyond its boiling point leads to decomposition, and purified TPA can be produced by recrystallization at process production scale. Currently, TPA is the monomer of choice for more than 70% of the PET production across the globe [48]. Where DMT is employed, purification is done by distillation to remove higher boiling constituents and light molecular weight esters. High-purity DMT is achieved by recrystallization. The quality and color of high-purity DMT are required to produce high molecular weight PET.

Industrially, PET is produced by the reaction of EG with the DMT or TPA in the presence of a catalyst (**Figure** 7) [48]. High conversions to condensed polymer in the first stage are established from a stoichiometric balance of reacting groups. Excess EG is employed in industrial processes, which is removed and recycled as part of the process. Early processes synthesized PET at industrial scale through the two-step polymerization reaction between DMT and a 30–50% excess of EG in the presence of a catalyst involving transesterification and polycondensation [49]. In the first step, transesterification of DMT with EG leads to the initial formation of BHET and a small number of assorted oligomers. This reaction is conducted at atmospheric pressure at a temperature range from 150 to 210°C under an inert atmosphere to minimize oxidative side reactions. Methanol is removed to force the reaction to completion with the formation of BHET and hydroxyethyl–terminated oligomers [49].

Polycondensation is the chemistry of the second step, which involves increasing the temperature to 270–280°C, exceeding the melting temperature of PET, and initiating the application of high vacuum (10–50 Pa). During this stage, EG is formed and removed as a by-product. The product PET has a degree of

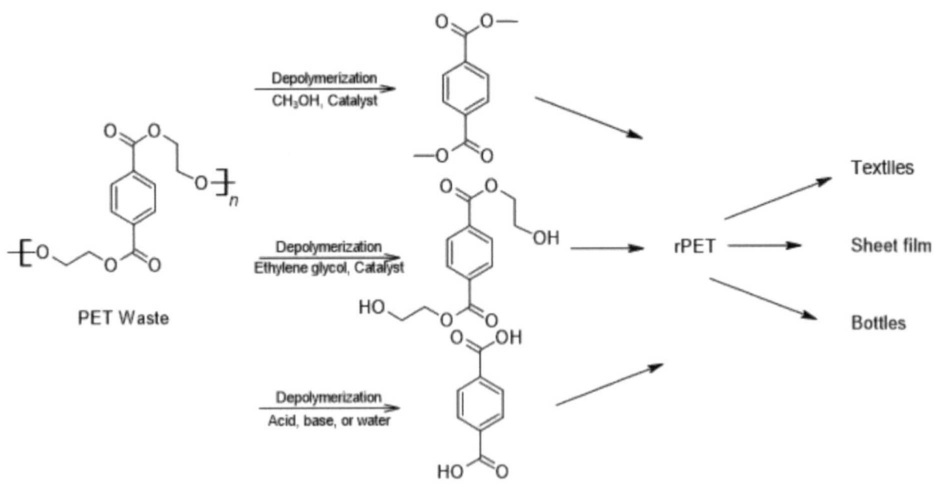

Figure 7.
rPET synthesis from PET waste using depolymerization.

polymerization of approximately 100. The equilibrium reaction of polycondensation requires the removal of excess EG through high vacuum and intensive mixing of the molten PET in the reactor. At the polycondensation stage, the intrinsic viscosity of the reaction mixture increases as the molar weight of the polymer increases. Removal of volatiles from the molten phase becomes a rate limiting process. Thermal degradation of the polymer occurs as heating more than 300°C is applied due to PET's limited thermal stability.

Judicious selection of polymerization catalyst and temperature control can be used to reduce the reaction time of the two steps. Significant investigation of the tertiary strata options for PET recycling has been conducted (**Figure 8**) leading to commercial opportunities for the recycle industry. Ammonia can be used to recycle PET to different materials using similar chemistry [50].

6.2.2 PET recycle

Globally, some 245 million tons of plastic are estimated to be in use annually [51]. The largest component is fibers with a quarter of this polymeric composition attributed to packaging [52]. The second largest component of the plastic mass is composed of PET bottles whose contribution expands annually. The oxygen content of PET identifies it to be a poor candidate for thermal recycling.

Secondary recycling based on mechanical technology enables the recycling of granulated plastic feedstock that has been processed from common waste plastic materials [15, 16, 53, 54]. The heated process of extrusion in a rotating screw configuration offers conditions that induce thermal softening of the plastic mass and plasticization. This process provides blending of the plastic melt, which permits the recycling of single, mixed, or blended plastic compositions. An example of the desired properties of vPET and rPET is displayed in (**Table 3**) [55]. The characteristics of rPET can vary widely with different sourced feedstocks. The components of a mechanical recycling process are shown in (**Figure 8**) targeting waste PET to provide the cleanest feedstock to the extrusion process. Thermal conditions of the extruder can lead to oxidative degradation of the plastic mass, and shear induced in the melt can result in undesirable polymer chain breaking, branching, and cross-linking. Polymer chain length degradation leads to reduced mechanical properties and lessens the features required for polymer processing into forms such as bottles, which require a higher molar weight polymeric mass.

The process conditions can be optimized to avoid some of these problematic issues. The chemical composition of the plastic mass in an extrusion process is important to the mechanical effects placed on the melt. The fluid mechanics of the extrusion is adjusted to allow the melt to begin to flow. The intrinsic viscosity of the melt rises with increased polymeric molecular weight. Heated extrusion processing is the primary technology employed in the mechanical recycling of plastics to form a granulated feedstock without the use of solvents.

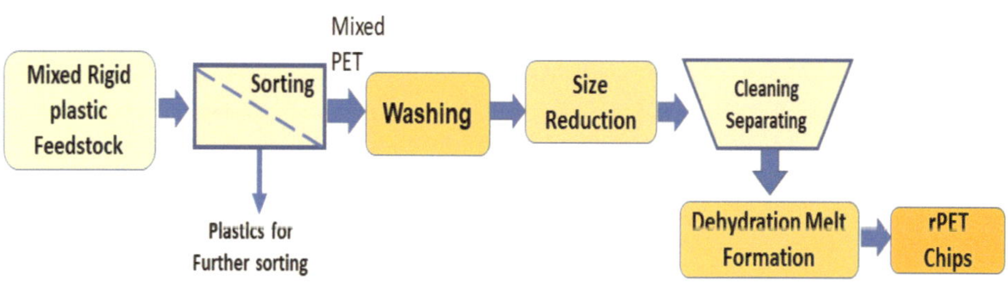

Figure 8.
Sequence of PET recycling process components.

Descriptor	vPET	rPET
Crystallinity		
Crystalline region (%)	37.2	28.5
Amorphous region (%)	62.8	71.5
Thermal properties		
Glass transition temperature (°C)	75.35	87.27
Melting point	242.72	244.15
Degradation temperature (°C)	420	419.6
Residue		
Strength properties		
Tensile strength (kg/cm^2)	75.4	87
Breaking strength (kg/cm^2)	82	42
Elongation strength (kg/cm^2)	7	5.0
Young's modulus (kg/cm^2)	5690	10,500
General		
Molar mass (g/mol)	19,342	15,812
Shrinkage % (150 denier)	9.31	6.18

Table 3.
Virgin vs. recycled polyester characteristic properties (Reconstructed after Vardicherla) [27].

Atmospheric oxygen can react with shear-induced radicals to form peroxy radicals, and this leads to radical-induced decomposition. Plastics with high oxygen permeability have been observed to exhibit enhanced thermo-oxidation rates within the melt. The chemical and physical forces accompanying extrusion leads to undesired changes in tensile strength, elongation, and many other properties for rPET. Inventive use of semi-closed or open-loop recycling involving the addition of vPET during the recycling process is employed to alleviate material properties decline experienced during extrusion.

6.2.3 Contamination effects

Contamination in the recycling feedstock occurs as a leftover from previous processing or desired modifications to the original feedstock but without proper treatment before recycling generally exerts significant negative influence on the quality and variability of the recycled plastic [30, 56, 57]. A list of contaminants (**Table 4**) found in PET feedstock have been identified for their significant contribution to quality decrease and performance variability increase of the regenerated polymer [57]. PET is hygroscopic and must be dried to remove water to reduce any chain length reduction through hydrolysis at the melt stage. Super cleaning techniques are used in some situations to ensure the optional flake quality for the later polymerization steps [58]. The label adhesives can release acetic acid, which will catalyze polymer chain reduction through hydrolysis, specifically during heat-related segments of the extrusion process. In the United States, the majority of virgin and rPET infrastructure are in the southeast and midwest (**Table 5**).

EG degradation and recombination products have been identified as the source of discoloration and clarity losses in the polymer product. Carry-over degradation products from thermal and oxidative conditions lead to yellowing and reduction of

Metal ion	Concentration (ppm)	Source
Antimony	240–260	Polycondensation
Cobalt	50–100	Polycondensation
Manganese	20–60	Transesterification
Titanium	0–80	Polycondensation
Iron	0–6	Washing
Sodium, magnesium, silicon	—	Food contamination

Table 4.
Metal contamination found in rPET.

Location	Throughput	Process
Tennessee	50 million lbs./yr. (2020)	Methanolysis of PET and glycolysis of PET
South Carolina	46 million pounds/yr. 1st yr. 88 million pounds/yr. (2020)	Low heat, pressure-less process
Illinois	Pilot plant	Glycolysis of highly colored recyclate
France	Lab scale/pilot plant	Depolymerization
France	Reactor development/pilot stage	Enzymatic Hydrolysis
Italy	Lab stage	Glycolysis
Switzerland	Lab stage	Microwave hydrolysis
Netherlands	Proof of concept	Glycolysis
Virginia	Startup stage	Subcritical water hydrolysis

Table 5.
Applications of chemical recycling with PET feedstock.

PET mechanical properties. Polyvinyl chloride (PVC) is used in bottle cap liners commonly connected with PET bottles and presents a major problem in the PET recycling process. Under thermal treatment conditions, PVC degrades to form hydrochloric acid, which in PET melt conditions leads to the reduction of the polymer chain length by hydrolysis reducing the value of the rPET [59, 60]. This problem is intensified, since PET and PVC have the same density and are difficult to separate.

Catalyst residues and processing additives such as trace metals (antimony, cobalt, and manganese) from the previous processing of consumer PET wastes promote transesterification and polycondensation reactions in the recycled PET as part of the heated extrusion process (**Table 6**). The recycled PET formed by the action of these trace metal has a chemically heterogeneous composition and may be affected by changes to the viscosity of the melt leading to batch–batch variability [60, 61]. Melt degradation can also be caused by the presence of pigments that are used to color plastics. Colored feedstock leads to the formation of an unattractive and lower commercial value gray color rPET product.

The mechanical recycling of complex and contaminated PET feedstock is difficult [62–64]. Mechanical recycled rPET is commonly characterized by intrinsic viscosities that are relatively low and heterogeneous. These properties have excluded rPET from direct incorporation in the production of bottle-grade PET, high-quality industrial fibers and films. Carpets, textile fibers for wearing apparel, and plastic containers designed for non-food applications have employed lower

Contaminants	Use/cause	Effects
Paper	Labels & adhesives	Incompatible with melt formation and releases minor amounts of acid that will catalyze polymer hydrolysis
Moisture	Residual water not removed by drying	Reduces the polymer molecular weight through hydrolysis
Degradation products	Products of thermal oxidation during extrusion	Reduces the PET mechanical properties and leads to yellowing
Dissimilar plastics	Poor feedstock sorting	Incompatibilities in the melt stage
PVC	Bottle cap incomplete removal	Releases HCl when treated thermally which catalyzes polymer chain hydrolysis
Colored feedstock	Inks and plastic dyes	Colored bodies exhibit undesirable gray color in product and may catalyze decomposition
Trace metals	Polymerization catalysts and processing additives	May catalyze transesterification or polycondensation reactions leading to batch–batch quality issues
Acetaldehyde	Thermal release of ethylene glycol and elimination of water	Easily oxidized to acetic acid which can catalyze polymer linkage hydrolysis

Table 6.
Contaminants to the PET recycling process and their effects.

quality rPET. Contaminant removal has been accomplished through a thorough cleaning step, which removes the spectrum of contaminants encountered. There is a significant difference in the performance of the cleaning step depending on the relative level of contamination. Heavily contaminated PET feedstock is much more difficult to recycle to high value products. An array of chemicals employed as necessary adjuvants to plastic production may exert significant roles to the recycling process and may be emitted throughout the plastic's life cycle [64, 65].

6.2.4 Market for PET recycling

Plastic recycling has become a crucial practice for the recovery of the fossil fuel resources embedded in the waste along with society's desire to control plastic use to avoid the detrimental littering of much of the annual plastic mass [61, 65]. A recent survey of U.S plastics recyclability paints a rather negative picture for the currently available technology [66, 67]. The recycling of plastics is at a significant growth stage begging for significant expansion. For certain segments of this expansion, an infrastructure development is necessary requiring considerable investment to a market that counts margins in pennies per pound.

A recognition of the resource value of this plastic mass continues to underscore the waste's value to create products of significant economic value [68]. Evaluations of the importance of this reutilization of plastic waste have been estimated in billions of dollars and have a transformative effect on the U.S. chemical industry and has led to an unforeseen boost to the economy [69, 70].

PET offers remarkable opportunities for recycling. At the mechanical recycling level, recycled material can be reprocessed multiple times. Feedstock is available in large quantities as a high PET content material. Dedicated collection systems for the recycling of PET bottles or separation from more complex feedstock can provide the required composition needed for recycling. Waste collection logistics is assisting the plastic recycling effort through the introduction of collection rate mandates, disposal bans, and quotas for reuse and reprocessing [71, 72]. Collection rates and plastic consumption in all markets continue to increase. Increased rPET content has been found to be highly desirable in the production of new bottles; this has been observed with the willingness of drink brands to pay slightly more for the rPET of the vPET. Furthermore, drink brands have agreed to increased rPET composition over the next few years.

6.2.5 rPET material flow

The collected feedstock for recycle is the result of manual or automated sorting to concentrate the PET fraction of a waste stream [61]. Post-consumer PET must be separated to reclaim the desired feed from contaminating non-PET. Items such as caps, lids, labels, adhesives, dirt, foreign plastics, and residual content reduce the quality of the final product and may significantly change the molar mass of the newly former polymer [61, 73]. Upon completion of the polymerization process, conversion to final product continues to the consumer and a new collection for recycle or disposal.

6.3 Pyrolytic and thermal processing

Pyrolysis is a chemical process using heat and pressure to depolymerize long-chain polymer molecules to form smaller, less complex molecules. Pyrolysis requires oxygen-free conditions for fast thermal degradation of plastic waste, which can be accomplished by using noncatalytic or catalytic conditions with zeolite, spent Fluid Catalytic Cracking catalyst, and MfO catalyst at temperatures varying from 300 to 800°C. Major products formed from the pyrolytic treatment of waste plastic feedstock are gas, oil, and char [23, 24]. Pyrolysis yields on average 45–50% of oil, 35–40% of gases, and 10–20% of tar. This conversion process can achieve up to 80% liquid oil at a moderate temperature of 500°C [27] and reaching 88 wt% liquid yields at 580°C. Plastics offer higher energy content mass than biofuels such as wood and coal.

A facility in Woodlands, Texas, USA initially was used to produce coal-derived products such as benzene, toluene, and xylene (BTX). It currently uses catalytic pyrolysis to convert post-consumer waste plastics into BTX. Similarly, a plastic-to-fuel facility in Virginia converts scrap plastic into fuel or chemical products [23]. The commonly used catalysts for pyrolysis conversion involve cracking, oligomerization, cyclization, aromatization, and isomerization reactions. Catalysts used in these processes include ZSM-5, zeolite, Y-zeolite, FCC, and MCM-41 [15, 28]. Light oil recovered from pyrolysis can be used as a chemical feedstock to produce new plastics or chemicals, whereas medium and heavy oils are commonly used as diesel fuel and heavy oils are used for power generation. For example, a large plant in Sapporo, Japan has a novel flexible system that receives not only PET, and PP, but also PVC, which is considered contaminants for PET recycling (**Figure 9**) [29].

7. Controlling elements for plastic recycling

Achieving a transition to a circular economy of plastics depends on the cost-effective and efficient reuse and recycling of plastics at large scales. A combination of end-of-life management approaches for plastics can reduce use, reuse, and repair, leading to a decreased energy and materials [74]. Recycling is a vital part of industrial ecology that can reduce environmental impacts and resource depletion. PET and HDPE are widely recycled when used in specific categories such as plastic bottles. However, this is an exception and is not seen for other plastic categories (**Table 1**).

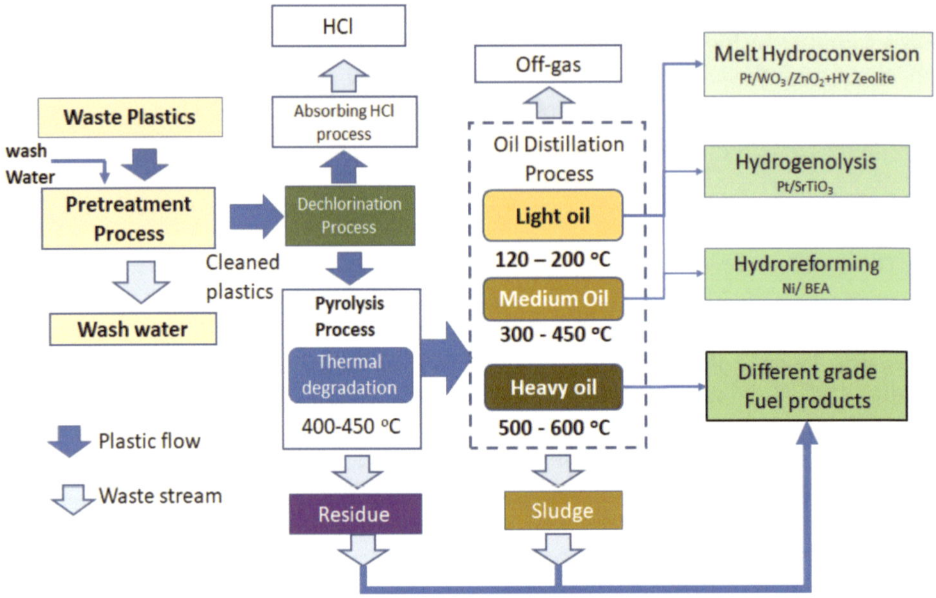

Figure 9.
Schematic diagram showing process steps for pyrolysis and other thermo-catalytic processing of plastic waste.

This could be due to the cost structure of recycled plastics, the challenges of sorting mixed plastics, and the variability of the cost of virgin plastics. Many post-consumer plastics may have to be converted to other products, and they are recycling a limited number of times. The controlling factors for recycling plastics include the effectiveness of collection and segregation of mixed materials, collection and storage of recyclables, geographical location of recycling facilities concerning the processing markets, and volume of the recyclable materials collected for the region.

The incompatible nature of most plastics due to their inherent immiscibility at a molecular level is one of the significant challenges for producing high-quality resins. For example, a small quantity of PVC in PET recycling degrades the quality of the recycled material. Hence, mixing recycled plastics with virgin plastics could reduce some of the attributes of the virgin materials, and hence, the recycled plastics are mainly used for non-critical applications. Hence, post-consumer recycling involves multiple steps: collection, sorting, washing, size reduction, and separation to minimize contamination by incompatible plastics [7]. Plastic additives such as plasticizers, color, and flame retardants used in manufacturing some plastics complicate the recycling process and present risks to human health and ecology.

8. Infrastructure development and access

The successful implementation of the plastic circular economy will require sufficient infrastructure to process-collected material. Due to the current lack of processing capacity, there is a need to modify or modernize existing recycling infrastructure [75].

8.1 Availability of suitable feedstock

The effective implementation of a CE for plastics is dependent upon the availability of feedstock, adequate recycling infrastructure, and technologies that can convert nonrecycled plastics to fuel or chemical feedstock. In 2016, only 16% of the inflow of polymers was collected for recycling, 40% were sent to landfills, and 25% incinerated. Recently, European countries have increased efforts to improve recycling rates. In 2018, 29.1 million tons of post-consumer plastic waste were collected in Europe. While less than a third of this was recycled, it represented a doubling of the quantity recycled and reduced plastic waste exports outside the European Union (EU) by 39% compared to what was recycled in 2006 [55].

The degree to which plastic recycling can be increased is limited by the availability of suitable feedstock. The ability to increase plastic recycling rates is inextricably linked to the collection rates, quality of the collected plastic, and the availability of recycling technologies to manage the post-use waste stream. Mechanical, or closed-loop processes are limited by the types of plastics that can be processed, resulting in sub-optimal yields. Plastic bottles are a highly desirable product for recyclers, but just about a third find their way into a recycling bin [76].

Chemical recycling enables the processing of mixed waste streams. However, just like mechanical recycling, advanced recycling facilities require a dependable flow of materials (feedstock) that fit the technology process [10]. Some chemical recyclers currently in operation rely on high-quality material, such as post-consumer bottles or recycled PET (rPET) polyethylene terephthalate flake as a feedstock to get a better quality of output. However, there is a concern in the recycling community that the new technologies are in danger of stripping the mechanical recycling industry of its much needed and increasingly scarce feedstock of high-quality post-consumer plastic waste. Because of the cost and challenge of sorting and separating, most chemical recyclers are not targeting low-quality waste but rather the same waste streams that would typically be used in mechanical recycling [10].

8.2 Availability of recycling facilities

The vision of a circular economy can only be realized by deeming used plastic as a resource and not as waste, and through employing advanced recycling technologies that "keep the molecule in play" and maintain materials at an economic value. Advanced recycling technologies have been used to turn plastics into fuels for decades; however, to meet the growing market demand for high-quality plastic recyclate, there is a need to improve domestic and global capabilities of waste management and recycling systems. (The future of plastics Sustainability: advanced recycling). A new era of advanced recycling for plastics, also referred to as "chemical recycling," has the potential to help the nation achieve global sustainability goals and a climate-neutral, circular economy. As much as half of all global plastics packaging could be recycled by 2040 if chemical recycling technologies were widely adopted [10].

Over the past 10 years, the capacity for processing post-consumer (rPET) has had its ups and downs. The industry recently lost 400 mm lbs. of capacity with the closure of some facilities. Capacity was expected to return to roughly 2 billion lbs./year by 2018, with at least 350 million lbs. of new PET-processing capacity becoming online. As of 2018, facilities that reprocess rPET were operating at 75% capacity [61, 73]. Brand container users are aggressively pursuing rPET for new container construction. Recently, the recycling market was unable to fill these demands for rPET.

The metropolitan area comprising Houston, Texas, USA, has the largest concentration of petrochemical manufacturing in the world. Significant tonnage quantities of BTX, synthetic rubber, insecticides, and fertilizers form the basis of this robust chemical industry. Chemical and thermal-based methods of breaking down plastics required for plastic recycling are currently available in the Houston area's chemical and petrochemical facilities to aid the utilization of a feedstock derived from plastic recycling. The metropolitan area comprising Houston, Texas, USA has the largest concentration of petrochemical manufacturing in the world. Significant tonnage quantities of BTX, synthetic rubber, insecticides, and fertilizers form the basis of this robust chemical industry.

Chemical and thermal-based methods of breaking down plastics required for plastic recycling are currently available in the Houston area's chemical and petrochemical facilities to aid the utilization of a feedstock derived from plastic recycling. This alternative based on the pyro-catalytic conversion of mixed waste polyolefins form pyrolysis oil, which can be inserted in the material flow stream of an olefin unit to be processed by a catalytic cracker to form a mixture of small molecules. This mixture contains a variety of small molecules including olefins, which can be separated and repolymerized into a recycled virgin version of the polymers. Tank cars and wagons can provide the transportation necessary to deliver the oil from a small pyrolytic oil plant to the refinery. Apart from the physical and technical assets of the Houston area, other contributions are evident in the intellectual capital of the energy industry value chain, which can provide leadership opportunities for chemical recycling advancements in the global economic, energy, and sustainability arenas [10]. The effort has been estimated to remove 10 million metric tons of CO_2, while supporting 100 advanced recycling facilities by 2020, each capable of processing 25,000 tons per year [76, 77]. Such recycling technology using facilities in Houston, TX area provides an example of how the circular plastic economy could be supported [78].

8.3 Availability of conversion facilities

The alternate approach under investigation would substitute plastics as feedstock material for oil. Even mixed plastic waste and difficult-to-recycle polymers can be used to make new, high-quality fuels and plastics—for the most demanding applications like food contact [79]. Conversion, or pyrolysis and depolymerization technologies, can convert waste plastic, which cannot be treated by mechanical recycling, into oil and gases. The compatibility of pyrolytic oil composition is important to the success of this technology. Avoidance of certain pyrolytic oil compositions will be critical to this recycling technology, since they can be detrimental to the material flow within the refinery. This is an attractive option for plastic products that are difficult to recycle mechanically due to their low-quality, composite nature, or low economic value. These monomers can be used as virgin material alternatives in manufacturing new polymers [80].

Although successful use of this technology has been demonstrated at PTF facilities worldwide, no commercial-scale systems have yet been developed in North America. There are several U.S.- and Canadian-based technology manufacturers that have operational pilot facilities. Several other global technology manufacturers have also emerged. Many of these companies have pilot-scale facilities that tend to be about one-fifth of the smallest recommended capacity for a commercial-scale facility [81].

9. Conclusion

The long-term solution for sustainable management of plastic requires stakeholder engagement from design to product across the value chain to achieve a circular economy. Although clean-up of the environment from plastic contamination is necessary for the short-term, sustainable solution demands a fundamental shift from the current practice of design, use, and recycling of plastics where plastic would not become a waste; instead, it contributes to the circular economy. The need for reliable plastic technology is imminent. Mechanical recycling offers technology that can be economically engaged to meet the clamoring requirements for highly increased plastic recycling in large quantities. The dispersion of plastic recycling technology across the U.S. is not responsive to the recycling needs of many population centers. Take-back and collection programs require attention to optimize the economics controlling the collection process. Incentive programs to enhance plastic waste collection efforts could provide a necessary support.

Competent technologies capable of converting plastic waste into useful materials have been developed to pilot scale or beyond. Performance data that is sufficiently transparent, describing the efficiency and environmental footprint for many of the technologies have not generally been available to assist the evaluation of a given technology. To the public, transparency of process operations is critical to the acceptance of the technology. This data may not be made available for technological and economic reasons in a very competitive market. The technology developer must offer the transparency of information regarding the performance of a conversion process to ensure public acceptance of new technology. Recent reviews have delineated that some advanced technologies in the thermal processing sector are not sustainable and should be avoided [82, 83]. Perhaps, this argument discards a viable technology without a replacement. Reviewers' expectations have shown an understanding of recycling process development and the time scales required for full-scale implementation. News of new plastic recycling technology often builds unachievable expectations in the minds of the public. Technologies described at a concept level may require a long and tortuous research path to a recycling technology that can be operated optimally at full-scale. The time scale for converting a recycling process at a concept level to a full-scale optimized process can be highly variable. A promising technology of chemical recycling still needs to be evaluated at full-scale to understand its performance and environmental footprint. New technologies will appear as developers can harness processing technology that is adaptable to the slight economic margins of operation currently found with plastic recycling.

The current availability and quality of waste plastic feedstock can be significant determining factors. The basic economics of plastic recycling impedes the desired expansion of recycling centers across the United States. The demonstration of plastic waste depolymerization processes has begun and requires optimization to ensure continued use against severe economic constraints. The success of these operations will be important to the expansion of the technology across the United States. The reduction of single-use plastics required innovative business models to develop novel packaging designs. There is also an urgent need to reduce the leaking of plastic waste into natural systems to become environmental waste. Increasing collection, sorting, and recycling rates are required to support circular economy to provide sustainable solutions to up-recycling, reducing plastic waste, and valorizing used plastics.

Author details

John A. Glaser*, Endalkachew Sahle-Demessie and Te'ri L. Richardson
U.S. Environmental Protection Agency, Office of Research and Development
Center for Environmental Solution and Emergency Response, 26 W King Drive, Cincinnati, Ohio, United States

*Address all correspondence to: glaser.john@epa.gov

References

[1] US EPA. National Overview: Facts and Figures on Materials, Wastes and Recycling. Washington, DC: US EPA; 2018

[2] Rosato MG, Rosato DV. Plastics Design Handbook. Berlin, Germany: Springer Science and Business Media; 2013

[3] Karlsson TM et al. The unaccountability case of plastic pellet pollution. Marine Pollution Bulletin. 2018;**129**(1):52-60

[4] Napper IE et al. Characterisation, quantity and sorptive properties of microplastics extracted from cosmetics. Marine Pollution Bulletin. 2015;**99**(1–2): 178-185

[5] Jambeck JR et al. Plastic waste inputs from land into the ocean. Science. 2015; **347**(6223):768-771

[6] Seidl LG et al. Concepts and sustainability assessment of the circular carbon economy: Chemical recycling for olefin production. Chemie Ingenieur Technik. 2021;**93**(3):421-437

[7] Hopewell J, Dvorak R, Kosior E. Plastics recycling: Challenges and opportunities. Philosophical Transactions of the Royal Society B: Biological Sciences. 2009;**364**(1526):2115-2126

[8] Silva ALP et al. Rethinking and optimising plastic waste management under COVID-19 pandemic: Policy solutions based on redesign and reduction of single-use plastics and personal protective equipment. Science of the Total Environment. 2020;**742**:140565

[9] World Economic Forum, Ellen MacArthur Foundation and McKinsey & Company. The New Plastics Economy: Rethinking the Future of Plastics & Catalyzing Action. 2016

[10] Baker Institute contributor. Breakthroughs In Advanced Plastic Recycling Will Help Deliver On Sustainability Goals, Forbes. Apr 19, 2021

[11] US EPA. Municipal Solid Waste Landfills. Washington, DC: US EPA; 2018

[12] Stahel WR. The product-life factor. In: Grinton Orr S editor. An Inquiry into the Nature of Sustainable Societies. Houston, TX: The Role of the Private Sector. HARC; 1981:72-96

[13] Bucknall DG. Plastics as a materials system in a circular economy. Philosophical Transactions of the Royal Society A. 2020;**378**(2176):20190268

[14] Pearce DW, Turner RK. Economics of Natural Resources and the Environment. Baltimore, MD: JHU Press; 1990

[15] Ragaert K, Delva L, Van Geem K. Mechanical and chemical recycling of solid plastic waste. Waste Management. 2017;**69**:24-58

[16] Maris J et al. Mechanical recycling: Compatibilization of mixed thermoplastic wastes. Polymer Degradation and Stability. 2018;**147**: 245-266

[17] Garcia JM, Robertson ML. The future of plastics recycling. Science. 2017;**358**(6365):870-872

[18] Van Berkel R et al. Industrial and urban symbiosis in Japan: Analysis of the Eco-Town program 1997–2006. Journal of Environmental Management. 2009;**90**(3):1544-1556

[19] ISO. ISO 15270 International Standard Organization, ISO 15270: Plastics-Guidelines for the Recovery and Recycling of plastic waste.

[20] Geyer R. Production, use, and fate of synthetic polymers. In: Plastic

Waste and Recycling. Amsterdam, Netherlands: Elsevier; 2020. pp. 13-32

[21] Hamad K, Kaseem M, Deri F. Recycling of waste from polymer materials: An overview of the recent works. Polymer Degradation and Stability. 2013;**98**(12):2801-2812

[22] Datta J, Kopczyńska P. From polymer waste to potential main industrial products: Actual state of recycling and recovering. Critical Reviews in Environmental Science and Technology. 2016;**46**(10):905-946

[23] Grigore ME. Methods of recycling, properties and applications of recycled thermoplastic polymers. Recycling. 2017;**2**(4):24

[24] Singh N et al. Recycling of plastic solid waste: A state of art review and future applications. Composites Part B: Engineering. 2017;**115**:409-422

[25] El Mehdi M et al. Recent advances in polymer recycling: A short review. Current Organic Synthesis. 2017;**14**(2): 171-185

[26] Vollmer I et al. Beyond mechanical recycling: Giving new life to plastic waste. Angewandte Chemie International Edition. 2020;**59**(36):15402-15423

[27] Awaja F, Pavel D. Recycling of PET. European Polymer Journal. 2005;**41**(7): 1453-1477

[28] Sinha V, Patel MR, Patel JV. PET waste management by chemical recycling: A review. Journal of Polymers and the Environment. 2010;**18**(1):8-25

[29] Aguado A et al. Chemical depolymerisation of PET complex waste: Hydrolysis vs. glycolysis. Journal of Material Cycles and Waste Management. 2014;**16**(2):201-210

[30] Vadicherla T, Saravanan D, Muthu SSK. Polyester recycling—Technologies, characterisation, and applications. In: Environmental Implications of Recycling and Recycled Products. Amsterdam, Netherlands: Springer; 2015. pp. 149-165

[31] Pohjakallio M, Vuorinen T, Oasmaa A. Chemical routes for recycling—Dissolving, catalytic, and thermochemical technologies. In: Plastic Waste and Recycling. Amsterdam, Netherlands: Elsevier; 2020. pp. 359-384

[32] Noreña L et al. Materials and Methods for the Chemical Catalytic Cracking of Plastic Waste. In Book: Material Recycling-trends and Perspective. Elsevier Science B.V; 2012. pp. 154-157

[33] Speight JG. Chemistry And Technology of Alternate Fuels. Singapore: World Scientific; 2020

[34] Qureshi MS et al. Pyrolysis of plastic waste: Opportunities and challenges. Journal of Analytical and Applied Pyrolysis. 2020;**152**:104804

[35] Dogu O et al. The chemistry of chemical recycling of solid plastic waste via pyrolysis and gasification: State-of-the-art, challenges, and future directions. Progress in Energy and Combustion Science. 2021;**84**:100901

[36] Armenise S, Syieluing, W, Ramírez-Velásquez JM, Launay F, Wuebben D, Ngadi N, et al. Plastic Waste Recycling via Pyrolysis: A Bibliometric Survey and Literature Review. Journal of Analytical and Applied Pyrolysis. 2021;**158**:105265

[37] Ágnes N, Rajmund K. The environmental impact of plastic waste incineration. Academic and Applied Research in Military and Public Management Science. 2016;**15**(3): 231-237

[38] Wong S et al. Current state and future prospects of plastic waste as source of fuel: A review. Renewable and

Sustainable Energy Reviews. 2015;**50**: 1167-1180

[39] Hita I, Sarathy SM, Castaño P. Polymeric waste valorization at a crossroads: ten ways to bridge the research on model and complex/real feedstock. Green Chemistry. 2021;**23**: 4656-4664

[40] Breilly D, Fadlallah S, Froidevaux V, Colas A, Allais F. Origin and industrial applications of lignosulfonates with a focus on their use as superplasticizers in concrete. Construction and Building Materials. Extraction. 2021;**301**:124065

[41] Jehanno C et al. Selective chemical upcycling of mixed plastics guided by a thermally stable organocatalyst. Angewandte Chemie International Edition. 2021;**60**(12):6710-6717

[42] Scheirs J, Long TE. Modern polyesters: Chemistry and technology of polyesters and copolyesters. Hoboken, NJ: John Wiley and Sons; 2005

[43] Barot AA et al. Polyester the workhorse of polymers: A review from synthesis to recycling. Archives of Applied Science Research. 2019;**11**(2):1-19

[44] McIntyre J. The historical development of polyesters. Modern Polyesters: Chemistry and Technology of Polyesters and Copolyesters. 2003;**2**:1

[45] Park SH, Kim SH. Poly (ethylene terephthalate) recycling for high value added textiles. Fashion and Textiles. 2014;**1**(1):1-17

[46] Sarioğlu E, Kaynak H.K. PET bottle recycling for sustainable textiles. In: Camlibel NO, editor. Polyester-production, Characterization and Innovative Applications. 2017. DOI: 10.57772/intechopen.69941

[47] Badia J et al. The role of crystalline, mobile amorphous and rigid amorphous fractions in the performance of recycled poly (ethylene terephthalate)(PET). Polymer Degradation and Stability. 2012;**97**(1):98-107

[48] Papaspyrides CD, Vouyiouka SN. Solid State Polymerization. Hoboken, NJ: John Wiley and Sons; 2009

[49] Pergal MV, Balaban M. Poly (ethylene terephthalate): Synthesis and physicochemical properties. In: Polyethylene Terephthalate: Uses, Properties and Degradation. New York: Nova Science Publishers; 2017. pp. 1-102

[50] Elamri A et al. Progress in Polyethylene Terephthalate Recycling. New York: Nova Science Publishers; 2017

[51] Gupta P, Bhandari S. Chemical depolymerization of PET bottles via ammonolysis and aminolysis. In: Recycling of Polyethylene Terephthalate Bottles. Amsterdam, Netherlands: Elsevier; 2019. pp. 109-134

[52] Heller MC, Mazor MH, Keoleian GA. Plastics in the US: Toward a material flow characterization of production, markets and end of life. Environmental Research Letters. 2020; **15**(9):094034

[53] Kosior E, Mitchell J. Current industry position on plastic production and recycling. In: Plastic Waste and Recycling. Amsterdam, Netherlands: Elsevier; 2020. pp. 133-162

[54] Feil A, Pretz T. Mechanical recycling of packaging waste. In: Plastic Waste and Recycling. Amsterdam, Netherlands: Elsevier; 2020. pp. 283-319

[55] Schyns ZO, Shaver MP. Mechanical recycling of packaging plastics: A review. Macromolecular Rapid Communications. 2021;**42**(3):2000415

[56] Han M. Depolymerization of PET bottle via methanolysis and hydrolysis.

In: Recycling of Polyethylene Terephthalate Bottles. Amsterdam, Netherlands: Elsevier; 2019. pp. 85-108

[57] Barnard E, Arias JJR, Thielemans W. Chemolytic depolymerisation of PET: a review. Green Chemistry. 2021;**23**:3765-3789. DOI: 10.1039/DIGC00887K

[58] Cabanes A, Valdés FJ, Fullana A. A review on VOCs from recycled plastics. Sustainable Materials and Technologies. 2020;**25**:e00179

[59] Welle F. Twenty years of PET bottle to bottle recycling—An overview. Resources, Conservation and Recycling. 2011;**55**(11):865-875

[60] Brouwer MT, Alvarado Chacon F, Thoden van Velzen EU. Effect of recycled content and rPET quality on the properties of PET bottles. Part III: Modelling of repetitive recycling. Packaging Technology and Science. 2020;**33**(9):373-383

[61] Di J et al. United States plastics: Large flows, short lifetimes, and negligible recycling. Resources, Conservation and Recycling. 2021;**167**:105440

[62] Chu J et al. Dynamic flow and pollution of antimony from polyethylene terephthalate (PET) fibers in China. Science of the Total Environment. 2021;**771**:144643

[63] López MMC et al. Assessing changes on poly (ethylene terephthalate) properties after recycling: Mechanical recycling in laboratory versus postconsumer recycled material. Materials Chemistry and Physics. 2014;**147**(3):884-894

[64] Geueke B, Groh K, Muncke J. Food packaging in the circular economy: Overview of chemical safety aspects for commonly used materials. Journal of Cleaner Production. 2018;**193**:491-505

[65] Pelzl B, Wolf R, Kaul BL. Plastics, Additives. Ullmann's Encyclopedia of Industrial Chemistry. Weinheim: Wiley-VCH Verlag GmbH & Co; 2000. p. 1-57. DOI: 10.1002/14356007.a20_459.pub2

[66] Bartolome L, Imran M, Cho BG. Recent developments in the chemical recycling of PET. Material recycling-trends and perspectives. 2012;**406**:65-85

[67] Greenpeace. Circular Claims Fall Flat: Comprehensive U.S. Survey of Plastics Recyclability. 2020. Available from: https://www.greenpeace.org/usa/wp-content/uploads/2020/02/Greenpeace-Report-Circular-Claims-Fall-Flat.pdf

[68] Weckhuysen BM. Creating value from plastic waste. Science. 2020;**370**(6515):400-401

[69] Hundertmark T et al. How plastics waste recycling could transform the chemical industry. McKinsey & Company. 2018;**12**:1-1

[70] Alessio DA et al. Plastics waste trade and the environment. Eionet Report-ETC/WMGE. ETC/WMGE, Copenhagen. 2019:50. Available from: http://hdl.handle.net/10807/146947

[71] Miller ZD et al. Identifying strategies to reduce visitor-generated waste in national parks of the United States: The zero landfill initiative. Applied Environmental Education and Communication. 2020;**19**(3):303-316

[72] Basuhi R et al. Environmental and economic implications of US postconsumer plastic waste management. Resources, Conservation and Recycling. 2021;**167**:105391

[73] Martin E. Closed Loop Partners, Cleaning the rPET Stream: How we scale post-consumer recycled PET in the US. 2017. p. 22. Available from: https://www.closedlooppartners.com/wp-content/uploads/2020/02/CLP-RPET-Report_Public-FINAL.pdf

[74] Nnorom IC, Osibanjo O. Overview of prospects in adopting remanufacturing of end-of-life electronic products in the developing countries. International Journal of Innovation, Management and Technology. 2010;**1**(3):328

[75] Trusts PC. System IQ, Breaking the Plastic Wave: A Comprehensive Assessment of Pathways towards Stopping Ocean Plastic Pollution, The Pew Charitable Trusts and SystemsIQ. 2020

[76] Billiet S, Trenor SR. 100th anniversary of macromolecular science viewpoint: needs for plastics packaging circularity. ACS Macro Letters. American Chemical Society. 2020;**9**(9): 1376-1390. DOI: 10.1021/acsmacrolett.0c00789

[77] McCauley D, Ramasar V, Heffron RJ, Sovacool BK, Mebratu D, Mundaca L. Energy justice in the transition to low carbon energy systems: Exploring key themes in interdisciplinary research. Applied Energy. 2019;**233–234**:916-921

[78] Morath S. Our plastic problem. Nat. Resources & Env't. University of Houston Law Center No. A-1. 2019:45-49. Available from: https://ssrn.com/abstract=3340072

[79] Tullo AH. Plastic has a problem; is chemical recycling the solution? Chemical & Engineering News. 2019;**97**: 39. ISSN 0009-2347

[80] Heller MC, Mazor MH, Keoleian GA. Plastics in the US: Toward a material flow characterization fo production, markets and end of life. Environmental Research Letters. 2020; **15**(9):094034

[81] 4R Sustainability, Inc. Conversion Technology: A complement to plastic recycling. Portland, 619 OR: American Chemistry Council; 2011

[82] Eunomia, CHEM Trust. Chemical Recycling: State of Play. 2020. Available from: https://www.eunomia.co.uk/reports-tools/final-report-chemical-recycling-state-of-play/

[83] Rollinson A, Oladejo J. Chemical recycling: Status, sustainability, and environmental impacts. Global Alliance for Incinerator Alternatives. 2020:1-45. Available from: www.no-burn.org/cr-technical -assessment. DOI: 10.46556/ONLS4535

Chapter 2

Recent Advances in Pre-Treatment of Plastic Packaging Waste

Rita Kol, Martijn Roosen, Sibel Ügdüler, Kevin M. Van Geem, Kim Ragaert, Dimitris S. Achilias and Steven De Meester

Abstract

There is an urgent need to close the loop of plastic waste. One of the main challenges towards plastic packaging waste recycling is the presence of a variety of contaminants. These contaminants include organic residues, additives, labels, inks and also other plastic types that can be present in the waste stream due to missorting or in multimaterial structures (e.g. multilayer films in packaging). In this context, pre-treatment processes are a promising route to tackle the difficulties that are encountered in mechanical and chemical recycling due to these contaminants. This chapter gives better insight on the already existing pre-treatment techniques and on the advances that are being developed and/or optimized in order to achieve closed-loop recycling. Some of these advanced pre-treatments include chemical washing to remove inks (deinking), extraction methods to remove undesired plastic additives and dissolution-based pre-treatments, such as delamination and dissolution-precipitation techniques.

Keywords: Recycling, Plastic Packaging, Circular Materials Economy, Pre-treatment, Chemical cleaning

1. Introduction

The largest share of post-consumer plastic waste is plastic packaging, [1, 2] which is typically composed of among others low-density polyethylene (LDPE), high-density polyethylene (HDPE), polypropylene (PP), poly(ethylene terephthalate) (PET) and polystyrene (PS) [2]. The European Commission has set recycling objectives to move towards a more sustainable plastic economy, namely: (i) achieve 55% recycling of plastic packaging by 2030, (ii) increase fourfold the sorting and recycling capacity (iii) and produce only reusable or recyclable plastic packaging [3].

The main challenges faced during the recycling processes of plastic materials is the complexity of this waste stream. Plastic packaging waste streams typically consist of a mixture of polymers and contaminants, such as paper, organic residues, odor constituents, adhesives and inks containing halogens and metals [4]. The presence of additives that are incorporated in the plastics during the manufacturing process also bring complications during mechanical and chemical recycling [1, 5]. The most used additives in packaging materials are: plasticizers, antioxidants, acid scavengers, light and heat stabilizers, lubricants, pigments, antistatic agents, slip compounds and thermal stabilizers [5]. However, during mechanical recycling these additives are blended in the recycled materials, which might be a potential health risk for consumers, especially in food contact materials [4]. For example, flame retardants such as persistent organic pollutants (POPs), phosphorous flame retardants and phthalates have been found in children toys [5]. Brominated flame retardants were also detected in food contact materials and household products [5]. Some additives can also have a direct impact on the recyclability of the plastics and even lead to the degradation of the plastics. This is the case for metal-containing additives, such as metals salts or oxides like Fe_2O_3, Cu_xO and ZnO, that will form pro-oxidants and photo-oxidation catalysts, promoting the degradation of the plastics during the reprocessing phases in mechanical recycling [5]. Furthermore, leaching to the environment of substances such as metals, volatile organic compounds, phthalates, polycyclic aromatic hydrocarbons, among others, is also a problem faced during different stages of mechanical recycling [5]. Another recycling issue is the presence of multilayer materials that typically consist of immiscible polymers (e.g. PET and polyethylene (PE)) that lead to recyclates with low mechanical properties [6]. Complications also arise during chemical recycling when e.g. halogenated compounds are released during the polymer decomposition, which can cause corrosion of the process equipment and reactors [4, 7]. Also for chemical recycling technologies, multilayer materials might cause complications as they typically contains polymers such as PET layers, which lead to the release of oxygenated compounds [4, 6].

With the currently applied pre-treatment technologies, such as sorting, washing, float-sink and grinding, there are still contaminants present in the post-consumer plastic waste, [4] as these techniques are not able to thoroughly clean the polymers and remove the impurities embedded in the polymer structure [8]. A more circular economy for plastics would need more advanced pre-treatments such as chemical washing technologies, deodorization, deinking, delamination and solvent-based extraction methods. Deodorization and deinking techniques enable the removal of odor constituents and inks present in the plastic waste, respectively. Delamination can tackle the issue of multilayer materials by selective decomposition of polymer layers and/or adhesives [6]. Solvent-based extraction methods are able to remove several additives from the polymer matrix. Solvent-based extraction methods can be divided into two groups: dissolution-precipitation and solid–liquid extraction methods, which include ultrasonic extraction, supercritical fluid extraction, microwave-assisted extraction and accelerated solvent extraction.

2. Pre-treatments

The aim of this chapter is to provide a review of the advances in pre-treatment proposed in literature for the purification of plastic waste. Thereto, technologies that are applied from lab-scale until industrial scale are discussed. The first section of this chapter focusses on the more conventional pre-treatment steps comprising sorting and washing of plastic packaging. Subsequently, the recent advances in deodorization, deinking and delamination of plastic packaging will be discussed in depth. In the last section, a detailed overview will be given on solvent-based methods, comprising dissolution-precipitation and extraction technologies.

2.1 Conventional pre-treatment steps

In Europe, extended producer responsibility (EPR) is set up for the collection and recycling of packaging waste. Via EPR a producer's responsibility for a product is extended to the post-consumer stage of a product's life cycle [9]. This is generally translated to an environmental fee that producers need to pay to have their products managed through the so-called producer responsibility organization (PRO). A PRO is a collective entity set up by producers or through legislation, which becomes responsible for meeting the recovery and recycling obligations of the individual producers [10]. Depending on the country, an individual or co-comingled kerbside collection system can be implemented, a deposit-refund system might be organized or combinations thereof. In Germany, for instance, there are two separate collection systems for plastic packaging, namely a deposit system for PET bottles and the Dual System, where packaging and non-packaging wastes made of plastic, paper, metals and composite material are disposed for commingled collection [11].

Depending on the particular collection scheme, the waste stream will comprise a whole range of different packaging products and contaminants [12]. Due this compositional complexity, sorting of plastics in a Material Recovery Facility (MRF) is an essential step to provide high quality materials for the recycling industry [13]. An MRF comprises sorting units such as waste screening, air separation, ballistic separation, magnetic separation, eddy current separation, sensor-based sorting such as near infrared technology (NIR) and manual sorting in a certain configuration in order to generate bales as pure as possible. The output materials achieve purity levels up to more than 97% [13]. However, even after sorting, the generated plastic bales still contain certain substances, such as polymers, paper, and organic residue, present in levels between 10% and 20% of the total mass of the end-of-life packaging product [4].

These bales are sent to recycling factories, currently mainly mechanical recyclers, that wash and regranulate the plastics. **Figure 1** illustrates a typical process chain applied at a recycler's plant.

The recycling process starts by reducing the size of the plastic products via a crude shredder. Henceforth, the shredder particles are transported via a conveyor belt to a first washing step. This is, for instance, a rotating drum washer where rocks, metals and glass are separated gravitationally; a water flow provides the washing [1]. This step is followed by a second washing by means of a friction washer. The plastic surfaces are here cleaned of organic residues, adhesives and glued-on labels that stick to the plastics by intensive mechanical agitation. The plastic material is transported via a screw, while dissolved impurities, fines, and process water are discharged through sieves [15]. The plastic particles are now fed into a miller to further reduce the particle size. From this point, they proceed to the float-sink installation, where a density separation of polymers is performed. Polymers with a density lower than 1 kg/L will float (e.g., PE and PP), whereas polymers with a density higher than 1 kg/L will sink (e.g., PET, PS and PVC). As

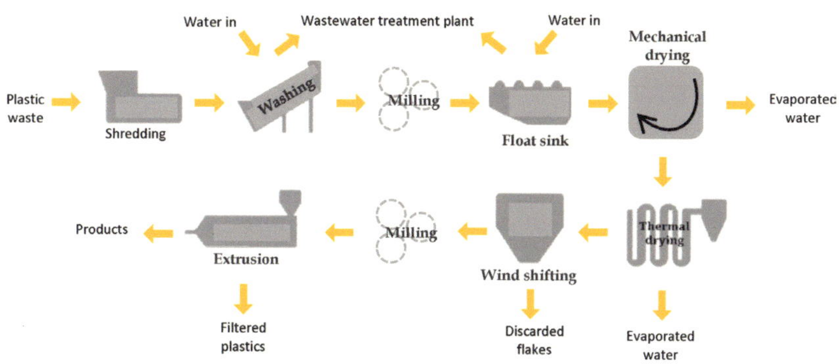

Figure 1.
Generic process flow of a recycling process. Adopted from [14].

this is a water-based technique, the flakes simultaneously undergo an extra washing [1]. After the float-sink step, the flakes are dried via a mechanical and/or a thermal dryer. Finally the plastics are sent to extrusion, which includes a melt filtration to remove fractions such as wood, paper, aged rubber particles and higher-melting polymers (e.g. PET in PP processed at 220°C) [16].

Most recycling plants use water to wash the plastics, whether or not at elevated temperature. In some cases sodium hydroxide (NaOH) or a detergent can be added as washing additives in order to reduce surface contamination such as dirt, labels and glue [17]. Logically, a water-based washing step cannot achieve 100% efficient removal of the heterogeneous substances that are present on a post-consumer plastic waste stream, even not when caustic or detergents are added. Therefore, more advanced purification steps are under developing at lab or pilot-scale or are being integrated before, during or after the extrusion step in order to provide recyclates with a higher purity. However, even for more advanced pre-treatment steps sorting and cleaning with an initial water-based washing step to remove organics, papers, and polymers e.g., with a friction washer or density separation, is probably still a key step within the plastic recycling chain.

2.2 Deodorization technologies

One of the main hurdles towards high-end recycling of plastics is the presence of odorous constituents. Due to the persistent odor that remains after washing with specialized equipment such as friction washers, large volumes of plastic packaging waste are currently only suitable for downcycled applications, such as plant trays, compost bins, street or garden benches, etc. [18]. In several scientific papers, the odor profile of different plastic packaging products is characterized. Especially the heterogeneity of odor components on packaging material in terms of physicochemical properties is described. This is illustrated by **Figure 2**, which gives an indication of the abundance of chemical groups that were detected on post-consumer waste plastics.

In total, over 400 volatile organic components (VOCs) are detected, which are here divided into 19 subcategories. With 60 and 56 components, respectively, esters and alkanes are two of the main functional groups found on plastics. However, their physicochemical properties are fundamentally different. This physicochemical heterogeneity of odor components makes efficient deodorization very challenging. Yet, most recycling plants only apply a water-based washing step, in best case with

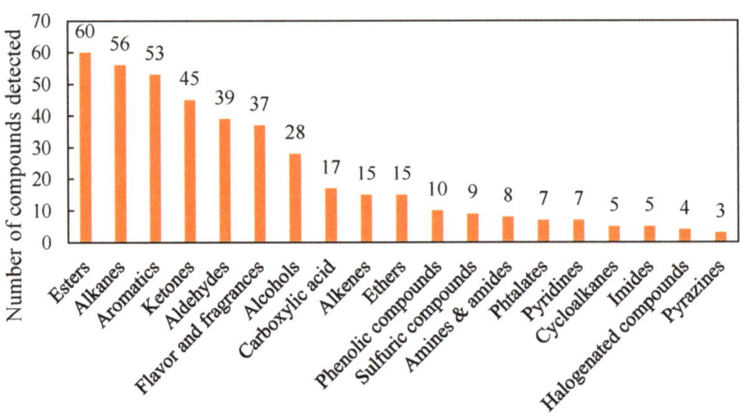

Figure 2.
Number of VOCs detected on plastic materials, divided into subcategories based on their chemical structure. Adopted from [19].

Material	Treatment	Result	Ref.
LDPE bags	Hot-water at lab-scale	Considerable odor intensity remained (6.3 out of 10)	[20]
MPO films	Industrial recycling process	No reduction observed	[21]
MPO films	Industrial recycling process	<65% reduction after washing Increase with a factor 2–6 after extrusion compared to the washed films	[18]
PP rigid	Industrial recycling process	Odor intensity ratings declined from 7.4 to 4.0	[22]
PE film	Cold (25°C) and hot (65°C) water at lab-scale Cold (25°C) and hot (65°C) caustic + detergent at lab scale	54% and 63% removal of VOCs 68% and 78% removal of VOCs	[8]

Table 1.
Overview of scientific literature to evaluate odor removal from plastics achieved by going through a full-scale water-based industrial mechanical recycling facility and their respective main results.

addition of a detergent. Logically, scientific and industrial experience have shown that the water-based washing media are insufficient to become recyclates with an acceptable odor threshold. **Table 1** shows an overview of different scientific studies that investigated the efficiency of industrial recycling plants in term of deodorization of plastic waste.

As water-based treatments are insufficient, the demand for new and improved deodorization technologies is increasing. Although, industrial application of such technologies remains relatively limited to date. An interesting option is the use of a solvent-based approach in order to remove the more hydrophobic constituents. A commercial solvent-based process applies hot ethyl acetate to clean polyolefin-based packaging products [23]. Scientific research at lab-scale has indicated that with a batch-wise extraction using ethyl acetate at 65°C, an average removal of analyzed odor components of 90% is feasible [8]. Taking into account the benefit that can be made with a continuous counter-current process, polymers with relatively high purity levels can be achieved.

Another solvent that is tested to remove, among others, odorous constituents from a HDPE waste stream is poly(ethylene glycol) (PEG). PEG is considered as a relatively eco-friendly solvent and is getting attention due to its low volatility and toxicity compared to conventional solvents, and its higher miscibility with organic compounds compared to water [24]. A batch lab-scale extraction with PEG has shown that the quantity of VOCs was reduced with 74% after PEG extraction at 100°C. Hence, PEG is considered to be a promising solvent towards deodorization [24]. A schematic representation of a deodorization by means of PEG can be found in **Figure 3a**.

Applying solvent-based technologies might be quite expensive, taking into account the extra capital expenditures (CAPEX) and operating expenditures (OPEX) that are typically linked to such technologies. This is often not preferable in plastic recycling, given the associated typically low profit margins [4]. Applying chemicals such as solvents and detergents on a relatively highly contaminated waste stream, comprising next to odorous constituents also glues, paper, inks, additives, degradation products, non-target polymers, etc., should preferably be able to remove a large range or even all of such substances in order to be interesting from economic perspective. In this perspective, such technologies are not only investigated towards removal of odor components, but also towards removal of a broader range of impurities. The application and development of more solvent-based

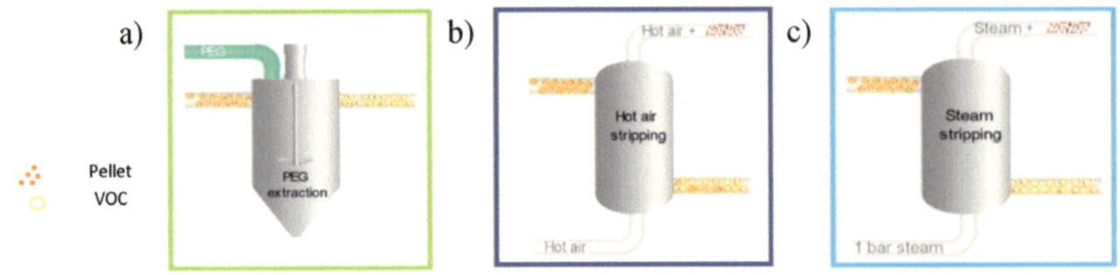

Figure 3.
Schematic representation of different deodorization strategies. a) Deodorization by means of a solvent, b) deodorization by means of hot air stripping, and c) deodorization by means of steam stripping. Adopted from [24].

technologies and their efficiency towards removal of different substances will be discussed more in detail in Section 2.5.

Besides applying extraction techniques, also the use of air to remove odorous constituents, as shown in **Figure 3b**. A commercial available example is the ReFresher technology with the INTAREMA® TVEplus® machine (EREMA Engineering Recycling Maschinen und Anlagen Ges.m.b.H., Ansfelden, Germany). This equipment applies heated air directly to the extruded pellets for flushing out volatile contaminants and simultaneously removes the air via a degassing unit. Applying a hot air stream during a few hours can significantly reduce the overall odor intensity of recycled HDPE pellets with an efficiency varying between 51.0 and 99.3% [25]. A disadvantage of this technique is the relatively long contact time that is needed to achieve the maximum feasible removal efficiencies, typically between 4 and 7 hours. Depending on its size, the ReFresher has a capacity between 350 kg/h and 4000 kg/h.

Likewise, steam can be applied to remove VOCs from plastic materials (see **Figure 3c**). A study has shown that an increased VOC reduction is achieved via steam stripping compared to hot air stripping [24]. Thereto, post-consumer HDPE was treated at a lab-scale distillation unit for 2 h. The produced vapor flowed through the plastic bed and left on the top, passing to a condenser. An overall reduction of volatile components above 70% was reported.

A similar approach is the use of a degassing system during extrusion is investigated towards the removal of VOCs [19]. Different methods are used e.g. degassing by vacuum, thermal degassing or degassing with the help of ultrasound [26]. A study to the removal of VOCs from plastics via a vacuum degassing system showed that the odor concentration was reduced with around 37% after three degassing steps. Devolatilization is considered to be a complex process as the correct choice of temperature and shear profiles, along with screw configuration and placement of venting influence the removal efficiencies of VOCs. For instance, a higher temperature and pressure during extrusion can increase the volatility of the moisture content and permit water and/or other volatile materials to be released [27]. **Figure 4** shows a typical set-up for devolatilization of LDPE.

Another investigated method to minimize odor on post-consumer plastic packaging waste is through the use of a probiotic bacteria solution during the recycling process. In a recent study, a commercial probiotics formulation was investigated at a pilot scale before the washing step [29]. Significant differences in the overall odor intensity of the untreated reference sample and the sample treated with probiotic bacterial cultures were obtained by applying the probiotic bacterial cultures to the input material, followed by 40 days of storage. Especially those substances that most likely originated from microbial degradation of organic matter are reduced up to 70%. However, further improvement and investigation to industrial implementation of probiotic treatment is required.

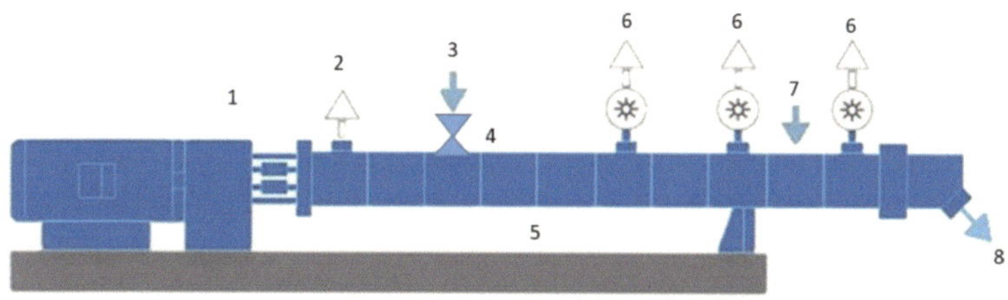

Figure 4.
A typical set-up for devolatilization of LDPE. 1) drive, 2) rear vent, 3) overheated solution with $p > p_{vapour}$, *4) flash valve, 5) kneading section, 6) vacuum, 7) stripping agent, and 8) discharge/pelletizing. Adopted from [28].*

A more established option to remove VOCs from plastic waste is the addition of high specific surface adsorbents during the extrusion process [21]. A great variety of adsorbents are available on the market. As they are added to a polymer in melting phase, VOCs can adsorb on the adsorbing agents' surfaces. It is stated that adding 0.30 wt% of a certain adsorbents such as zeolite and activated silicate, can significantly reduce the amount of VOCs coming from post-consumer HDPE with approximately 50% [30]. A similar more recent development is the addition of reactive additives that undergoes a chemical reaction with the functional groups of odor-causing substances and, hence, convert them into non-volatile components [31]. Examples of such a commercial available additives are zinc ricinoleate (Tego Sorb PY 88; Evonik Industries AG, Essen, Germany) and Recycloblend 660 (PolyAd Services). However, scientific studies to quantify the effectiveness of such additives are scarce.

2.3 Deinking technologies

Plastic packaging is typically heavily printed with inks for functional benefits such as including information about composition, presence of allergens and nutritional details, etc., but also for marketing purposes to make them more appealing to consumers. The main constituents of printing inks are resins, solvents, colorants and additives [32]. Resins are high molecular weight polymers constituting 15 to 50% of the composition of the ink and they act as binder for colorant stabilization [32, 33]. Solvent constitutes the largest part of the ink composition (up to 65%) [32]. Solvents are used to dissolve the resins and also to keep the ink liquid for supporting ink transfer [34]. Colorants used to give desired color to plastic packaging, constitute 5–30% of the ink composition [32, 33]. Colorants can be used as pigments or dyes. Pigments are insoluble solid fine particles which are dispersed in the binder, while dyes are substances that are completely soluble in the binder [35, 36]. In addition to colorants, lacquers or overprint varnishes are uncoloured forms of printing inks, which can be used to provide gloss and protective properties to the print [33]. Lastly, additives are generally used up to 10% in order to improve physicochemical properties of inks such as among, others, adhesion, slip and scratch resistance [37, 38]. The composition of these ink components can considerably differ depending on the printing process and also on the substrate. For example, for most substrates solvent-based inks in which the resin is dissolved in a suitable solvent are used as they allow sufficient wetting and adhesion [34]. Compared to solvents, the evaporation rate of water is much slower, making the drying process of the ink energy intensive [34]. Therefore, water-based inks are generally used for substrates which can promote absorption mechanism such as paper and board [39]. In addition to solvent and water-based inks where drying of inks is performed through evaporation of the liquid medium, in UV-based inks UV radiation is used for drying, which allows the ink to immediately form a three-dimensionally cross-linked film. UV-based inks require reactive resins such as acrylates which can react with free radicals created by UV radiation [33, 34].

Although inks are one of the necessary components of plastic packaging, they are a significant source of contamination in plastic recycling. As all printed plastic films are generally collected and processed together, a low quality brownish, grayish or black recyclate is obtained, making it only suitable for downcycled products [40]. The presence of ink also causes recycled films to be less stiff, weaker, and denser compared to the original material, thus its price is considerably lower than the price of films free of ink. Furthermore, during the processing or reprocessing, residual ink can also decompose and produce gases causing rancid odor formation and also decrease the physical properties of a raw material [41]. In order to eliminate these problems and obtain high quality recyclates, interest in deinking technologies is increasing. However, only a few deinking technologies are so far active in deinking of plastic packaging. For example, in the patented process known as known as the Nordenia Extraction and Cleaning process or NorEC (DE19651571A1), ethyl acetate was used as a solvent-based extraction medium to remove broad range of inks [42]. The NorEC process has currently been applied in an industrial packaging plant in the North of Germany with a capacity of 15000 tons [42]. In this plant industrial PE film waste is being shredded and treated with this extraction technology. It is stated that the NorEC process requires lower amount of energy compared to the conventional wet processing [43]. Furthermore, surfactants are extensively studied as a potential deinking medium [44–48]. Deinking mechanism by using a surfactant consists of four main steps: [1] adsorption of the surfactant on plastic surfaces, [2] solubilization of the binder in the surfactant aggregates so called micelles, [3] detachment of ink particles from the surface, and [4] stabilization of the detached ink particles (**Figure 5**) [48]. According to the study of Chotipong et al. [48], cationic surfactants such as cetyl trimethylammonium bromide (CTAB) were more effective to remove both water- and solvent-based inks. In addition, it is shown that critical micelle concentrations (CMC), pH of the medium, temperature and stirring are important parameters on deinking efficiency [48].

Furthermore, in the patented method of the University of Alicante (EP2832459B1), a surfactant is used to remove inks from plastic packaging [49]. This method has also a semi-industrial demonstration deinking plant with a treatment capacity of 100 kg/h [49]. In this closed-loop recycling plant, printed plastic films pass through several treatments such as grinding, deinking, washing, drying, and pelletizing in order to obtain ink free plastics with high optical quality [41]. As a water-based medium is used, the medium can go to wastewater treatment. The use of surfactants to deink plastic films was also described in another patent filed by Duchenaud Uniflexo (EP1419829A1) [50]. In this method, the deinking medium contains organic solvents and a non-ionic surfactant. Although high deinking efficiencies can be achieved via this method, it has been stated that physico-mechanical properties of recycling plastic substantially decrease [41]. Furthermore, the use of

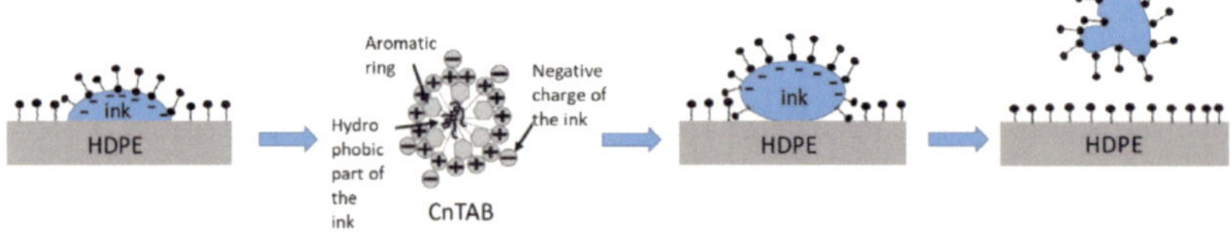

Figure 5.
Four-step mechanism for removal of solvent-based ink from HDPE surface. Adopted from [48].

dangerous products and the high cost of the process limit the potential scaling-up [41]. Similarly, the Italian company Gamma Meccanica uses mechanical brushes for deinking purposes, but again deinking is limited to packaging having the ink layer on the surface [51].

2.4 Delamination technologies

Plastic packaging can be laminated with different polymer layers in order to increase their physico-chemical properties. Combination of functionality of each polymer provides plastic film with superior preservation performances such as among others, extended shelf life, high mechanical strength, good sealability (**Figure 6**) [52]. For example, sealing properties of PET are poor, thus it is often laminated with polyolefins [53]. Similarly, use of aluminum layer provides protection against UV light, as such nutritional value of products are preserved for a longer time period [53]. Although the combination of different polymer layers extends the functionality and application area of plastic packaging, they make the recycling of multilayer packaging more complex. For example, during mechanical recycling, incompatibility issues may arise in the polymer blends due to their difference in physico-chemical properties, such as PE and PET [54]. Similarly, in thermochemical recycling heterogeneous polymers such as PET, polyamide (PA), polycarbonate (PC) contaminate polyolefinic plastic waste [1]. Therefore, multilayer plastic film fractions are still mainly incinerated or landfilled to date [55]. Recently, there is a growing interest towards single layer plastic films in order to eliminate those complexities encountered in multilayer packaging [55]. However, it is not always feasible to achieve functionalities that combination of different plastics can provide, thus in order to enhance circularity of flexible packaging, delamination of multilayer packaging might still be needed.

There are different option to separate the layers of multilayer plastic packaging. One of them is selective dissolution of a certain polymer layer. For example, in the study of Samorì et al. [56] switchable hydrophilic solvents were used to selectively dissolve the LDPE layer and recover Al from food packaging. Similarly, in the study of Mumladze et al. [52] adhesive polymers were dissolved by using N, N-Dimethylcyclohexylamine (DMCHA) as a solvent and also during dissolution ultrasonic treatment was employed to speed up the separation of multilayers. Furthermore, in the patented method of Nauman and Lynch, pure polymer fractions were obtained through sequential dissolution of multilayer structures by gradually increasing temperature in the presence of a single solvent system [57]. There are several studies focusing on selective dissolution-precipitation of polymers to delaminate multilayer components, but they are mainly focused on recovering

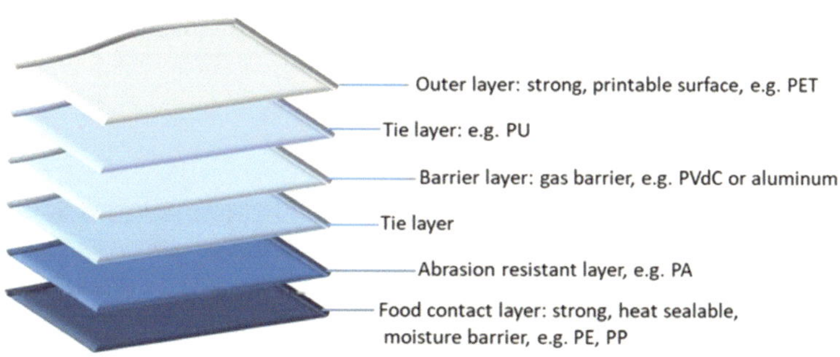

Figure 6.
An example of a multilayer flexible packaging film structure. Adopted from [6].

the polyolefins [58–60]. On the other hand, solvent-targeted recovery and precipitation (STRAP) process was studied to recover all the constituent polymer layers of multilayer plastic packaging [61]. This process would become competitive to design solvent systems for recycling of multilayer packaging. Another option to delaminate multilayer packaging is through selective decomposition of polymer layers [62]. For example, in the study of Kulkarni et al. [63], the aluminum layer was recovered from multilayer packaging structures by depolymerizing PET and PA in the presence of sub and supercritical water. In another study, sulfuric acid was used to degrade PET and recover the PE layer from multilayer structures consisting of PE and PET layers [64]. Although selective degradation of polymer layers are promising in terms of polyolefin recovery, degraded polymers affect the medium recovery adversely. It is shown in a study that during selective PET degradation, energy consumption for the solvent and product recovery contributes to a major part of the greenhouse gas emissions [62]. Multilayer packaging can also be separated through dissolution of tie layers such as among others, polyurethanes (PU), acrylates, acid anhydrides, or others which are used to laminate dissimilar polymer layers. There are several studies focusing on the dissolution of tie layers for separation of polymer-aluminum multilayer packaging by using organic solvent systems [65, 66]. For example, in the patented method of Panagiotis et al. [67] the cured composite laminate material was preconditioned to delaminate composite laminate materials by using organic solvents such as water, benzyl alcohol, acetone, methyl ethyl ketone (MEK), or a combination of one or more thereof. As an alternative to solvents, acids are also used dissolve tie layers towards delamination of a broader range of multilayer structures [6, 68, 69]. For example, in the patented method of Massura et al. [69] polymer, aluminum and/or paper were separated by using protonic carboxylic acids such as acetic acid together with organic solvents to increase the solubility of adhesives. In the study of Ügdüler et al. [6], it has been proven that diffusion rate of formic is faster compared to other longer chain carboxylic acids e.g. hexanoic acid, decanoic acid, thus formic acid was selected as a superior medium to delaminate different types multilayer packagings. Similarly, in various patents inorganic acids such as nitric acid and phosphoric acid are also used for delamination of composite packaging or industrial refuse containing aluminum layers [70, 71].

To date industrial delamination technologies have mainly focused on tetra pack recycling. For instance, in China various companies are performing acid-based delamination. In this approach, recycling of composite packaging waste is carried out in a continuous industrial scale through separation of PE-Al by using formic acid and nitric acid. Afterwards, both materials are recovered by for instance a sink-float separation [72, 73]. In 2015, the German company Saperatec GmbH patented a method for recovery of saleable products from composite waste by using micro-emulsion comprising swelling agents, carboxylic acids, water, and surfactants [74]. Based on this patent, Saperatec has launched a project to build a recycling plant with a capacity of approximately 17000 t/a input [75]. The discussed delamination technologies are summarized in **Table 2**.

2.5 Solvent-based extraction methods

Solvent-based extraction methods can be applied to remove target additives from the polymer matrix without dissolving (e.g. Soxhlet extraction method, ultrasonic extraction method, etc.) or alternatively, (selective) dissolution can be applied to recover a certain polymer (from a mixture), which is typically called the dissolution-precipitation technique. In this chapter, solvent-based extraction methods will be divided into two groups: dissolution-precipitation and solid–liquid extraction methods.

Technology	Principle	Current state	Ref.
Deodorization / deinking	Extraction with ethyl acetate	Patented, industrial scale (RPC-BPI)	[42]
Deinking	Micelle formation through surfactants	Pilot-scale stage (Cadel Deinking Industries, Gamma Meccanica, and Duchenaud Uniflexo)	[49, 50, 76]
Delamination	Acid based delamination using diluted organic acids	Full industrial stage (China)	[72]
Delamination	Acid based delamination using a mixture of swelling agents, carboxylic acids, water, and surfactants	Pilot-scale stage (Saperatec GmbH)	[74]
Delamination	Switchable hydrophilicity solvents to break the chemical and mechanical bonds between different layers	Research stage	[52]

Table 2.
Overview of the state of the art chemical pre-treatment steps, their main principle, and current state.

2.5.1 Dissolution-precipitation technique

The dissolution-precipitation technique allows the removal of contaminants and additives from polymers. The principle of this technique is the dissolution of the polymer in a suitable solvent followed by solid–liquid separation steps for removal of contaminants. Next, the polymer can be recovered by rapid evaporation of the solvent or by adding a proper antisolvent that will make the polymer to precipitate [77]. Drying of the polymer grains and recovery of the solvent and antisolvents are the final steps [78]. Advantages of the dissolution of plastics in a suitable solvents are (i) the decrease in the bulk volume of the plastics, (ii) the precipitated polymer is in a more acceptable form for reuse, (iii) insoluble contaminants can be removed with a solid–liquid separation process and (iv) the final product might be competitive with virgin material in terms of quality [79, 80].

Furthermore, dissolution-based processes can also be used for the separation of mixed plastic wastes based on the selective dissolution (**Figure 7**) [78]. Selective dissolution-precipitation has been applied in a laboratory scale to different plastics, namely PP pipes, PVC bottles, PS waste foam, LDPE film and HDPE bottles from agrochemical packaging [77]. Pappa et al. [77] studied the selective dissolution-precipitation techniques at a laboratory scale and pilot-scale for the two-component mixture LDPE/PP. The solvent/antisolvent used was xylene/i-propanol in a 3:1 ratio and the dissolution was performed at different temperatures in the range of 85 to 135°C, depending on the polymer. The recovery of the two polymers was higher than 99%.

The choice of the solvent (S) and antisolvent (AS) is dependent on the solubility of the polymer and/or polymers. The amount of solvent used for dissolution also plays an important role, as low concentrated solution will lead to low viscous fluids, but will require higher amounts of antisolvent, since normally the added ratio of S/AS is 3:1 [81, 82]. On the other hand, concentrated polymer solutions lead to very high viscous fluids, which are hard to process [82]. Therefore, typical recommended concentrations are in the range of 5–15 wt% [81]. Papaspyrides et al. [82] studied the dissolution-precipitation technique for LDPE pellets using xylene and toluene as solvents at 85°C. Toluene proved to be the most suitable solvent, as it permitted to

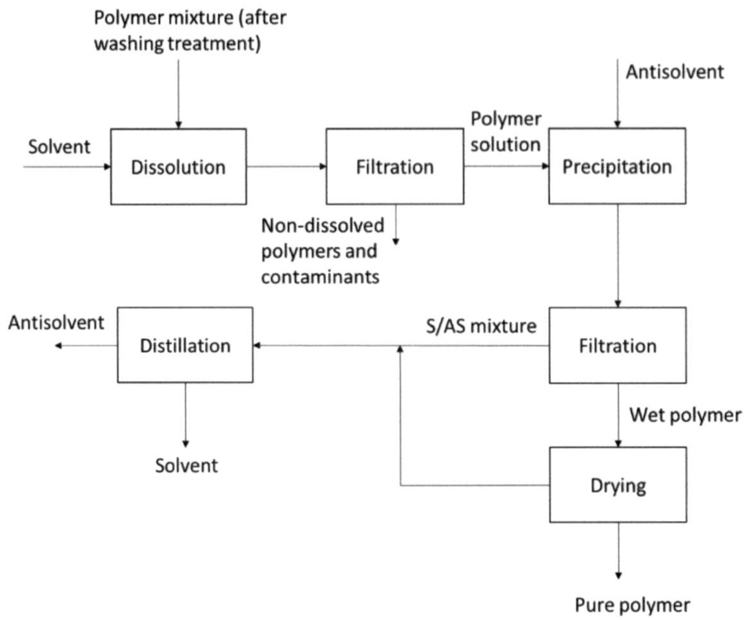

Figure 7.
Example of the selective dissolution-precipitation process. Redrawn from [77].

achieve higher concentrations (0.30 kg/L) and remaining within the limit of viscosity. The choice of the antisolvent is also important since it influences the form of the precipitated polymer. Some antisolvent may lead to gelly polymers while other permit the precipitation in the form of powder or grains [78, 82]. Papaspyrides et al. [82] concluded that acetone as antisolvent was the most successful one for the LDPE-toluene solution as the polymer was precipitated in the form of powder without forming gelling lumps. **Table 3** summarizes solvent/antisolvent systems that have been proposed in literature for different polymer types.

Currently, there are some pilot plants on the market for plastics. The CreaSolv® technology, patented by Fraunhofer Institute IVV, consists of a dissolution-precipitation technique that is able to remove additives, for instance plasticizers, from different polymers, such as polyolefins, PS and PET in scrap packaging, among others [87]. The APK's Newcycling® technology is designed to separate different polymer types such as PE and polyamide (PA) from multilayer plastic waste via a chemical dissolution process [38, 88]. The PureCycle Technologies^{SM} process is designed to remove contaminants and purify PP, [89] and Polystyvert developed a dissolution process for recycling all types of PS [90].

2.5.2 Solid – liquid extraction methods

Next to the conventional solid–liquid extraction (SLE) methods such as the shake-flask and Soxhlet extraction method, [91] alternative techniques like ultrasonic extraction, supercritical fluid extraction (SFE), microwave-assisted extraction (MAE) and accelerated solvent extraction (ASE®) have been developed [91, 92]. Compared to the traditional SLE, these techniques reduce the amount of solvent and shorten the residence time [92]. Further advantages of increasing the temperature and pressure during extraction is the performance enhancement due to the increase of solubility, mass transfer effects and disruption of the surface equilibria [91]. **Table 4** summarizes the advantages and disadvantages of these techniques.

Polymer	S/AS	Concentration	T (°C)	S/AS ratio	Polymer recovery (%)
PET	*n*-Methyl-2-pyrrolidone/*n*-octane [78]	0.20 kg/L	165	1:2	100
	Benzyl alcohol/methanol [83]	5%w/v	180	1:3	99
PP	Xylene/*n*-hexane [83]	5%w/v	140	1:3	98.7
PP	Xylene/acetone [84]	0.15 kg/L	135	1:3	100
LDPE	Xylene/*n*-hexane [83]	5%w/v	100	1:3	98.9
	Toluene/Acetone [82]	0.30 kg/L	85	1:7	~100
HDPE	Toluene/acetone [85]	0.10 kg/L	110	1:4	100
	Xylene/*n*-hexane [83]	5%w/v	100	1:3	98.6
	Xylene/methanol [83]	5%w/v	100	1:3	97
PS	Toluene/*n*-hexane [83]	5%w/v	25	1:3	87.7
	Toluene/*n*-hexane [83]	5%w/v	50	1:3	92.1
	Toluene/*n*-hexane [83]	5%w/v	100	1:3	94.5
	Xylene/methanol [83]	5%w/v	25	1:3	89.2
	Xylene/methanol [83]	5%w/v	50	1:3	95.8
	Xylene/methanol [83]	5%w/v	100	1:3	97.9
	Toluene/Water [80]	0.50 kg/L	75	1:3	~100
PVC	Cyclohexanone/n-hexane [86]	0.30 kg/L	25	1:7	95
	Dichloromethane/methanol [83]	5%w/v	25	1:3	91.1
	Dichloromethane/methanol [83]	5%w/v	40	1:3	98.2
	Toluene/methanol [83]	5%w/v	25	1:3	94.1
	Toluene/methanol [83]	5%w/v	50	1:3	94.6

Table 3.
Solvent/Antisolvent systems proposed in literature for several polymers typically found in plastic packaging waste streams.

2.5.2.1 Conventional SLE: shake-flask extraction and Soxhlet

In a shake-flask extraction the polymer is mixed with the extraction solvent. The solvent will penetrate through the pores of the solid matrix and dissolve the polymer [38]. To speed up the extraction methods, the sample can be heated or refluxed [91]. After the extraction, the insoluble solid matrix can be separated from the solvent-containing analytes via decantation, filtration or centrifugation [38, 91]. The shake-flask method can be performed in batches and multiple extractions can be carried out [91]. The extraction efficiency is dependent on the type of solvent, extraction time, temperature, particle and pore size of the solid compound [38]. For example, Spell and Eddy [93] used the shake-flask technique to extract antioxidants (Ionol and Santonox) and a slip agent (oleamide) from PE at room temperature. For the extraction of these antioxidants with carbon disulphide, a time of 1–3 hours for Ionol and 50–72 h for Santonox at room temperature were necessary to achieve recoveries around 98%. Carbon tetrachloride was used for the extraction of the slip agent. After 21 h at room temperature, recoveries of 93 to 100% were obtained [38, 94]. The shake-flask method (which is basically a batch solid–liquid extraction) is a simple method, but it requires high residence

Method	Principle	Advantage	Disadvantage	Industrial application
Shake-flask extraction [38, 91]	Extraction with solvent	Low investment Simple equipment	Needs high residence time Generally produces the lowest yields compared to the other extraction methods	Not yet proven for plastics, but is commonly applied in other sectors [38, 95].
Soxhlet extraction [38, 91, 97]	Extraction with solvent at high T	Low investment Simple equipment Automation High recovery Excellent reproducibility No reverse diffusion of additives back into polymer	Time and solvent consuming Relative selectivity Requires more energy (compared to shake-flask) Potential for solvent loss Clean-up strategy required Less suitable for thermolabile analytes	Not yet available on the market for plastics, but commonly applied in other sectors, such as agriculture [38, 100].
Ultrasonic extraction [38, 91]	Ultrasonic frequencies	Economical Simple equipment	Labour intensive Clean-up needed No automation	Still a challenge to scale up to industrial scale due to the high plastic volumes that might create engineering problems [38]. Not yet on the market for plastics but applied in other sectors, such as extraction of caffeine. [38, 100, 101]
Microwave assisted extraction [91]	Solvent with microwave absorbing component	Rapid Automation Low solvent use Simultaneous extraction	Modest investment Safety precautions Dielectric solvents Requires optimization	Not yet available on the market for plastics, but commonly applied in other sectors, such as agriculture [38, 104].
Supercritical fluids extraction [91, 97]	Use of supercritical fluids, e.g. $scCO_2$	Reduced usage of organic solvent Shorter extraction time Adjustable solvent strength, Compared to Soxhlet, SCF can easily penetrate into a porous matrix	High investment Not eco-friendly Requires high-purity extractants Requires optimization	Not yet on the market for plastics, but applied to extract bioactive compounds in food industry [38, 104].

Method	Principle	Advantage	Disadvantage	Industrial application
Accelerated solvent extraction [91]	Pressurized fluid extraction	Rapid User friendly Automation Not analyst labour intensive Limited solvent use	Modest investment Pressure needed Requires optimization	Not yet on the market for plastics, but applied in other market,. e.g. food and agriculture industries [38, 104].

Table 4.
Summary of solid–liquid extraction methods.

times [38, 91]. Currently, there are no such industrial plants within the plastic recycling sector, but it has been applied in other sectors [38, 95].

The Soxhlet extraction method has been traditionally performed to extract polymer additives and residues with strong solvents at high temperatures [96]. The principle of the Soxhlet method is the continuous extraction of the solid target compound using a suitable solvent with repeated boiling-condensation cycles [91]. The extraction temperature is limited by the boiling point of the solvent used [97]. One of the disadvantages of this method is the time needed to extract the additives, that sometimes can go higher than 12 h [98]. For example, hindered amine light stabilizers (HALS) (Tinuvin 770 and Hostavin TMN 20) were extracted from a polyolefin with chloroform at 60°C for 16 h and around 96% recovery was obtained [38, 99]. The Soxhlet extraction method is not yet available on the market for plastics, but has been commonly applied in other sectors, such as agriculture [38, 100].

2.5.2.2 Advanced solvent extraction methods: ultrasonic extraction, supercritical fluid extraction (SFE), microwave-assisted extraction (MAE) and accelerated solvent extraction (ASE®)

The principle of ultrasonic extraction is to detach the target compound by agitating and creating cavitation in the solid matrix using ultrasonic frequencies in the range of 20 to 2000 kHz [38]. Ultrasonic extraction is most typically performed using direct sonification by using an ultrasonic bath [38]. The parameters that are expected to influence the performance of ultrasonic extraction are temperature, particle size, solvent type and concentration, among others [38]. Haider and Karlsson [98] used the ultrasonic method to extract a HALS (Chimassorb 944) and two antioxidants (Irganox 1010 and Irgafos 168) from LDPE with chloroform at different temperatures (30 to 60°C) and times (5 to 60 min), obtaining recoveries of around 100% [38, 98]. This technique is not yet on the market for plastics but is applied in other sectors, such as extraction of caffeine [38, 100, 101].

Supercritical fluids extraction enables the extraction of components from the polymer matrix by using a supercritical fluid, i.e. a substance above its critical temperature and pressure [96, 102]. A commonly used fluid is supercritical carbon dioxide (scCO2) [96, 102]. The SFE process usually involves three steps: (i) diffusion of the solute from the core of the polymer to the surface, (ii) transfer of the compounds from the surface to the extraction fluid and (iii) elution of the compounds to the supercritical extractant [97]. The supercritical fluid extraction has several advantages over the Soxhlet extraction, these are (i) the use of less amount of solvent, (ii) shorter extraction time, (iii) adjustable solvent strength and (iv) a wider range of extraction temperatures, as it will not be limited by the boiling point of the solvent [97]. Bermúdez et al. [96] used $scCO_2$ to extract 4,4 – dibutyl azobenzene from polystyrene matrices at 60–65°C and 22.4–24.5 MPa. The extraction yield improved by 41% using the SFE compared to the Soxhlet extraction [96]. Garde et al. [103] extracted phenolic antioxidants from PP using CO2 with hexane or methanol as a modifier at a temperature range of 30 to 60°C for 90 to 180 min and obtained recoveries higher than 80% [38, 103]. The supercritical fluid extraction is not yet on the market for plastics, but has been applied to extract bioactive compounds in food industry [38, 104].

Microwave assisted extraction is based on the principle of using microwave energy in the range of 30 to GHz to 300 MHz to heat the solvent that is in contact with the solid sample [91]. The microwave energy is also used to partition the target compounds from the solid matrix into the solvent [91]. The use of polar solvents, such as water and acetone, is more advised for MAE compared to nonpolar solvents due to the low permittivity and thus, these would not be affected by the microwave energy [38]. Nielson [105] compared the extraction of antioxidants (BHT, Irganox 1010, Irganox 1076) and slip agents (erucamide) from PP, HDPE and LDPE with the Soxhlet and MAE technique. With the Soxhlet method, CYHA: 2-propanol and DCM: 2-propanol solvent mixtures at room temperature for 30 to 60 min were used and recoveries higher than 90% were obtained [38, 105]. With MAE at 48°C for 20 min Nielson [105] obtained recoveries higher than 90% [38, 105]. Costley et al. [106] used the Soxhlet method to extract cyclic trimer and other low molecular weight oligomers from PET with different solvents (xylene, DCM, acetone, water and hexane) at 140°C for 24 h. The recovery rates were around 60% [38, 106]. With MAE, Costley et al. [106] were also able to obtain recoveries around 60% at temperatures between 70 and 140°C for 30 to 120 min. Microwave assisted extraction is not yet available on the market for plastics, but has been commonly applied in other sectors, such as the agriculture sector [38, 104].

Accelerated solvent extraction is a pressurized fluid extraction method [38, 91]. This technique is performed at elevated temperatures, usually between 50 and 200°C and pressure of 6.9 and 13.8 MPa, for a short period of time (5–10 min) and using low amounts of solvents (max 100 mL) [91]. The principle of ASE® is the swelling of the polymer matrix by the solvent, followed by the desorption of the compounds from the solid matrix. Next, the diffusion through the solvent placed inside a particle core takes place and finally the transfer of the bulk of flowing fluid [38]. Garrido-López and Tena [107] extracted HALS and phenolic antioxidants from PE using 2-propanol with THF and/or CYHA swelling solvent at 80–110°C and 10.3 MPa for 2 to 22 min, obtaining recoveries higher than 97% [38, 107]. Vandenburg et al. [108] extracted a phenolic antioxidant from PP using 2-propanol at 150°C and 13.8 MPa for 5 min, obtaining a recovery of around 90% [38, 108]. The accelerated solvent extraction is not yet on the market for plastics, but has been applied in other sectors, e.g. in food and agriculture industries [38, 104].

3. Conclusions

Plastic packaging waste streams are complex, typically containing several contaminants, additives and multilayer materials that complicate the recycling processes. Cleaning of plastic waste through advanced pre-treatment processes is a promising route to tackle the difficulties that are encountered in mechanical and chemical recycling due to contaminants.

Conventional pre-treatment steps that are currently applied in industries, which comprise sorting and water based washing, have shown to be not 100% efficient towards removal of the broad range of heterogeneous substances that are present on a post-consumer plastic waste stream. Therefore, advances in pre-treatments have been proposed in literature, some already applied at pilot or industrial scale, whilst other are not yet on the market for the plastic recycling sector. Techniques such as deodorization and deinking have shown to be suitable for the removal of odors constituents and inks from plastics waste, respectively. Deodorization and deinking pilot plants are already available on the market but there are still some limitations to overcome. Delamination allows the separation of the different layers present in multilayer plastic packaging and currently, industrial delamination technologies have mainly focused on recycling of multilayers containing aluminum. Finally, solvent-based extraction techniques such as the dissolution-precipitation technique have shown to be able to remove target additives from the polymer matrix as well as to selectively recover different polymers. The dissolution-precipitation is already applied on the market for the removal of several contaminants, whilst other solvent based extraction techniques are not yet well established in the plastic recycling industry.

To conclude, there is still the need for further improvement and investigation to industrial implementation of several techniques for the plastic recycling sector. Nonetheless, advanced pre-treatment processes show a great potential towards upcycling and closed-loop recycling of plastics.

Acknowledgements

This research has received funding via the C-PlaNeT (Circular Plastics Network for Training) project from the European Union's Horizon 2020 research and innovation program under the Marie Sklodowska-Curie grant agreement No. 859885. This work has also received support from the European Regional Development Fund (ERDF) via the PSYCHE project (Interreg France-Wallonie-Vlaanderen) with co-financing from the provinces of East-Flanders and West-Flanders.

Conflict of interest

The authors declare no conflict of interest.

Abbreviations

Al	Aluminum
CAPEX	Capital expenditures.
CMC	Critical micelle concentrations.
CTAB	Cetyl trimethylammonium bromide.
DMCHA	N, N-Dimethylcyclohexylamine.
EPR	Extended producer responsibility.
HALS	Hindered amine light stabilizer.
HDPE	High-density polyethylene.
LDPE	Low density polyethylene.
MEK	Methyl ethyl ketone.
MRF	Material recovery facility.
NIR	Near infrared.
OPEX	Operating expenditures.
PA	Polyamide.
PC	Polycarbonate.

PE	Polyethylene.
PEG	Poly(ethylene glycol).
PET	Poly(ethylene terephthalate).
POPs	Persistent organic pollutants.
PP	Polypropylene.
PRO	Producer responsibility organization.
PS	Polystyrene.
PU	Polyurethane.
PVC	Poly(vinyl chloride).
VOC	Volatile organic component.

Author details

Rita Kol[1,2], Martijn Roosen[1], Sibel Ügdüler[1,3], Kevin M. Van Geem[3], Kim Ragaert[4], Dimitris S. Achilias[2] and Steven De Meester[1]*

1 Laboratory for Circular Process Engineering (LCPE), Department of Green Chemistry and Technology, Ghent University, Kortrijk, Belgium

2 Laboratory of Polymer Chemistry and Technology, Department of Chemistry, Aristotle University of Thessaloniki, Thessaloniki, Greece

3 Laboratory for Chemical Technology (LCT), Department of Materials, Textiles and Chemical Engineering, Faculty of Engineering and Architecture, Ghent University, Zwijnaarde, Belgium

4 Department of Materials, Textiles and Chemical Engineering, Faculty of Engineering and Architecture, Center for Polymer and Material Technologies (CPMT), Ghent University, Zwijnaarde, Belgium

*Address all correspondence to: steven.demeester@ugent.be

References

[1] Ragaert K, Delva L, Van Geem K. Mechanical and chemical recycling of solid plastic waste. Waste Manag. 2017;69:24-58.

[2] PlasticsEurope. Plastics – the Facts 2020 [Internet]. 2020. Available from: https://www.plasticseurope.org/en/resources/publications/4312-plastics-facts-2020

[3] European Commission. A European Strategy for Plastics in a Circular Economy [Internet]. 2018. Available from: https://eur-lex.europa.eu/legal-content/EN/TXT/HTML/?uri=CELEX:52018DC0028&from=PT

[4] Roosen M, Mys N, Kusenberg M, Billen P, Dumoulin A, Dewulf J, et al. Detailed Analysis of the Composition of Selected Plastic Packaging Waste Products and Its Implications for Mechanical and Thermochemical Recycling. Environ Sci Technol. 2020;54(20):13282-13293.

[5] Hahladakis JN, Velis CA, Weber R, Iacovidou E, Purnell P. An overview of chemical additives present in plastics: Migration, release, fate and environmental impact during their use, disposal and recycling. J Hazard Mater. 2018;344:179-199.

[6] Ügdüler S, De Somer T, Van Geem KM, Roosen M, Kulawig A, Leineweber R, et al. Towards a Better Understanding of Delamination of Multilayer Flexible Packaging Films by Carboxylic Acids. ChemSusChem. 2021;1-17.

[7] Achilias DS, Andriotis L, Koutsidis IA, Louka DA, Nianias NP, Siafaka P, et al. Recent Advances in the Chemical Recycling of Polymers (PP, PS, LDPE, HDPE, PVC, PC, Nylon, PMMA). In: Material Recycling - Trends and Perspectives. 2012.

[8] Roosen M, De Somer T, Demets R, Ügdüler S, Meesseman V, Van Gorp B, et al. Towards a better understanding of odor removal from post-consumer plastic film waste: A kinetic study on deodorization efficiencies with different washing media. Waste Manag. 2020;120:564-575.

[9] OECD. Extended Producer Responsibility. 2001. 164 p.

[10] Andreasi Bassi S, Boldrin A, Faraca G, Astrup TF. Extended producer responsibility: How to unlock the environmental and economic potential of plastic packaging waste? Resour Conserv Recycl. 2020;162:105030.

[11] Picuno C, Alassali A, Chong ZK, Kuchta K. Flows of post-consumer plastic packaging in Germany: An MFA-aided case study. Resour Conserv Recycl. 2021;169:105515.

[12] Brouwer MT, Alvarado Chacon F, Thoden van Velzen EU. Effect of recycled content and rPET quality on the properties of PET bottles, part III: Modelling of repetitive recycling. Packag Technol Sci. 2020 Sep;33(9):373-383.

[13] Kleinhans K, Hallemans M, Huysveld S, Thomassen G, Ragaert K, Van Geem KM, et al. Development and application of a predictive modelling approach for household packaging waste flows in sorting facilities. Waste Manag. 2021;120:290-302.

[14] Larrain M, Van Passel S, Thomassen G, Van Gorp B, Nhu TT, Huysveld S, et al. Techno-economic assessment of mechanical recycling of challenging post-consumer plastic packaging waste. Resour Conserv Recycl. 2021;170:105607.

[15] Feil A, Pretz T. Chapter 11 - Mechanical recycling of packaging

waste. In: Letcher TMBT-PW and R, editor. Academic Press; 2020. p. 283-319.

[16] Brouwer MT, Thoden van Velzen EU, Augustinus A, Soethoudt H, De Meester S, Ragaert K. Predictive model for the Dutch post-consumer plastic packaging recycling system and implications for the circular economy. Waste Manag. 2018;71:62-85.

[17] Welle F. Twenty years of PET bottle to bottle recycling - An overview. Vol. 55, Resources, Conservation and Recycling. 2011. p. 865-875.

[18] Demets R, Roosen M, Vandermeersch L, Ragaert K, Walgraeve C, De Meester S. Development and application of an analytical method to quantify odour removal in plastic waste recycling processes. Resour Conserv Recycl. 2020;161.

[19] Cabanes A, Valdés FJ, Fullana A. A review on VOCs from recycled plastics. Sustain Mater Technol. 2020;e00179.

[20] Cabanes A, Strangl M, Ortner E, Fullana A, Buettner A. Odorant composition of post-consumer LDPE bags originating from different collection systems. Waste Manag. 2020;104:228-239.

[21] Strangl M, Ortner E, Fell T, Ginzinger T, Buettner A. Odor characterization along the recycling process of post-consumer plastic film fractions. J Clean Prod. 2020;260.

[22] Strangl M, Lok B, Breunig P, Ortner E, Buettner A. The challenge of deodorizing post-consumer polypropylene packaging: Screening of the effect of washing, color-sorting and heat exposure. Resour Conserv Recycl. 2021;164:105143.

[23] Velzen EUT Van, Jansen M. Solvent extraction as additional purification method for post- consumer plastic packaging waste. 2011;

[24] Cabanes A, Fullana A. New methods to remove volatile organic compounds from post-consumer plastic waste. Sci Total Environ. 2021;

[25] Strangl M, Ortner E, Buettner A. Evaluation of the efficiency of odor removal from recycled HDPE using a modified recycling process. Resour Conserv Recycl. 2019;146:89-97.

[26] Hirschfeld S, Wünsch O. Mass transfer during bubble-free polymer devolatilization: a systematic study of surface renewal and mixing effects. Heat Mass Transf. 2020;56(1):25-36.

[27] Alshahrani SM, Morott JT, Alshetaili AS, Tiwari R V, Majumdar S, Repka MA. Influence of degassing on hot-melt extrusion process. Eur J Pharm Sci. 2015;80:43-52.

[28] Coperion GmBH. Integrated devolatilization solutions [Internet]. 2021 [cited 2021 May 16]. Available from: https://www.coperion.com/en/know-how/technology/extrusion-technology/devolatilization-processes

[29] Lok B, Buettner A, Denk P, Ortner E, Fell T. Exploring Odor Minimization in Post-Consumer Plastic Packaging Waste through the Use of Probiotic Bacteria. Vol. 12, Sustainability. 2020.

[30] Bledzki AK, Kessler A, Lieser J. Odour reduction on plastics and its measurement. Polym Test. 1999;18(1):63-71.

[31] Pfaendner R. Improving the Quality of Recycled Materials. Kunststoffe international. 2015;41-44.

[32] PCC Group. Printing Inks [Internet]. 2018 [cited 2021 May 16]. Available from: https://www.products.pcc.eu/en/pcc-group/

[33] Leach RH, Pierce RJ, Hickman EP, Mackenzie MJ, Simth HG. The printing

ink manual. 5th ed. Dordrecht, The Netherlands: Springer; 2007.

[34] Sutter J, Dudler V, Meuwly R. Packaging Materials 8. Printing Inks for Food Packaging Composition and Properties of Printing Inks. 2011.

[35] Pedersen GA, Carlson E, Ekroth S, Kostamo P, Nordström AL, Olafsson G, et al. Food contact materials and articles: Printing Inks. Norden, Nordic Council of Ministers; 2012.

[36] Forrest MJ. Coatings and Inks for Food Contact Materials. Volme 16,. Rapra Review Reports; 2005.

[37] Gürses A, Açıkyıldız M, Güneş K, Gürses MS. Dyes and Pigments: Their Structure and Properties. In 2016.

[38] Ügdüler S, Van Geem KM, Roosen M, Delbeke EIP, De Meester S. Challenges and opportunities of solvent-based additive extraction methods for plastic recycling. Waste Manag. 2020;104:148-182.

[39] Niry Group. Printing Inks [Internet]. 2020 [cited 2021 May 16]. Available from: https://www.niry-group.com/en/

[40] Gabriel DS, Maulana J. Impact of plastic labelling, coloring and printing on material value conservation in the products of secondary recycling. In: Hu JW, editor. Applied Engineering, Materials and Mechanics II. Trans Tech Publications Ltd; 2018. p. 384--389.

[41] Horodytska O, Valdés FJ, Fullana A. Plastic flexible films waste management – A state of art review. Waste Manag. 2018 Jul;77:413-425.

[42] RPC bpi group. Norec [Internet]. 2018 [cited 2021 May 16]. Available from: https://www.bpipackagingsolutions.com/

[43] Narr T, Giefing W. Recycling process treating wastes, plastic mixtures and plastic composites. DE19651571A1, 1996.

[44] Gecol H, Scamehorn JF, Christian SD, Grady BP, Riddell F. Use of surfactants to remove water based inks from plastic films. Colloids Surfaces A Physicochem Eng Asp. 2001;

[45] Gecol H, Scamehorn JF, Christian SD, Grady BP, Riddell FE. Deinking of water-based ink printing from plastic film using nonionic surfactants. J Surfactants Deterg. 2002;5(4):363-374.

[46] Gecol H, Scamehorn JF, Christian SD, Riddell FE. Use of surfactants to remove solvent-based inks from plastic films. Colloid Polym Sci. 2003;281(12):1172-1177.

[47] Domeño C, Aznar M, Nerín C, Isella F, Fedeli M, Bosetti O. Safety by design of printed multilayer materials intended for food packaging. Food Addit Contam - Part A Chem Anal Control Expo Risk Assess. 2017;34(7):1239-1250.

[48] Chotipong A, Scamehorn JF, Rirksomboon T, Chavadej S, Supaphol P. Removal of solvent-based ink from printed surface of high-density polyethylene bottles by alkyltrimethylammonium bromides: Effects of pH, temperature, and salinity. Colloids Surfaces A Physicochem Eng Asp. 2007;297(1-3):163-71.

[49] Cadel Deinking Industries. Cadel Deinking [Internet]. 2020 [cited 2021 May 17]. Available from: http://cadeldeinking.com/en/#_process

[50] Piolat A. Method and apparatus for recycling printed plastic films. EP1419829A1, 2002.

[51] Gamma Meccanica. Gamma Meccanica [Internet]. 2020 [cited 2021 May 17]. Available from: https://www.gamma-meccanica.it/

[52] Mumladze T, Yousef S, Tatariants M, Kriukiene R, Makarevicius V, Lukošiute SI, et al. Sustainable approach to recycling of multilayer flexible packaging using switchable hydrophilicity solvents. Green Chem. 2018;

[53] Ebnesajjad S. Plastic films in food packaging: Materials, technology and applications. In: Ebnesajjad S, editor. Waltham, USA: Elsevier; 2013. p. 78-93.

[54] Schyns ZOG, Shaver MP. Mechanical Recycling of Packaging Plastics: A Review. Macromol Rapid Commun. 2020;1-27.

[55] Kaiser K, Schmid M, Schlummer M. Recycling of polymer-based multilayer packaging: A review. Recycling. 2017;3(1):1.

[56] Samorì C, Cespi D, Blair P, Galletti P, Malferrari D, Passarini F, et al. Application of switchable hydrophilicity solvents for recycling multilayer packaging materials. Green Chem. 2017;19(7):1714-1720.

[57] Lynch CJ, Nauman EB. Polymer recycling by selective dissolution. US; US5278282A, 1994.

[58] Thome RD, Kraus S, Schubert J. Process for treating or working up composite and plastic materials. EU; EP0644230A1, 1993.

[59] Kippenhahn R, Knauf U, Luck T, MÄURER A, Schlummer M, Wolz G. Method for separating and recovering target polymers and their additives from a material containing polymers. EU; EP1311599A1, 2003.

[60] Mäurer A, Schlummer M. Good as new. Recycling plastics from WEEE and packaging wastes. Waste Manag World. 2004;4:33-34.

[61] Walker TW, Frelka N, Shen Z, Chew AK, Banick J, Grey S, et al. Recycling of multilayer plastic packaging materials by solvent-targeted recovery and precipitation. Sci Adv. 2020 Nov;6(47):eaba7599.

[62] Ügdüler S, Van Geem KM, Denolf R, Roosen M, Mys N, Ragaert K, et al. Towards closed-loop recycling of multilayer and coloured PET plastic waste by alkaline hydrolysis. Green Chem. 2020;22(16):5376-5394.

[63] Kulkarni AK, Daneshvarhosseini S, Yoshida H. Effective recovery of pure aluminum from waste composite laminates by sub- and super-critical water. J Supercrit Fluids. 2011;55:992-997.

[64] Patel KM, Patel MM, Vaviya MH. Process for recovering low-density polyethylene from flexible packaging material. WO2015159301A3, 2016.

[65] Gartzen J, HEIL G, Mang T. Process for treating materials adhesively secured together by an adhesive agent. WO1993004116A1, 1993.

[66] Kirk TE. Method and apparatus for separation and recovery of the components from foil-containing laminates. US; US5246116A, 1992.

[67] George PE, Hurley K, Carter E. Methods and Systems for the Recycling of Laminated Materials. US; EP2591900B1, 2011

[68] Xianghui D. Plastic-aluminum separating agent and plastic-aluminum separating method. China; CN100535039C, 2007.

[69] Massura AC, Marçal De Souza EA, Crochemore GB. Process for separation of multilayered films used for packagings. WO2003104315A1, 2002.

[70] Huang C-K, Shao C-H. Separating method for recycling foil-laminated material. US; US20040129372A1, 2002.

[71] Mukhopadhyay A. Process of de-lamination of multi-layer laminated packaging industrial refuse. US; US20040054018A1, 2001.

[72] Zawadiak J. Tetra Pak Recycling – Current Trends and New Developments. Am J Chem Eng. 2017;

[73] Yan D, Peng Z, Liu Y, Li L, Huang Q, Xie M, et al. Optimizing and developing a continuous separation system for the wet process separation of aluminum and polyethylene in aseptic composite packaging waste. Waste Manag. 2015;35:21-28.

[74] Lovis F. Method and apparatus for recycling packaging material. United Kingdom; GB2557682A, 2016.

[75] Saperatec. Industrial pilot plant [Internet]. 2020 [cited 2021 May 16]. Available from: https://www.saperatec.de/en/plant-service.html

[76] Gamma Meccanica. Gamma Meccanica [Internet]. 2020 [cited 2021 May 16]. Available from: https://www.gamma-meccanica.it/en/plastic-recycling/eco-clean/

[77] Pappa G, Boukouvalas C, Giannaris C, Ntaras N, Zografos V, Magoulas K, et al. The selective dissolution/precipitation technique for polymer recycling: A pilot unit application. Resour Conserv Recycl. 2001;34(1):33-44.

[78] Poulakis JG, Papaspyrides CD. Dissolution/reprecipitation: A model process for PET bottle recycling. J Appl Polym Sci. 2001;81(1):91-95.

[79] Kampouris EM, Papaspyrides CD, Lekakou CN. A model recovery process for scrap polystyrene foam by means of solvent systems. Conserv Recycl. 1987;10(4):315-319.

[80] Kampouris EM, Papaspyrides CD, Lekakou CN. A model process for the solvent recycling of polystyrene. Polym Eng Sci. 1988;28(8):534-537.

[81] Notari M, Rivetti F. Use of Dialkyl Carbonates as Solvents for Expanded Polystyrene. Brindisi; USOO7745503B2, 2010.

[82] Papaspyrides CD, Poulakis JG, Varelides PC. A model recycling process for low density polyethylene. Resour Conserv Recycl. 1994;12(3-4):177-184.

[83] Achilias DS, Giannoulis A, Papageorgiou GZ. Recycling of polymers from plastic packaging materials using the dissolution-reprecipitation technique. Polym Bull. 2009;63(3):449-465.

[84] Poulakis JG, Papaspyrides CD. Recycling of polypropylene by the dissolution/reprecipitation technique: I. A model study. Resour Conserv Recycl. 1997;20(1):31-41.

[85] Poulakis JG, Papaspyrides CD. The dissolution/reprecipitation technique applied on high-density polyethylene: I. Model recycling experiments. Adv Polym Technol. 1995;14(3):237-242.

[86] Kampouris EM, Diakoulaki DC, Papaspyrides CD. Solvent recycling of Rigid Poly(Vinyl Chloride) Bottles. J Vin. 1986;8:79-82.

[87] Fraunhofer IVV. The Role of Chemistry in Plastics Recycling. Kunstoffe international. Munich; 2020.

[88] APK AG. Newcycling® - how packaging waste becomes high-quality plastic granules [Internet]. [cited 2021 May 16]. Available from: https://www.apk-ag.de/en/newcycling/process/

[89] PureCycle. PureCycle Technologies [Internet]. [cited 2021 May 16]. Available from: https://purecycletech.com/

[90] Polystyvert. Polystyrene Recycling [Internet]. [cited 2021 May 16].

Available from: http://www.polystyvert.com/en/

[91] Bart JCJ. Additives in Polymers - Industrail Analysis and Applications. Chichester: John Wiley & Sons, Ltd; 2005.

[92] Lou X, Janssen HG, Cramers CA. Parameters Affecting the Accelerated Solvent Extraction of Polymeric Samples. Anal Chem. 1997;69(8):1598-1603.

[93] Spell HL, Eddy RD. Determination of Additives in Polyethylene by Absorption Spectroscopy. Anal Chem. 1960;32(13):1811-1814.

[94] Spell HL, Eddy RD. Determination of Additives in Polyethylene by Absorption Spectroscopy. Anal Chem. 1960;32(13):1811-1814.

[95] Yang Y, Zhang L, Han X, Zhang S, Li B, Peng Y. Determine the operational boundary of a pilot-scale single-stage partial nitritation/anammox system with granular sludge. Water Sci Technol. 2016;73(9):2085-2092.

[96] Bermúdez MD, Carrión-Vilches FJ, Martínez-Nicolás G, Pagán M. Supercritical carbon dioxide extraction of a liquid crystalline additive from polystyrene matrices. J Supercrit Fluids. 2002;23(1):59-63.

[97] Lou X, Janssen H -G, Cramers CA. Investigation of parameters affecting the supercritical fluid extraction of polymer additives from polyethylene. J Microcolumn Sep. 1995;7(4):303-317.

[98] Haider N, Karlsson S. A rapid ultrasonic extraction technique to identify and quantify additives in poly(ethylene). Analyst. 1999;124(5):797-800.

[99] Sevini F, Marcato B. Chromatohraphic determination of some hindered amine light stabilizers in polyolefins. 1983;260:507-512.

[100] Poirot R, Prat L, Gourdon C, Diard C, Autret JM. Fast batch to continuous solid-liquid extraction from plants in continuous industrial extractor. Chem Eng Technol. 2007;30(1):46-51.

[101] Preece KE, Hooshyar N, Krijgsman AJ, Fryer PJ, Zuidam NJ. Pilot-scale ultrasound-assisted extraction of protein from soybean processing materials shows it is not recommended for industrial usage. J Food Eng. 2017;206:1-12.

[102] Cotton NJ, Bartle KD, Clifford AA, Dowle CJ. Rate and extent of supercritical fluid extraction of additives from polypropylene: Diffusion, solubility, and matrix effects. J Appl Polym Sci. 1993;48(9):1607-1619.

[103] Garde JA, Catalá R, Gavara R. Analysis of antioxidants extracted from polypropylene by supercritical fluid extraction. Food Addit Contam. 1998;15(6):701-708.

[104] Turner C. Overview of modern extraction techniques for food and agricultural samples. ACS Symp Ser. 2006;926:3-19.

[105] Nielson RC. Extraction and quantitation of polyolefin additives. J Liq Chromatogr. 1991;14(3):503-519.

[106] Costley CT, Dean JR, Newton I, Carroll J. Extraction of Oligomers from Poly(ethylene terephthalate) by Microwave-assisted Extraction. Anal Commun. 1997;34(3):89-91.

[107] Garrido-López Á, Tena MT. Experimental design approach for the optimisation of pressurised fluid extraction of additives from polyethylene films. J Chromatogr A. 2005;1099(1-2):75-83.

[108] Vandenburg HJ, Clifford AA, Bartle KD, Zhu SA, Carroll J, Newton ID, et al. Factors Affecting High-Pressure Solvent Extraction (Accelerated Solvent Extraction) of Additives from Polymers. Anal Chem. 1998;70(9):1943-1948.

Chapter 3

Current Topics in Plastic Recycling

Maria Anna Charitopoulou, Eleni Alexopoulou, Panagiotis Alexiou and Dimitris S. Achilias

Abstract

This chapter presents an overview of current trends in plastic recycling and focuses on specific topics of interest. Firstly, there are presented all methods used for plastic recycling, along with the advantages and disadvantages of each method. Extra attention is paid to chemical recycling and especially, pyrolysis (thermal and catalytic), which is an environmentally friendly method that results in the formation of value-added products. Emphasis is given on three case studies where there are difficulties as regards the recycling of the plastic part: *polymeric blends*, since the existence of mixed plastic wastes may be challenging for their recycling; *plastics originating in multilayer packaging*, since the multilayer packaging consists of various materials, including plastics, paper, and metals that may be an obstacle for the recycling of the plastic part; and *brominated flame-retarded plastics from waste electric and electronic equipment (WEEE)*, since in this case safe handling is required in order to avoid environmental contamination and a pretreatment step before recycling may be of paramount importance. These three case studies along with the mentioned difficulties and suggestions in order to overcome them are presented here, with the aim of offering insights for future studies on the management of plastic materials.

Keywords: plastics recycling, pyrolysis, polymeric blends, multilayer packaging, WEEE, BFR

1. Introduction

Undoubtedly, plastics play a major role in our everyday life, since plastic parts are used in numerous applications, such as packaging (for instance, food containers), automotive industry, electric and electronic equipment (EEE), etc., due to their unique properties [1]. Some of their most important characteristics that necessitate their use in these applications are lightness, ease of processing, resistance to corrosion, transparency, and others. Nevertheless, their wide use in various applications in combination with the short life span of many plastic products leads to large amounts of end-of-life plastics. Taking all these into account, along with plastic nonbiodegradability, research has focused on exploring environmentally friendly approaches for their safe disposal [2]. Plastic handling involves collection, treatment, and afterward recycling. Unfortunately, finding environmentally friendly approaches for their disposal is no mean feat (**Figure 1**); due to the variation in types of plastics, which are often of unknown composition, the existence of polymer blends, or composites, multilayer structures with other materials apart from polymers, as well as the wide range

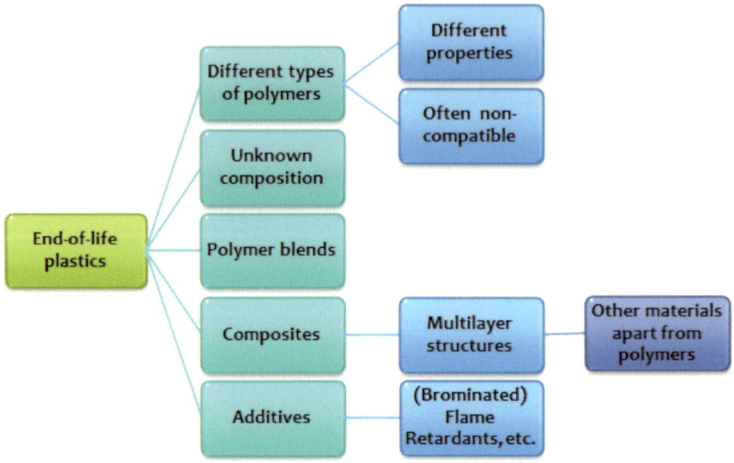

Figure 1.
Difficulties encountered during end-of-life plastic handling.

of additives (such as UV and thermal stabilizers, antistatic agents, (brominated) flame retardants, colorants, plasticizers, etc.) they may contain [3, 4].

The disposal of post-consumer plastics occurs via landfilling, primary recycling, energy recovery, mechanical recycling, and chemical recycling [2]. Although landfilling is an undesirable, non-recycling method, since it results in serious environmental problems, such as soil and groundwater contamination, until now large amounts of end-of-life plastics still end up in landfilling [5, 6]. With a view to eliminating plastic landfilling, research has focused on recycling methods (**Figure 2**) that can be applied, which are primary recycling, recycling without quality losses, energy recovery-quaternary, mechanical or secondary recycling-downcycling into lower qualities and chemical or tertiary recycling-recovery of chemical constituents [7]:

- In primary recycling (re-extrusion), the plastic scrap is reinserted in the heating cycle of the processing line in order to increase the production [8]. It remains a very popular method, because of its simplicity and low cost. However, it can be applied only in case of clean, uncontaminated single-type waste [2].

- Mechanical recycling involves reprocessing and modification of plastic waste using mechanical-physical means with the aim of forming similar, plastic

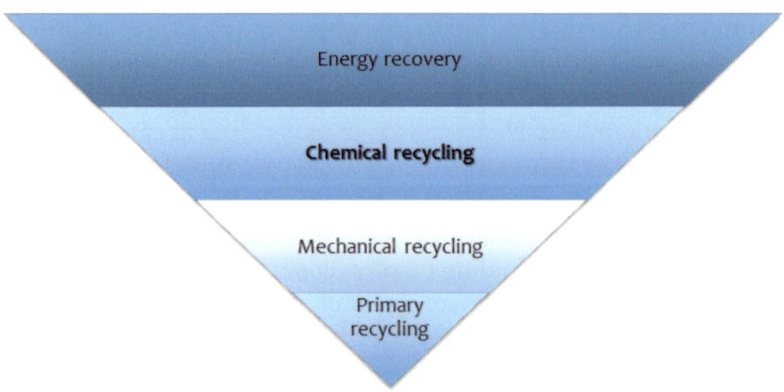

Figure 2.
Recycling methods for post-consumer plastics.

products, at nearly the same or lower performance level when compared with the original products [6]. Since mechanical recycling can be used only in case of homogeneous plastics, heterogeneous plastics require sorting and separation before their recycling. In mechanical recycling, the presence of brominated flame retardant (BFR) incorporated in plastics must be identified before its application, in order to avoid the possible formation of toxic substances, such as polybrominated dibenzo-p-dioxins/furans (PBDD/Fs) [9, 10]. Its main drawback is the fact that product's properties are deteriorated during every cycle [2]; and it should be underlined that each polymer can endure only a limited number of reprocessing cycles [11]. An additional challenge is the existence of mixed plastic waste (polymer blends), since different polymer types have different melting points and processing temperatures. In such cases, the processing temperature is usually set to the highest melting component. Nevertheless, this may result in overheating and possible degradation of the lower melting components and so, in reduced final properties [12].

- In chemical or feedstock recycling, plastic wastes are converted into lower-molecular-weight products, such as: fuels, monomers, or secondary valuable products that can be used as feedstock for refineries. Conversion takes place through chemical reactions in the presence of solvents and reagents [10]. It is an environmentally friendly method, since, as mentioned previously, it results in the formation of valuable products or monomers [9].

- During energy recovery, plastics are incinerated in a boiler or in other industrial equipment, taking advantage of their high energy value; for energy production in the form of heat and electricity. Nevertheless, if incomplete incineration takes place, then toxic substances, such as dioxins, furans, and others, may be formed and released into the atmosphere, resulting in environmental issues [2, 8–10].

In conclusion, during chemical recycling, plastics are converted into smaller molecules (mainly liquids and gases), which can be used for the production of new, valuable products; and that is why it is considered as an environmentally friendly and economically feasible technique. Furthermore, chemical recycling seems to be more advantageous than the other existing methods; taking into account, for instance, the fact that during chemical recycling, both heterogeneous and contaminated polymers can be treated, only with a limited pretreatment. Moreover, the energy consumption of the process is very low, if compared with that of mechanical recycling or energy recovery [6].

Chemical recycling comprises two processes: solvolysis and thermolysis. During solvolysis, polymers are dissolved in a solvent and treated with or without catalysts and initiators. *Solvolysis* can also be applied as a pretreatment before thermochemical processes (such as pyrolysis). During *thermolysis*, polymers are heated in an inert atmosphere (e.g., N_2 atmosphere) in the absence of air or oxygen. It consists of various processes including (thermal and catalytic) pyrolysis, gasification, and hydrogenation (**Figure 3**) [13–14].

Thermal pyrolysis involves polymer cracking in an inert atmosphere (usually nitrogen atmosphere), at high temperatures, and in the absence of catalysts. During this, plastic waste is converted into liquids (pyrolysis oil), gases, and solid residues (chars) [6]. Various temperatures within the range of 300–900°C as well as different heating rates varying from 4 to 25°C/min and different retention times have been investigated in literature in order to find the optimal conditions [15]. When pyrolysis of brominated flame-retarded plastics occurs, the liquid fraction usually

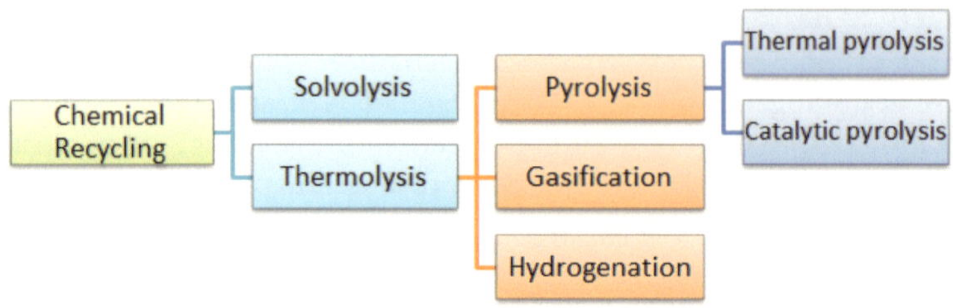

Figure 3.
Chemical recycling routes.

contains many brominated compounds that inhibit their reuse. In such cases, a pretreatment step before or during pyrolysis is of paramount importance, in order to obtain bromine-free products.

Catalytic pyrolysis involves polymer cracking in an inert atmosphere (usually nitrogen atmosphere) and in the presence of catalysts. It offers many advantages if compared with thermal pyrolysis, such as the fact that there are required lower temperatures and shorter reaction times; and so, in this case, less energy is consumed. Furthermore, the selectivity of the products is increased, since catalysts enhance the formation of high commercial value and quality products; and in the meantime, the formation of undesired products (e.g., brominated compounds) can be suppressed [15, 16]. As a consequence, various catalysts have been explored for pyrolysis of various types of plastics, including silica-alumina, zeolites (HZSM-5, etc.), mesoporous catalysts (MCM-41), metal-based catalysts, fluid catalytic cracking (FCC) catalysts, and minerals [9]. Among them, zeolites are the most widely investigated in case of nitrogen-containing polymers such as poly(acrylonitrile-butadiene-styrene) (ABS), since they promote the formation of aromatics [13]; but of course, their properties vary depending on the zeolite type.

Gasification includes partial oxidation or indirect combustion of polymers at high temperatures (up to 1600°C) and in the presence of oxygen. It results in the formation of two main products: CO and H_2 (synthesis gas – syngas). Syngas can be used either in order to run a gas engine or it can be converted into hydrocarbon fuels via the Fischer-Tropsch process. More often than not, it is preferable to gain condensable liquids or petrochemicals as the main products; and that is the reason why pyrolysis is favored over gasification, since the latter requires multiple steps in order to obtain liquid products [13, 14].

Hydrogenation entails the conversion of large hydrocarbon molecules into lower-molecular-weight products. It takes place in hydrogen atmosphere, high pressure (approximately 100 atm), and at moderate temperatures between 150 and 400°C [14].

Generally, it should be underlined that pyrolysis can be considered as one of the best options for plastics recycling, since its advantages are aplenty. Specifically, pyrolysis enables material and energy recovery from polymer waste, as a very small amount of the energy content of waste is consumed for its conversion into valuable hydrocarbons. Furthermore, pyrolysis products are valuable, since they can be used as fuels or chemical feedstock. Last but not least, in case that flame retardants are present in plastic waste, via pyrolysis the formation of toxic substances may be restricted, due to the fact that it takes place in the absence of oxygen [17]. Of course, catalyst's presence, as mentioned previously, plays a vital role. Apart from catalysts, various other parameters, including temperature, heating rate, residence time, operating pressure, etc., can strongly affect the quality and distribution of pyrolysis products [6].

2. Difficulties in plastics recycling

As mentioned previously, many obstacles can be found during the end-of-life plastic recycling. In this unit there are presented in detail three case studies, including: polymeric blends (difficulties due to the coexistence of mixed plastic wastes), plastics originating in multilayer packaging (challenging because of the coexistence of different materials, such as plastics, paper, and metals), and brominated flame-retarded plastics from WEEE (possible formation of undesirable, toxic substances due to the BFR's presence), along with suggestions on how to overcome these difficulties.

2.1 Polymeric blends

Polymer blends are mixtures of two or more polymers in concentration greater than 2%wt. The blends can be miscible or immiscible, a parameter that depends on the thermodynamics of the system and molecular structure, weight, and polymer concentration. More information on the complicated thermodynamics that govern polymer blend miscibility can be found in the Polymer Blends Handbook [18, 19]. Miscible polymer blends are also known as homogeneous blends and are monophasic while immiscible blends with morphologies that differ such as, spheres, cylinders, fibers, or sheets (**Figure 4**) [12].

Subject to polymer compatibility, polymer blends can exhibit synergistic, antagonistic, or additive behavior. A common method used to assuage the immiscibility of polymers blends is the inclusion of compatibilizers—a polymeric surface tension reduction agent that promotes interfacial adherence—in the blend. The three most common types of compatibilizers are reactive functionalized polymers, nonreactive polymers containing polar groups, and block or graft polymers [12, 19, 20].

The difficulty during polymer blend recycling lies in the different properties presented by its component parts such as melting points and processing temperatures between polymers [12]. Most recycling efforts are concentrated on the procedure of pyrolysis to extract energy through the oils, wax, char, and gasses produced. Furthermore, research in recent years has focused on the use of various, different catalysts in order to lower the energy consumption of the whole process and increase the exploitable yield. Along with those some novel methods of polymer blend recycling will be explored.

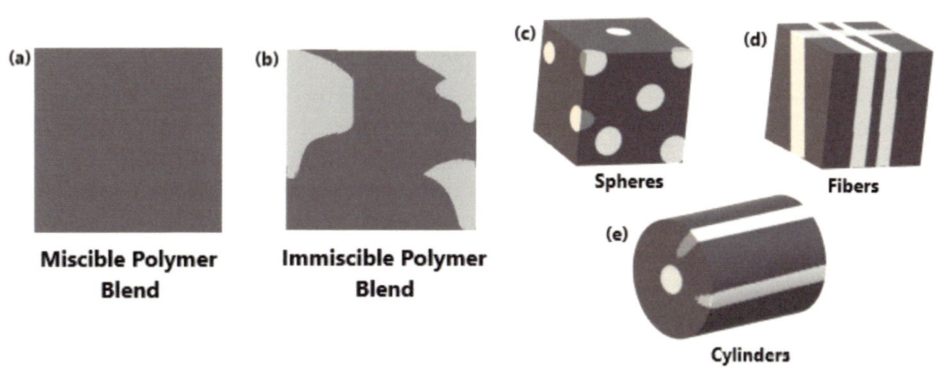

Figure 4.
(a) and (b) are a visual representation of the differences between miscible and immiscible Polymer Blends. Images (c), (d), and (e) show the spherical, fibrous, and cylindrical morphologies of immiscible Polymer Blends, respectively. Image inspired by Ragaert et al. [12].

2.1.1 Production of composite materials

Polymer composites are made up of two or more elements resulting in a multiphase, multicomponent system that exhibits superior properties compared with the constituent materials due to a synergistic effect. It comprises two parts:

- A polymeric matrix that can be either thermoplastic polymers such as polypropylene (PP), polycarbonate (PC), acrylonitrile-butadiene-styrene (ABS) and poly(ethylene terephthalate) (PET) or thermoset polymers such as epoxy, vinyl ester, and polyester.

- A reinforcing filler such as glass, carbon, and aramid [21].

One way that polymer blend can be recycled is by acting as the matrix for secondary elements creating composite materials. In this way it is possible to unite the two components in a form that reinforces the secondary materials and reuses the polymer blends. This method can be adapted to use natural fillers or fibers as the reinforcing fillers. Those can be added along with a coupling agent to optimize the interaction of the fillers with the matrix further and have the positive side effect of making the whole process environmentally friendly. It is important, however, that these fillers have the capacity to be chemically treated.

In a research conducted by Choudory et al., [22], Low-density polyethylene (LDPE)/Linear low-density polyethylene (LLDPE) blend extracted from milk pouches was used as a matrix for coir fibers. The result was composites with properties only slightly lacking from the virgin material ones. In case a maleated styrene pretreatment was applied, the mechanical properties and thermooxidative stability were drastically increased [23].

In another research conducted by Lou et al., [24], PET/PP blend and bamboo charcoal were used to create extruded or injection-molded composite materials. A great increase in mechanical properties was observed in the injection-molded composites, which maintained their mechanical properties even after three rounds of processing. The percentage of total mass of PET in the blend plays a particularly significant role in the product's final behavior [23].

2.1.2 Pyrolysis

Pyrolysis is a promising choice as regards the recycling of polymer blends. With pyrolysis, high levels of conversion of the polymer blend into oil and gas with high calorific values can be attained. These can be used afterward to either fuel the process, or they can be utilized elsewhere [25]. This can be an invaluable asset to the petrochemical industry and a green way for the recycling of plastic waste [26].

Another advantage of pyrolysis is that a sorting process is not needed in contrast to other recycling methods that are extremely susceptible to contamination. This can of course save money and time when recycling polymer blends. Lastly, with the use of the pyrolysis procedure, waste management becomes easier as it is a cheap and environmentally friendly method. In the meanwhile, it allows for minimization of landfill capacity—a serious contemporary difficulty [5]. As the combination of polymers that make up polymer blends is wide, with every blend presenting different properties and pyrolysis behavior, it would be impractical to analyze each one of them. Instead, this chapter will focus on the pyrolysis route taken for the most common polymer blends by examining the research conducted by scientists in the field.

In general, the pyrolysis process can be either thermal or catalytic. In practice, however, the latter is widely preferred by the industry as it demands lower operating temperatures—and thus cost is minimized—that produce a more satisfactory yield of pyrolytic oils, if the correct catalyst has been elected [5].

In a study conducted by Vasile et al., [26], a blend with a composition similar to that originating in municipal waste—24%wt high-density polyethylene (HDPE), 39%wt LDPE, 21.5%wt isotactic polypropylene (IPP), 10%wt PS, 4%wt ABS, and 1.5%wt PET—was investigated. The blend underwent the process of catalytic pyrolysis two separate times each with a different catalyst—HZSM-5 in the first batch and PZSM-5 zeolite catalyst in the second batch, in order to find which catalyst led to better results. It was concluded that the PZSM zeolitic catalyst was characterized by higher selectivity and stability. The optimal temperature for the pyrolysis was found to be 450–480°C, and the gas produced increased sixfold in comparison to the non-catalytic process. Furthermore, the liquid products were found to contain high concentrations of aromatic hydrocarbons. As such, both the liquid and the gas phase can be utilized by the petrochemical industry. Lastly, the pyrolysis oil could be useful as petrochemical feedstock [26].

A novel research conducted by Bober et al. [27] proposed a way to produce hydrogen gas from the catalytic pyrolysis of different consistency HDPE/poly(methyl methacrylate) PMMA polymer blends. After trial and error, the optimal temperature for maximum hydrogen production was found to be 815°C, a temperature where the catalyst used, Ni/Co, operated the best for hydrogen production. It was also found that, the higher the HDPE content in the blend, the bigger the hydrogen output. In contrast, when PMMA was the dominant polymer in the blend, CO was produced at a greater rate than the previous procedure. The research team proposed that the best ratio for HDPE/PMMA in the blend is 4:1 [27].

It must also be noted that concerning the production of hydrogen from pyrolysis of polymer blends, a popular option is the co-pyrolysis of the polymer blends with biomass [28].

A largely untapped potential of Polymer Blends is their recycling as feedstock for the chemical industry. A study presented by Plastics Europe [29], displays that only 2–3% of the collected plastic waste in Europe is utilized as feedstock (**Figure 5**).

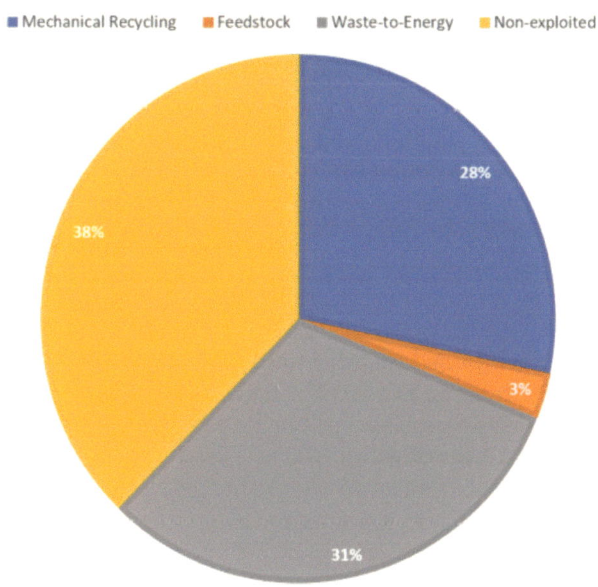

Figure 5.
The fate of the European collected plastic waste. Image inspired by Donaj et al. [30].

A possible procedure for the creation of feedstock from pyrolysis of Polymer Blends on the group of polyolefins was suggested by Donaj et al., [30]. For the purposes of the process, the researchers used a blend of polyolefins—46% LDPE, 30% HDPE, 24% PP- taken from MSW/plastic waste. The collected material was firstly reduced in size to about 3 mm pieces and then pyrolysis ensued under temperatures of 600–700°C in a fluidized bed reactor and with the use of steam and a catalyst if that was deemed feasible as the latter materials increase the yield of olefines. To optimize the procedure, Ziegler-Natta catalyst was used.

The research noted that after the procedure's conclusion, plastic pyrolysis had directly yielded 15–30% gaseous olefins that can then be channeled directly into a polymerization plant. The residue produced consists of a naphtha-like consistency. To be used, this residue must undergo reformation via petrochemical technologies to be upgraded into olefins. Also, as in the previous cases of pyrolysis, the products of the process can be used to fuel the procedure itself. However, work still needs to be done on this field as the process described is not as cost-effective as desired [30].

2.1.3 Melt processing

A last noteworthy method for the utilization of immiscible Polymer Blends is their direct melting processing into fibers with good mechanical properties proposed by Shi et al. [31]. The blend used in this research was PS/PP while fibers were chosen due to two distinct reasons: (a) The fiber spinning technique is known to endow improved properties to polymer blends. (b) Fibers from polymer blends may display new properties in comparison to pure polymers. This method is widely cost-effective for preparing strong fibers for the industry, and it is expected to see great development in the coming years [31].

2.2 Multilayer packaging

In this age of climate change and overall pollution, it has been the priority of policymakers to ensure the viable and sustainable future of human development. An example of this is the EU with the European Plastic Strategy dictating that all packaging used should be reusable or recyclable by 2030 [32].

A prime example of the challenges the industry faces to reach this standard is Tetra Pak, a multilayer packaging used mostly in the food, medicine, chemical, and commodities industry. Tetra Pak most usually consists of three elements: paper cardboard, aluminum, and LDPE.

As stated by the Tetra Pak company, its composition is as follows: (a) 71% paperboard, (b) 24% plastics, and (c) 5% aluminum foil (**Figures 6** and 7).

These three make up the six layers that combined make Tetra Pak. Each layer has a particular use elaborated on below:

However, this is not an absolute rule. For example, certain products with a short shelf life have no need for the protection given by the aluminum layer. On the other hand, when the aforementioned shelf life needs to be extended, the LDPE layers can be substituted by PP providing a chance for further heat treatment of the product. HDPE, PET, and PA are also possible options for replacing the LDPE layers. Lastly, polyurethanes and EMAA are often utilized as adhesives between layers [34] while the Tetra Pak carton may also contain various chemical additives such as plasticizers, stabilizers, lubricants, fillers, foaming agents, colorants, flame retardants, and antistatic agents [35].

As Tetra Pak cartons are composed of mainly paper, the removal and recycling of the carboard layer are of much significance. As such there are two main processing routes: recycling without hydropulping and recycling with hydropulping. The initial

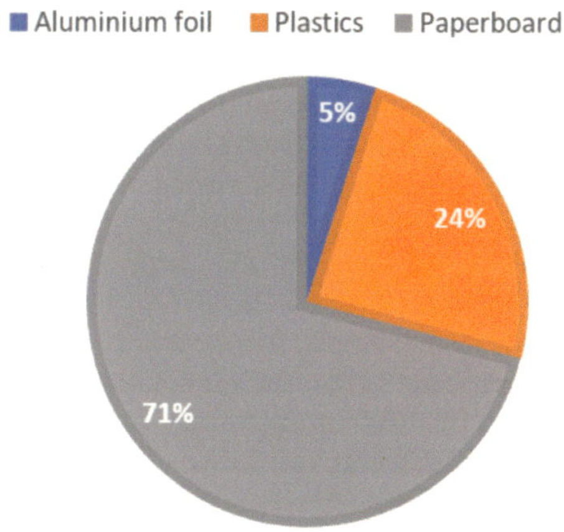

Figure 6.
Raw materials used to produce Tetra Pak. Image inspired by the Tetra Pak site information.

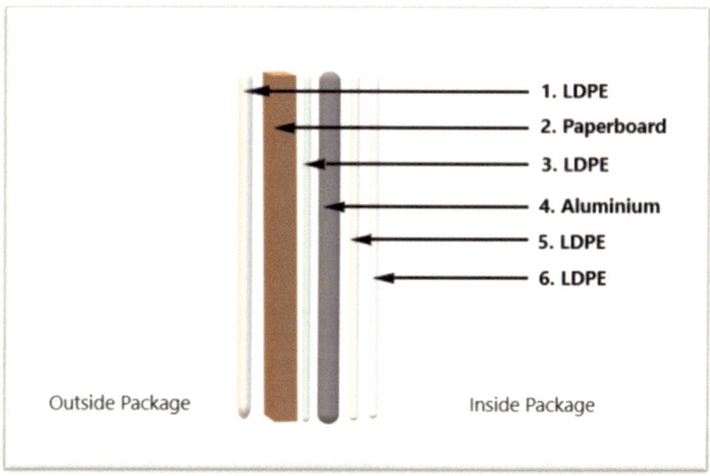

1.	LDPE	Shields the Tetra Pak and the information printed on it from moisture and impacts
2.	Paper Cardboard	The layer upon which the manufacturer can print. It is rigid and is the main source of stability and endurance of the packaging.
3.	LDPE	Acts as an adhesive layer between the paper cardboard and the aluminum layer
4.	Aluminum	Protects product against UV radiation, oxygen, odors and microorganisms
5.	LDPE	Acts as an adhesive layer between the aluminum and the LDPE layer
6.	LDPE	Encloses and isolates the liquid product

Figure 7.
The layers of Tetra Pak. Image inspired by Georgiopoulou et al. [33].

procedure processes the cartons as a whole, while the latter uses the technique of hydropulping to first separate the cellulosic fibers from the Al-LDPE laminate.

2.2.1 Recycling without hydropulping

The main aim of those following this route is energy recovery or downcycling. Energy recovery is attained in combination with solid municipal waste through means of pyrolysis, gasification, or incineration. However, this method comes with many downsides. Paper—the main ingredient of Tetra Pak cartons—has a low heat combustion (16 MJ/Kg), high moisture content, and a significantly high ash value. This makes the entire process inefficient, and thus it is in general not widely used [34].

2.2.2 Recycling with hydropulping

Before proceeding with the options in this category, it would be useful to briefly go over the hydropulping process. When the soon-to-be recycled material first arrives into the recycle unit, the hydropulper breaks apart the paper with rotating blades that use high pressure water and a slurry of fibers is produced. Further processing ensues in centrifugal cleaners that remove heavy materials such as sand, adhesives, staples, etc. [36]. The end result of this procedure is a pulp of cellulosic fibers and can be used as a substitute for wood pulp, in the production of brown paper and pulp board [37]. What remains after the process is the external LDPE layer and the Al-LDPE laminates. However, residual cellulosic fibers can account for up to 5% of the finished products (**Figure 8**).

2.2.3 Pyrolysis

The appeal of this method lies in its simplicity and cost-effectiveness. The pyrolysis procedure has two steps: (a) the degradation of paper (200–400°C) and (b) the devolatilization of LDPE (420–515°C) [38–40]. It should be noted that the temperature plays an important role in the composition of the final products. For example, the production of char is minimized with higher temperatures, and the opposite is true for wax.

The solid products that follow the process are aluminum, char caused by paper degradation, wax from LDPE degradation and tar. A great deal of gaseous products are also formed that mainly consist of CO_2, CO, H_2, CH_4, C2–6 hydrocarbons, and volatile matter. Lastly, there is an aqueous phase consisting of water and phenols.

Many uses have been proposed for those pyrolytic products. The produced gases could be used to sustain the pyrolysis procedure itself or used elsewhere entirely, the char and tar can be exploited as a solid and oil fuel, respectively, while char can also act as a primal resource for the production of carbon-based materials. Lastly, the wax and aqueous phase can readily be utilized as a raw material for the chemical industry [39, 40].

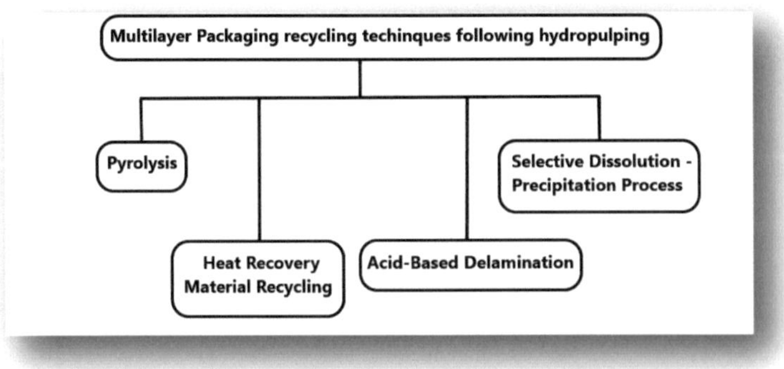

Figure 8.
The main recycling routes.

A novel approach has been taken by researchers in Mexico and Spain who have used the char and the aluminum from the pyrolysis to have them act as absorbents of mercury in aqueous solutions. By means of trial and error and using thermodynamical analyses, they did conclude that char obtained from pyrolysis at 600°C at a 3 h procedure demonstrated the most promising mercury adsorption capacity at 21.0 mg/g. The field of char absorbents is still expanding with hopes of Tetra Pak pyrolysis chars acting as major absorbents for industry in the future [41].

2.2.4 Selective dissolution-precipitation process (SDP)

The basic principle of this approach is the immersion of the Al-PE laminate in a carefully selected solvent and under specific temperature conditions with the aim of the dissolution of the LDPE in the solvent. What follows is the removal by means of filtration of additives and impurities. Lastly an antisolvent is added, and as a result precipitation of the dissolved polymer follows. To maximize LDPE and pure aluminum recovery, the SDP process is repeated three times.

The LDPE produced is of quality that matches that of the virgin product while the aluminum collected is also of high purity. Along with the hydropulping process, this is a very promising option for Tetra Pak recycling. However, the procedure is not without drawbacks: firstly, because of the cost-effective energy consumption needed to separate the solvent-antisolvent mixture and secondly, due to its high environmental impact. The economic viability of this technique rests upon whether the solvent-antisolvent mixture can be separated cheaply (**Figure 9**) [33].

2.2.5 Acid-based delamination

This technique has been developed by researchers in China and focuses on the separation of LDPE and aluminum by means of a separation reagent, mostly aqueous solutions of organic acids or even mixtures of acids. The procedure works by breaking the mechanical bonds holding the laminate together and as such allows for recovery of the products.

The yield of the process is highly dependent on the conditions of the reaction. In the process some of the aluminum is dissolved by the acid—which is also

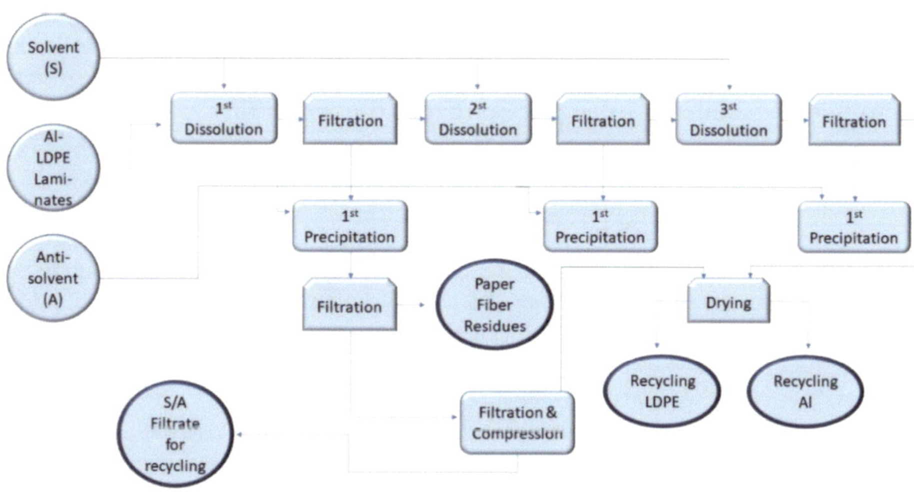

Figure 9.
The SDP process. Image inspired by Georgiopoulou et al. [33].

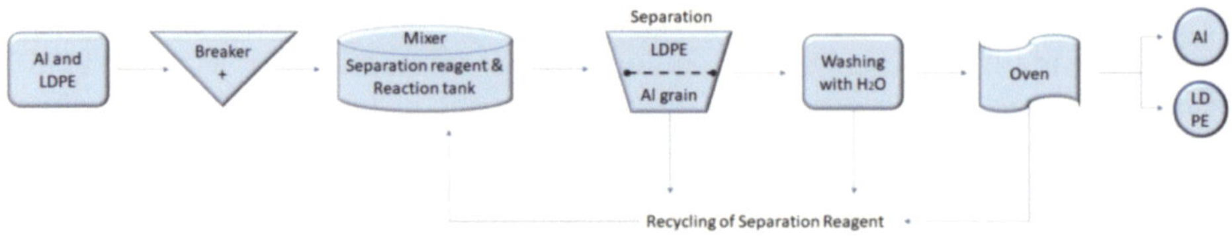

Figure 10.
The acid-based delamination process. Image inspired by Zhang Ji-fei et al. [37].

consumed—and thus losses are to be expected. However, this depends on many factors such as acid used, temperature, etc. Product purity is also correlated with those factors.

After trial and error, it has been found that methanoic acid is the best separation reagent for Tetra Pak. Lastly, there seems to be a high correlation between the separation rate, the temperature the reaction is taking place at, and the concentration of the reagent. More specifically, reaction time decreases with the rise of reagent concentration and temperature (**Figure 10**) [37].

2.2.6 Heat recovery and material recycling

Thanks to the high heating value of the Al-LDPE laminate (40 MJ/Kg), it can be used as a sufficient fuel source. This has taken precedent especially in Europe. Although the laminate can be used directly after the hydropulping process, it is most usually used in conjunction with other fuel sources. This recycling route can be considered environmentally friendly as the LDPE of Tetra Pak burns cleanly without producing fumes containing elements such as sulfur, nitrogen, or halogens.

Also, the Al_2O_3 produced during pyrolysis, by the reaction between Al and moisture in high heat conditions, is in big part exploited by the cement industry, which uses it as a desired component of cement production [36]. Lastly there is the choice of forming finished products directly by using the laminates in roof tile production, injection and rotational molding, and PE-Al agglomeration and pulverization [42, 43].

In these times that society demands a more environmental way of thinking from the industry, recycling of multilayer packaging becomes a priority for many scientists. They have developed a plethora of ways to recycle such packaging, from using it as a fuel to using its pyrolysis products as a mercury absorbent. It is most likely that this field will keep on expanding with ever more innovative and cost-effective ways to fully exploit, reuse, and transform the Tetra Pak multilayer packaging as human development is going into the future.

2.3 Brominated flame-retarded plastics originating in WEEE

The rapid technological advances along with people's need for better living conditions resulted in a global rise in the consumption of EEE over the last years and so in huge amounts of WEEE [44]. Plastics in WEEE account for ~30% of WEEE and in most cases contain BFR that necessitates careful handling [9], since BFR's presence in plastics leads to the formation of various, toxic brominated substances in the liquid fraction obtained after pyrolysis, inhibiting its further use. In such cases a pretreatment step before or during the recycling is necessary in order to receive bromine-free products. So, due to the fact that *brominated plastics from WEEE* are increasing more and more and the BFR enhances the difficulties in their recycling, this unit focuses on pretreatment methods that can be applied either before or during their recycling.

One very common pretreatment method for the removal of BFR applied before pyrolysis is *solvent extraction*. "Traditionally" it takes place using a *soxhlet* extraction apparatus. It is a very popular method until now, due to its low cost and simplicity, although large amounts of solvents and much time are usually required [45]. For instance, Evangelopoulos et al. applied *solvent extraction* as a pretreatment before pyrolysis, via a soxhlet extraction apparatus, with the aim of reducing tetrabromobisphenol A (TBBPA) from real WEEE samples. They tried two different solvents, isopropanol and toluene, due to their different properties; and they found that isopropanol was more efficient in removing bromine from the solid fraction, whereas toluene was more efficient in removing TBBPA from the liquid fraction [46].

Apart from the typical soxhlet extraction, many advanced solvent extraction techniques have been explored over the years, including supercritical fluid extraction (SFE), pressurized liquid extraction (PLE), ultrasonic-assisted extraction (UAE), and microwave-assisted extraction (MAE). These techniques require less time and volumes of solvents than those during soxhlet extraction [47]. Vilaplana et al. applied MAE for the removal of TBBPA and decabromodiphenyl ether (Deca-BDE) from virgin high-impact polystyrene (HIPS) and standard samples from real WEEE. They found that complete extraction of TBBPA took place when they used a combination of polar/nonpolar solvent system (isopropanol/n-hexane) and high extraction temperatures (130°C). On the other hand, in case of Deca-BDE, there were obtained lower extraction yields due to its high molecular weight and its nonpolar nature [47].

In another study [48], UAE and MAE were investigated for the recovery of TBBPA from real WEEE samples that consisted of ABS, polypropylene (PP), polycarbonate (PC), and blends of ABS/PC. From the results obtained it was proved that MAE was more efficient in extracting TBBPA than UAE, especially in case of ABS polymers. The optimal solvent media was isopropanol: n-hexane (1:1), which is a binary mixture of a polar –nonpolar solvent, whereas pure isopropanol, as a solvent, could not result in complete extraction of TBBPA [48].

As mentioned previously, SFE has also attracted a lot of attention as regards the degradation of brominated flame-retarded plastics from WEEE, because of the supercritical fluids' unique properties, such as high density, low viscosity, varied permittivity related to pressure, and high mass transfer, as well as the fact that their viscosity, density, and diffusion coefficient are very sensitive to changes in temperature and pressure. Supercritical fluids appear at temperature and pressure higher than their critical state. Supercritical CO_2 is the most widely used fluid in SFE, since it presents remarkable advantages, including: low critical point, low cost, ease of availability, nontoxicity, recyclability, and simplicity as regards its operation. Water is also, a cheap, nontoxic, and easily available fluid, but it has a relatively high supercritical point [49].

Onwudili and Williams [50] studied supercritical water ($T > 374°C$ and $P > 22.1$ MPa) due to the fact that it presents different characteristics in comparison with organic solvents. They focused on ABS and HIPS, since they are some of the most representative brominated plastics in WEEE and degraded them in supercritical water (up to 450°C and 31 MPa) in a batch reactor. Furthermore, they investigated the effect of alkaline additives, NaOH and $Ca(OH)_2$, by treating the plastics both in the absence and in the presence of them. They noticed that oils, which were the main reaction products, had almost zero bromine and antimony content in the presence of NaOH additive [50]. In another work, [51] there was used subcritical water for the debromination of printed circuit boards (PCB) that contained BFR in a high-pressure batch reactor. They applied three different temperatures, 225, 250, and 275°C, and noticed that debromination increased with increase in temperature. After the debromination of the samples, they applied recycling methods, such as pyrolysis.

Apart from water, organic solvents such as acetone, methanol, and ethanol can also be used as supercritical fluids in chemical recycling of plastics from WEEE [52]. For instance, Wang and Zhang [52] used various supercritical fluids: acetone, methanol, isopropanol, and water with a view to studying the degradation of waste computer housing plastics that contained BFR. They came to the conclusion that supercritical fluid process was efficient for the debromination and decomposition of brominated flame-retarded plastics enabling the recycling of bromine-free oil. As for solvent's efficiency in debromination, the order was the following: water > methanol > isopropanol > acetone.

It should be highlighted here that although SFE technology is considered as a green choice for resource recovery, it has some important drawbacks as well. One of the main obstacles in such technology is the fact that only equipment able to withstand high pressures and temperatures and very resistant to corrosion can be used. These demands, however, increase the cost a lot, and along with the large amount of energy that is required, prevent its industrial implementation [49].

To avoid the latter difficulties, there are other approaches that can be applied in case of flame-retarded plastics. One such approach is that of *two-step pyrolysis.* In this case pyrolysis steps affect the obtained products and by controlling the pyrolysis parameters, the formation of brominated products can be suppressed, without requiring resistant equipment. For instance, according to Ma et al. [53], who applied single- and two-step pyrolysis of waste computer casing plastics, two-step pyrolysis led to the transfer of the biggest part of brominated compounds into the liquid fraction of the first step, in comparison with that of the second step. This observation showed that high-quality oils with low bromine content can be obtained when applying two-step pyrolysis [53].

Co-pyrolysis is another worth mentioning process, in which two or more materials are pyrolyzed together, with the aim of improving the quality and quantity of the liquid fraction, without the need of a pretreatment step prior to pyrolysis. Co-pyrolysis is based on the synergistic effect of different materials that can react together during pyrolysis and leads to a reduction of the total volume of waste, since more waste (e.g., polymers) is consumed as feedstock. The mechanisms of co-pyrolysis and pyrolysis are almost the same, and it is performed at moderate operating temperatures and in the absence of oxygen [54]. Ma et al. [55] examined co-pyrolysis of HIPS, which contained decabromodiphenyl oxide (DDO) as the BFR and antimony trioxide (Sb_2O_3) as a synergist, in the presence of PP (at three different mass ratios) in order to investigate PP's effect on the bromine reduction. From the results obtained it was proved that PP's presence not only increased the yield of various, valuable products, such as toluene, styrene, etc., in the pyrolysis oil, but also led to a reduction of the bromine content [55].

As described above, during co-pyrolysis, the end-of-life brominated plastics along with other (plastic) waste are pyrolyzed together and result in bromine reduction in the derived pyrolysis oil, without any kind of pretreatment before the pyrolysis process. Another idea, in order to reduce bromine while avoiding the extra pretreatment step, is that of the *use of additives or catalysts* during pyrolysis. According to current literature data, many types of *additives*, such as NaOH, $Ca(OH)_2$, CaO, scallop shell, and others, have been investigated for their effect on the reduction of bromine [56, 57]; but of course the degree of debromination depends on the types of the polymers and additives used.

During *catalytic pyrolysis*, as mentioned in the introduction, catalysts influence products' distribution. This has to do not only with favoring the formation of valuable products but also with reducing the formation of the undesirable ones, such as the brominated compounds. Here there are given some representative examples of catalysts that were examined for their debromination effect. In a recent work of Ma et al. [58], there were examined three zeolite catalysts: HY, Hβ, and HZSM-5 along with two mesoporous catalysts: all-silica MCM-41 and active Al_2O_3, for their influence on products distribution. They carried out catalytic pyrolysis of brominated flame-retarded HIPS and observed that catalysts enhanced the formation of valuable, aromatic compounds, such as toluene, styrene, etc., and, in the meantime, enhanced the debromination of the liquid fraction [58].

In another study [59], there was investigated activated Al_2O_3 for catalytic pyrolysis of waste PCB examining three different temperatures: 400, 500, and 600°C, as well as different ratios of PCB: Al_2O_3. They noticed that higher temperatures improved the oil production; and the optimal results as regards the production of light oil and the debromination were obtained at 600°C. The catalyst's presence increased the formation of light hydrocarbons and in the meantime the debromination. Wu et al. [60] carried out catalytic pyrolysis of brominated HIPS that also contained Sb_2O_3, in the presence of red mud, limestone, and natural zeolite, with a view to eliminating bromine and antimony from the pyrolysis oil. They found that in their presence, the total amount of bromine (and antimony) in the oil was reduced. Nevertheless, red mud was the most efficient catalyst in reducing bromine, since Fe_2O_3 present in red mud reacted with HBr that was formed during the degradation of the BFR and hindered the formation of the volatile $SbBr_3$; in the meanwhile, its zeolite property catalytically destroyed the organobromine compounds [60].

Co-pyrolysis can also take place in the presence of catalysts, known as *catalytic co-pyrolysis*. For instance, in [61], they applied catalytic co-pyrolysis of PCB in the presence of (more waste) high-density polyethylene (HDPE) and PP, with a view to reducing the brominated compounds formed. Apart from using other waste polymers as co-feeding for catalytic pyrolysis, there have been reported studies (e.g., [62]) where additives such as $CaCO_3$ and Fe_3O_4 were investigated along with the catalysts for their debromination efficiency in the pyrolysis oil. A two-step process (pyrolysis and catalytic upgrading) can also occur when catalysts are used and enable the conversion of e-waste plastics into high-value materials. In such cases the first step involves the pyrolysis of brominated plastics so as to decompose them; and the second one involves the catalytic upgrading of their products into valuable and bromine-free products. This two-step process is very useful when dealing with real WEEE plastics that contain impurities, etc., that may result in the catalysts' deactivation if direct catalytic pyrolysis occurs. It can be divided into two categories, based on which pyrolysis products are used as raw material; the first category includes pyrolysis vapors as raw material, and the second one includes pyrolysis oil [9].

An example that belongs in the first category is [63], in which they examined a small-scale two-stage pyrolysis and catalytic reforming of brominated flame-retarded HIPS at 500°C using four zeolites: natural zeolite (NZ), iron oxide loaded natural zeolite (Fe-NZ), HY zeolite (YZ), and iron oxide–loaded HY zeolite (Fe-YZ). They observed that the bromine content in the oil was reduced in the presence of catalysts; however, Fe-NZ and Fe-YZ showed better debromination results, due to the reactions between the iron oxide that was loaded and the derived HBr. Compared with Fe-YZ, Fe-NZ did not greatly change the pyrolysis products and so preserved the valuable single-ring aromatic compounds. As a result, Fe-NZ was more effective and feasible for the feedstock recycling of brominated HIPS via the pyrolysis process.

Areeprasert and Khaobang [64] studied pyrolysis and catalytic reforming of a polymer blend (ABS/PC) and PCB, at 500°C, using some conventional catalysts: Y-zeolite (YZ), ZSM-5, iron oxide–loaded Y-zeolite (Fe/YZ), and iron oxide–loaded ZSM-5 (Fe/ZSM-5), as well as some alternative, green catalysts: biochar (BC), electronic waste char (EWC), iron oxide–loaded biochar (Fe/BC), and iron oxide–loaded electronic waste char (Fe/EWC). They found that all catalysts increased the single-ring hydrocarbon products of the liquid fraction. As for the debromination, it was noticed that in case of ABS/PC, the most effective catalyst was Fe/BC, whereas in case of PCB, it was Fe/EWC. Also, they concluded that the green-renewable catalysts could be a promising choice for removing bromine from the liquid fraction [64]. Ma et al. [65] investigated pyrolysis-catalytic upgrading of brominated flame-retarded ABS. The process took place in a two-stage fixed bed reactor; and the second stage included the catalytic upgrading of the vapor intermediates that were obtained from pyrolysis (first stage). The examined catalysts were: HZSM-5 and Fe/ZSM-5. Both catalysts had high catalytic cracking activities that led to an increased yield of oil and to a reduction of the bromine in the liquid fraction.

3. Conclusions

This chapter briefly presents all methods that are used nowadays for plastic recycling, including primary recycling, energy recovery, mechanical recycling, and chemical recycling. The advantages and disadvantages of each method are discussed. Emphasis though is given on chemical recycling and mainly, pyrolysis, due to its many benefits, which are fully described. Furthermore, three case studies that involve some difficulties in plastic recycling are thoroughly investigated. The first one includes the case of polymeric blends, where the coexistence of different plastic materials makes their recycling more difficult. The second one is focused on the recycling of plastics that come from multilayer packaging. The main obstacle in this case lies in the fact that multilayer packaging comprises various, different materials, such as paper and metals, apart from the plastics, so extra attention is required for their separation and recycling. The last case study that is presented here is that of brominated flame-retarded plastics from WEEE, since in such cases direct recycling is not that easy due to the formation of undesirable brominated compounds and more often than not a pretreatment step prior to their recycling is necessary. Taking into account the mentioned difficulties, the aim of this chapter is to present and analyze various recent literature data along with suggestions on how to overcome the mentioned problems.

Acknowledgements

The research work was supported by the Hellenic Foundation for Research and Innovation (HFRI) under the HFRI PhD Fellowship grant (Fellowship Number: 853).

Author details

Maria Anna Charitopoulou, Eleni Alexopoulou, Panagiotis Alexiou and Dimitris S. Achilias*
Laboratory of Polymer and Colors Chemistry and Technology, Department of Chemistry, Aristotle University of Thessaloniki, Thessaloniki, Macedonia, Greece

*Address all correspondence to: axilias@chem.auth.gr

References

[1] Charitopoulou MA, Vouvoudi EC, Achilias DS. Recycling of engineering plastics (PC, HIPS, ABS, PA), their blends and composites. In: Parameswaranpillai J, Sanjay MR, Arpitha GRI, Siengchin HS, editors. Recent Developments in Plastic Recycling. Berlin, Germany: Springer; 2021

[2] Achilias D, Andriotis L, Koutsidis IA, Louka DA, Nianias NP, Siafaka P, et al. Recent advances in the chemical recycling of polymers (PP, PS, LDPE, HDPE, PVC, PC, nylon, PMMA). In: Achilias D, editor. Material Recycling—Trends and Perspectives. London: InTech Open; 2012. DOI: 10.5772/33457

[3] Crawford RJ, Martin PJ. Plastics Engineering. 4th ed. Amsterdam, Netherlands: Butterworth-Heinemann–Elsevier; 2020

[4] Buekens A, Yang J. Recycling of WEEE plastics: A review. Journal of Material Cycles and Waste Management. 2014;**16**:415-434. DOI: 10.1007/s10163-014-0241-2

[5] Anuar Sharuddin SD, Abnisa F, Wan Daud WMA, Aroua MK. A review on pyrolysis of plastic wastes. Energy Conversion and Management. 2016;**115**:308-326. DOI: 10.1016/j.enconman.2016.02.037

[6] Charitopoulou MA, Kalogiannis KG, Lappas AA, Achilias DS. Novel trends in the thermo-chemical recycling of plastics from WEEE containing brominated flame retardants. Environmental Science and Pollution Research. 2021;**28**:59190-59213. DOI: 10.1007/s11356-020-09932-5

[7] Wagner F, Peeters JR, De Keyzer J, Janssens K, Duflou JR, Dewulf W. Towards a more circular economy for WEEE plastics—Part A: Development of innovative recycling strategies. Waste Management. 2019;**100**:269-277. DOI: 10.1016/j.wasman.2019.09.026

[8] Al-Salem SM, Antelava A, Constantinou A, Manos G, Dutta A. A review on thermal and catalytic pyrolysis of plastic solid waste (PSW). Journal of Environmental Management. 2017;**197**:177-198. DOI: 10.1016/j.jenvman.2017.03.084

[9] Ma C, Yu J, Wang B, Song Z, Xiang J, Hu S, et al. Chemical recycling of brominated flame retarded plastics from e-waste for clean fuels production: A review. Renewable and Sustainable Energy Reviews. 2016;**61**:433-450. DOI: 10.1016/j.rser.2016.04.020

[10] Yang X, Sun L, Xiang J, Hu S, Su S. Pyrolysis and dehalogenation of plastics from waste electrical and electronic equipment (WEEE): A review. Waste Management. 2013;**33**:462-473. DOI: 10.1016/j.wasman.2012.07.025

[11] Rahimi A, García JM. Chemical recycling of waste plastics for new materials production. Nature Reviews Chemistry. 2017;**1**(46):1-11. DOI: 10.1038/s41570-017-0046

[12] Ragaert K, Delva L, Van Geem K. Mechanical and chemical recycling of solid plastic waste. Waste Management. 2017;**69**:24-58

[13] Vinu R, Ojha DK, Nair V. Polymer pyrolysis for resource recovery. In: Reedijk J, editor. Reference module in chemistry, molecular sciences and chemical engineering. Amsterdam, Netherlands: Elsevier Inc; 2016. DOI: 10.1016/B978-0-12-409547-2.11641-5

[14] Sahin O, Kirim Y. Material recycling. In: Dincer I, editor. Comprehensive Energy Systems. Amsterdam, Netherlands: Elsevier; 2018. pp. 1018-1042

[15] Miandad R, Barakat MA, Aburiazaiza AS, Rehan M, Nizami AS. Catalytic pyrolysis of plastic waste: A

review. Process Safety and Environment Protection. 2016;**102**:822-838. DOI: 10.1016/j.psep.2016.06.022

[16] Aguado J, Serrano DP, Escola JM. Catalytic upgrading of plastic wastes. In: Scheirs J, Kaminsky W, editors. Feedstock recycling and pyrolysis of waste plastics: Converting waste plastics into diesel and other fuels. New York: Wiley; 2006. pp. 73-110

[17] Liu W-J, Tian K, Jiang H, Yu H-Q. Lab-scale thermal analysis of electronic waste plastics. Journal of Hazardous Materials. 2016;**310**:217-225

[18] Manias E, Utracki LA. Thermodynamics of Polymer Blends. In: Utracki LA, Wilkie C, editors. Polymer Blends Handbook. Berlin, Germany, Springer; 2014. pp. 171-289. DOI: 10.1007/978-94-007-6064-6_4

[19] Utracki LA. Polymer Blends Handbook. Dordrecht, Netherlands: Kluwer Academic Publishers; 2002. DOI: 10.1016/j.radphyschem.2004.12.001

[20] Koning C, van Duin M, Pagnoulle C, Jerome R. Strategies for compatibilization of polymer blends. Progress in Polymer Science. 1998;**23**:707-757. DOI: 10.1016/S0079-6700(97)00054-3

[21] Francis R. Recycling of Polymers Methods, Characterization and Applications. Vol. 12. Weinheim, Germany: Wiley-VCH Verlag GmbH and Co; 2017. pp. 209-219

[22] Choudhury A, Kumar S, Adhikari B. Recycled milk pouch and virgin low-density polyethylene/linear low-density polyethylene based coir fiber composites. Journal of Applied Polymer Science. 2007;**106**:775-785. DOI: 10.1002/app.26522

[23] La Mantia FP, Scaffaro R. Recycling Polymer Blends. Berlin, Germany: Springer Science+Business Media Dordrecht; 2014. pp. 1886-1910. DOI: 10.1007/978-94-007-6064-6_23

[24] Lou C-W, Lin C-W, Lei C-H, Su K-H, Hsu C-H, Liu Z-H, et al. PET/PP blend with bamboo charcoal to produce functional composites. Journal of Materials Processing Technology. 2007;**192-193**:428-433. DOI: 10.1016/j.jmatprotec.2007.04.018

[25] Kaminsky W, Schlesselmann B, Simon CM. Thermal Degradation of Mixed Plastic Waste to Aromatics and Gas. Amsterdam, Netherlands: Elsevier; 1996. pp. 189-197. DOI: 10.1016/0141-3910(96)00087-0

[26] Vasile HP, Mihai B, Petru Onu H, Darie SC. Thermal and catalytic decomposition of mixed plastics Cornelia. Journal of Analytical and Applied Pyrolysis. 2001;**57**:287-303

[27] Bober P, Orinák A, Orinaková R, Zamostny P, Ladomersky J, Fedorková A. Hydrogen production by catalysed pyrolysis of polymer blends. Fuel. 2011;**90**(6):2334-2339. DOI: 10.1016/j.fuel.2011.02.001

[28] Sharypov VI, Marin N, Beregovtsova NG, Baryshnikov SV, Kuznetsov BN, Cebolla VL, et al. Co-pyrolysis of wood biomass and synthetic polymer mixtures. Part I: Influence of experimental conditions on the evolution of solids, liquids and gases. Journal of Analytical and Applied Pyrolysis. 2002;**64**(1):15-28. DOI: 10.1016/S0165-2370(01)00167-X

[29] Plasticeurope. Plastics—The Facts, 2010, An analysis of European plastics production, demand, and recovery for 2009. 2010. Available from: plasticeurope.org

[30] Donaj PJ, Kaminsky W, Buzeto F, Yang W. Pyrolysis of polyolefins for increasing the yield of monomers' recovery. Amsterdam, Netherlands:

Elsevier; 2012. DOI: 10.1016/j.wasman.2011.10.009

[31] Shi J, Fang X, Maffe AP, Yao D. An Effective Method of Processing Immiscible Polymer Blends into Strong Fiber. New York: Wiley; 2019. DOI: 10.1002/pen.25205

[32] European Comission. A European Strategy for Plastics in a Circular Economy; 2018

[33] Georgiopoulou I, Pappa GD, Vouyiouka SN, Magoulas K. Recycling of post-consumer multilayer Tetra-Pak® packaging with the selective dissolution-precipitation process. Resources, Conservation and Recycling. 2021;**165**:105268. DOI: 10.1016/j.resconrec.2020.105268

[34] Zawadiak J, Wojciechowski S, Piotrowski T, Krypa A. Tetra Pak recycling–Current trends and new developments. American Journal of Chemical Engineering. 2017;**5**(3):37-42. DOI: 10.11648/j.ajche.20170503.12

[35] Groh KJ, Backhaus T, Carney-Almroth B, Geueke B, Inostroza PA, Lennquist A, et al. Overview of known plastic packaging-associated chemicals and their hazards. Science of the Total Environment. 2019;**651**:3253-3268. DOI: 10.1016/j.scitotenv.2018.10.015

[36] Selke SEM. Recycling of paper products. Reference Module in Materials Science and Materials Engineering. Elsevier; 2016. DOI: 10.1016/B978-0-12-803581-8.03651-1.

[37] Ji-fei ZHANG, Da-hai YAN, Zhong-he LI. The recycling of the Tetra-Pak packages research on the wet process separation conditions of aluminum and polythene in the Tetra-Pak packages. 3rd International Conference on Bioinformatics and Biomedical Engineerin IEEE Xplore. 2009;1-6. DOI: 10.1109/ICBBE.2009.5163649

[38] Wu CH, Chang HS. Pyrolysis of Tetra pack in municipal solid waste. Journal of Chemical Technology & Biotechnology. 2001;**76**:779-792. DOI: 10.1002/jctb.404

[39] Korkmaz A, Yanik J, Brebu M, Vasile C. Pyrolysis of the Tetra Pak. Waste Management. 2009;**29**:2836-2841. DOI: 10.1016/j.wasman.2009.07.008

[40] Haydary J, Susa D, Dudáš J. Pyrolysis of aseptic packages (Tetra Pak) in a laboratory screw type reactor and secondary thermal/catalytic tar decomposition. Waste Management. 2013;**33**(5):1136-1141. DOI: 10.1016/j.wasman.2013.01.031

[41] Zúniga-Muro NM, Bonilla-Petriciolet A, Mendoza-Castillo DI, Duran-Valle CJ, Silvestre-Albero J, Reynel-Avila HE, et al. Recycling of Tetra Pak wastes via pyrolysis: Characterization of solid products and application of the resulting char in the adsorption of mercury from water. Journal of Cleaner Production. 2021;**291**:125219. DOI: 10.1016/j.jclepro.2020.125219

[42] Lopes C, Carmo Gonçalves M, Felisberti M. Blends of poly (ethylene terephthalate) and low density polyethylene containing aluminium: A material obtained from packaging recycling. Journal of Applied Polymer Science. 2007;**106**(4):2524-2535. DOI: 10.1002/app.26769

[43] Hidalgo M. Manufacturing Rigid Board by Packaging Waste Containing Aluminum and Polyethylene. New Delhi: NISCAIR-CSIR; 2011

[44] Parajuly K, Habib K, Liu G. Waste electrical and electronic equipment (WEEE) in Denmark: Flows, quantities and management. Resources, Conservation and Recycling. 2017;**123**:85-92. DOI: 10.1016/j.resconrec.2016.08.004

[45] Covaci A, Voorspoels S, de Boer J. Determination of brominated flame retardants, with emphasis on polybrominated diphenyl ethers (PBDEs) in environmental and human samples—A review. Environment International. 2003;**29**:735-756. DOI: 10.1016/S0160-4120(03)00114-4

[46] Evangelopoulos P, Arato S, Persson H, Kantarelis E, Yang W. Reduction of brominated flame retardants (BFRs) in plastics from waste electrical and electronic equipment (WEEE) by solvent extraction and the influence on their thermal decomposition. Waste Management. 2019;**94**:165-171. DOI: 10.1016/j.wasman.2018.06.018

[47] Vilaplana F, Ribes-Greus A, Karlsson S. Microwave-assisted extraction for qualitative and quantitative determination of brominated flame retardants in styrenic plastic fractions from waste electrical and electronic equipment (WEEE). Talanta. 2009;**78**:33-39. DOI: 10.1016/j.talanta.2008.10.038

[48] Kousaiti A, Hahladakis JN, Savvilotidou V, Pivnenko K, Tyrovola K, Xekoukoulotakis N, et al. Assessment of tetrabromo-bisphenol-A (TBBPA) content in plastic waste recovered from WEEE. Journal of Hazardous Materials. 2020;**390**:121641. DOI: 10.1016/j.jhazmat.2019.121641

[49] Li K, Xu Z. A review of current progress of supercritical fluid technologies fore-waste treatment. Journal of Cleaner Production. 2019;**227**:794-809. DOI: 10.1016/j.jclepro.2019.04.104

[50] Onwudili JA, Williams PT. Degradation of brominated flame-retarded plastics (Br-ABS and Br-HIPS) in supercritical water. Journal of Supercritical Fluids. 2009;**49**:356-368. DOI: 10.1016/j.supflu.2009.03.006

[51] Soler A, Conesa JA, Ortuno N. Emissions of brominated compounds and polycyclic aromatic hydrocarbons during pyrolysis of E-waste debrominated in subcritical water. Chemosphere. 2017;**186**:167-176. DOI: 10.1016/j.chemosphere.2017.07.146

[52] Wang Y, Zhang FS. Degradation of brominated flame retardant in computer housing plastic by supercritical fluids. Journal of Hazardous Materials. 2012;**205-206**:156-163. DOI: 10.1016/j.jhazmat.2011.12.055

[53] Ma C, Yan Q, Yu J, Chen T, Wang D, Liu S, et al. The behavior of heteroatom compounds during the pyrolysis of waste computer casing plastic under various heating conditions. Journal of Cleaner Production. 2019;**219**:461-470. DOI: 10.1016/j.jclepro.2019.02.100

[54] Abnisa F, Wan Daud WMA. A review on co-pyrolysis of biomass: An optional technique to obtain a high-grade pyrolysis oil. Energy Conversion and Management. 2014;**87**:71-85. DOI: 10.1016/j.enconman.2014.07.007

[55] Ma C, Sun L, Jin L, Zhou C, Xiang J, Hu S, et al. Effect of polypropylene on the pyrolysis of flame retarded high impact polystyrene. Fuel Processing Technology. 2015;**135**:150-156. DOI: 10.1016/j.fuproc.2014.12.01

[56] Jung S-H, Kim S-J, Kim J-S. Fast pyrolysis of a waste fraction of high impact polystyrene (HIPS) containing brominated flame retardants in a fluidized bed reactor: The effects of various Ca-based additives (CaO, $Ca(OH)_2$ and oyster shells) on the removal of bromine. Fuel. 2012;**95**:514-520. DOI: 10.1016/j.fuel.2011.11.048

[57] Hlaing ZZ, Wajima T, Uchiyama S, Nakagome H. Reduction of bromine compounds in the pyrolysis oil of computer casing plastics using shell, $Ca(OH)_2$ and NaOH. APCBEE Procedia. 2014;**10**:193-197. DOI: 10.1016/j.apcbee.2014.10.037

[58] Ma C, Yu J, Yan Q, Song Z, Wang K, Wang B, et al. Pyrolysis-catalytic upgrading of brominated high impact polystyrene over Fe and Ni modified catalysts: Influence of HZSM-5 and MCM-41 catalysts. Polymer Degradation and Stability. 2017;**146**:1-12. DOI: 10.1016/j.polymdegradstab.2017.09.005

[59] Wang Y, Sun S, Yang F, Li S, Wu J, Liu J, et al. The effects of activated Al_2O_3on the recycling of light oil from the catalytic pyrolysis of waste printed circuit boards. Process Safety and Environment Protection. 2015;**98**:276-284. DOI: 10.1016/j.psep.2015.07.007

[60] Wu H, Shen Y, Harada N, An Q, Yoshikawa K. Production of pyrolysis oil with low bromine and antimony contents from plastic material containing brominated flame retardants and antimony trioxide. Energy and Environmental Research. 2014;**4**(3):105-118. DOI: 10.5539/eer.v4n3p105

[61] Kim Y-M, Han TU, Kim S, Jae J, Jeon J-K, Jung S-C, et al. Catalytic co-pyrolysis of epoxy-printed circuit board and plastics over HZSM-5 and HY. Journal of Cleaner Production. 2017;**168**:366-374. DOI: 10.1016/j.jclepro.2017.08.224

[62] Ye Z, Yang F, Qiu Y, Chen N, Lin W, Sun S. The dibrominated and lightweight oil generated from two stage pyrolysis of WPCBs by using compound chemical additives. Process Safety and Environment Protection. 2018;**116**:654-662. DOI: 10.1016/j.psep.2018.03.025

[63] Wu H, Shen Y, Ma D, An Q, Harada N, Yoshikawa K. Fuel oil production from two-stage pyrolysis-catalytic reforming of brominated high impact polystyrene using zeolite and iron oxide loaded zeolite catalysts. Open Journal of Ecology. 2015;**5**:136-146. DOI: 10.4236/oje.2015.54012

[64] Areeprasert C, Khaobang C. Pyrolysis and catalytic reforming of ABS/PC and PCB using biochar and e-waste char as alternative green catalysts for oil and metal recovery. Fuel Processing Technology. 2018;**182**:26-36. DOI: 10.1016/j.fuproc.2018.10.006

[65] Ma C, Yu J, Chen T, Yan Q, Song Z, Wang B, et al. Influence of Fe based ZSM-5 catalysts on the vapor intermediates from the pyrolysis of brominated acrylonitrile-butadiene-styrene copolymer (Br-ABS). Fuel. 2018;**230**:390-396. DOI: 10.1016/j.fuel.2018.05.077

Chapter 4

Chemical Recycling of Polyolefins (PE, PP): Modern Technologies and Products

Daria Frączak

Abstract

Chemical recycling is one of the most intensively developed potential solutions for the global plastic waste issue. This broadly defined term covers several different technologies that lead to many diverse products. Polyolefins (polyethylene and polypropylene) can be chemically recycled by pyrolysis (cracking) or gasification. These polymers' chemical composition and structure make them a great potential source of valuable hydrocarbons or carbon atoms for syngas production. Thermal and catalytic cracking of polyethylene and polypropylene can be optimised to maximise specific types of hydrocarbons that, after optional additional processing, such as hydrotreatment, steam cracking or distillation, can be used as intermediates in petrochemical plants, fuels or fuel components, monomers for polymerisation of new, virgin polymers or as specialty chemicals (final market products). Gasification of plastic waste transforms polymers into a mixture of hydrogen, carbon monoxide and carbon dioxide, which can be further used as a source of these gasses, transformed into chemicals and fuels, or used directly to produce energy. This chapter presents all of these process paths with examples of existing technologies and their level of technology readiness and perspectives for scale-up.

Keywords: polyolefins, polyethylene, polypropylene, chemical recycling, pyrolysis, thermal cracking, catalytic cracking, gasification

1. Introduction

Reduce, reuse, recycle – this is the current strategy to prevent the world from being flooded by plastic waste. There are many initiatives led by governments, regulators, and also companies and entrepreneurs to follow these guidelines: the so-called "Single Use Plastics" Directive (SUP) [1]. The Circular Plastics Alliance, and The Alliance to End Plastic Waste – to name a few. Incineration and energy recovery of plastics seem to no longer be the promoted solution due to increasing carbon dioxide emissions and causing the loss of the potential hidden within polymers. Although landfilling of plastics practically does not emit CO_2, it can lead to microplastic release into the environment [2].

Chemical recycling is now more recognised as a potential solution to recycling or ending the life of plastic, as it can potentially treat the waste that cannot be mechanically recycled, keeping "carbon" in the industry loop. But there is no one official understanding of the term "chemical recycling" worldwide. European regulation defines recycling only at a very general level as any operation that takes waste and makes products, materials, and substances, except fuel components [3]. ISO 15270 is even more precise, limiting chemical (feedstock) recycling to cracking, gasification and depolymerisation processes to convert plastic waste into monomer of new raw materials, excluding energy recovery and incineration [4]. On the other hand, American Chemistry Council (ACC) defines chemical (advanced) recycling as any technology that converts post-use plastics into monomers, specialty polymers, feedstocks and fuels [5]. What is more, the process of waste to value-added chemicals is also known as upcycling [6].

Recycling itself, both mechanical or chemical, can produce two types of products. Suppose the properties of the recycled material are not considerably different from those of the virgin material and can be used in the same application. In that case, the recycling process is called "closed-loop recycling". This approach is difficult for the mechanical recycling of some polymers for specific applications (like food packaging or specialty applications) as, in many cases, the processing of plastic waste causes partial degradation of the polymer structure and a change to its mechanical properties. When recycled material has different properties and is used in different applications to the original one, the recycling process is called "open-loop".

In the current chapter, both types of chemical recycling technologies of polyolefins (polyethylene and polypropylene) – closed-loop and open-loop, will be described.

2. Structure of polyolefins

Polyethylene (PE) and polypropylene (PP) are polymers called polyolefins. These are the two largest plastic resins based on production worldwide. One of the reasons for this is the variety of applications where these plastics are used: mainly in packaging but also in toys, piping, cable covers, automotive parts, ship ropes or even bulletproof vests. Polyolefins are inert, have low thermal conductivity (are good insulators) and are not subject to attack by most chemicals. As thermoplastic resins, most polyolefins can be mechanically recycled; however, thanks to the chemical and hydrocarbon structure, they are also proper materials for chemical recycling via cracking (pyrolysis) and gasification.

Polyethylene is a product of ethylene polymerisation. Depending on the production process and consequent chemical structure and properties, many types of PE are produced. The major ones are:

- low-density polyethylene (LDPE),
- high-density polyethylene (HDPE),
- linear low-density polyethylene (LLDPE),
- crosslinked polyethylene (PEX).

LDPE is a type of polyethylene with the most branched structure and a density lower than that of HDPE. HDPE is a polymer with a linear structure with a low degree of branching. LLDPE is produced by the polymerisation of ethylene with other olefins with longer hydrocarbon chains, like 1-butene, 1-hexene or 1-octene. The result, a linear backbone with short and uniform branches. Finally, PEX is a polymer obtained by crosslinking the process of polyethylene. This crosslinking

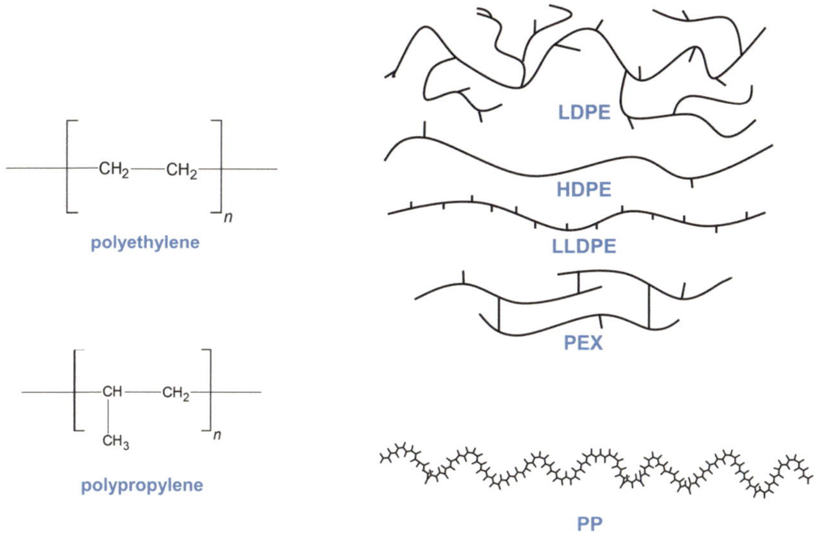

Figure 1.
Structures of polyethylene and polypropylene.

changes the properties of polyethylene significantly – increased temperature, pressure, or corrosion resistance. Thermoplastic polyethylene becomes thermoset which limits the possibility of mechanical recycling of this material significantly.

Polypropylene is a product of the polymerisation of propylene. It means that every monomer has a methyl side group. Due to the presence of asymmetric carbon atoms in the chain, stereoisomerism is observed for this polymer.

Differences in the structures are presented in **Figure 1**. These differences influence the polymer's mechanical properties and the cracking or gasification process conditions, and the composition of products from chemical recycling processes.

It should be noted that many different additives are used to change the properties of the polymer during the production of plastic items. For example, pigments and dyes are added to change the colour; glass fibres can be added to alter the mechanical properties of the polymer, and talk is sometimes added to reduce the price of the final goods. Stabilisers (like UV stabilisers), flame retardants, lubricants and plasticisers are other types of additives commonly used in the plastic industry. The quantity of these additives may be vast, from parts of a percent up to 60–70%. What is more, in many cases, polyolefins are used together with other polymers. For example, in multilayer films used for packaging, PE is used together with poly(ethylene terephthalate) (PET). What is more, flexible packaging is often highly printed. The final yield and composition of chemical recycling products will depend on the type and quantity of all of the impurities.

3. Cracking (pyrolysis)

The cracking of polymers is a process where long polymer chains are broken into products with shorter chains. During heating (at around 350–600°C), molecules start to vibrate until the vibrations are intense enough to overcome van der Waals forces. The short-chain molecules then evaporate. When the energy of the van der Waals force is higher than the enthalpy between carbon atoms in the molecule, the bond in the molecule will break instead of evaporating. In polyolefins' chains, the most unstable bonds will be the first to break, leading to the creation of radicals.

The dissociation energy needed to break the bond between carbon atoms lowers with the order of carbon classification: primary > secondary > tertiary, which means that the first place in a polymer molecule where the bond will break is at the branching. This already indicates that polypropylene cracks at lower temperatures than polyethylene because all carbons in the structure (apart from terminal ones) are tertiary. It also means that LDPE cracks easier than HDPE because of higher branching. All types of polyethylene and polypropylene can be cracked, including those that are challenging for mechanical recycling, like PEX.

3.1 Mechanism of cracking reaction

In general, the following steps of the cracking reaction can be identified:

- initiation – free-radical creation,
- propagation – products and secondary radicals creation, isomerisation,
- termination – recombination, disproportionation, cyclisation, arenes creation, polycondensation.

When the bond between hydrogen and carbon atom is broken, a radical is created (initiation step). In the second step of the process, polymer bond breaks at the β position, creating one unsaturated molecule and another radical. Depending on the type of polymer, this cracking can follow one of these types of reactions:

- unzipping – the chain break at the β position of the molecule (β-scission) with monomer's creation;
- backbiting – intramolecular transfer of a hydrogen atom from one part of the molecule to the other part of the same molecule and then β-scission in the middle part of the molecule with the creation of specific molecules with shorter chains;
- random scission – intermolecular transfer of hydrogen atom and then β-scission in the middle part of the molecule with the creation of products with shorter chains (**Figure 2**) [8].

In the case of polyolefins, cracking occurs mainly by random scission and backbiting reactions.

In the last step of the process, the radical can react with another radical, creating saturated alkane (recombination), one alkane and one alkene (disproportionation), create cyclic structure (cyclisation), dehydrogenate and condensate and take part in other reactions. In the end, a mixture of different types of hydrocarbons with varying chain lengths is created. When condensation and dehydrogenation reaction progress, polycyclic aromatic hydrocarbons (PAH) are created, which still react and finally create a coke, that is reach in carbon.

3.2 Product composition

As mentioned before, the cracking of polyolefin chains produces different hydrocarbons. In general, three types of streams are created during the pyrolysis of polyethylene and polypropylene: gas fraction that consists of hydrocarbons with the lowest molecular weights, a liquid or semisolid fraction (pyrolysis oil) that consists

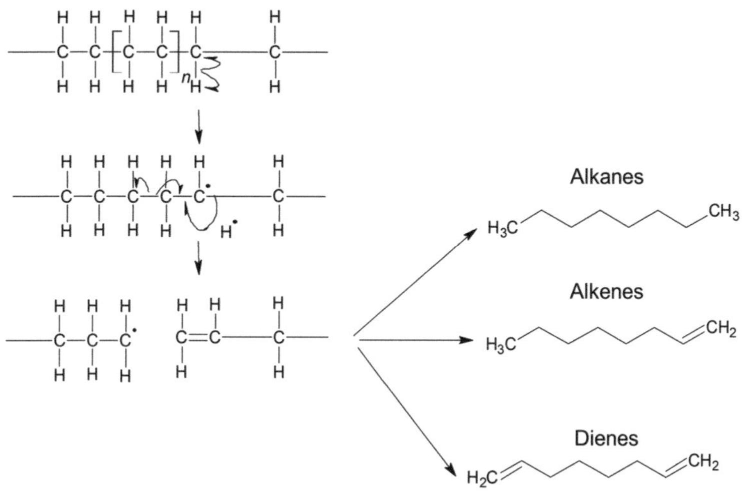

Figure 2.
Cracking of polyolefins by random scission reaction, redrawn from [7].

of hydrocarbons that were created in the form of vapours which after cooling were condensed and the process residue (char), that consists of plastic's additives and coke, which was created during the process. The composition and yields of each of these products depend on the feedstock's composition and process conditions such as temperature, pressure, residence time, and catalyst used.

For example, during the process in a rotary kiln reactor with quartz sand used as a heat carrier, pyrolysis oil consisting of 43,1% of aliphatic hydrocarbons and 55,5% of aromatic hydrocarbons were obtained from polyethylene. Pyrolysis oil from polypropylene consisted of 44,7% aliphatic and 52,9% aromatic hydrocarbons [9]. On the other hand, during thermal cracking of PE and PP in a microreactor at different temperatures gave products consisting of 59,7% alkanes, 31,90% alkenes, 8,40% cycloalkanes and 66,55% alkanes, 25,7% alkenes and 7,58% cycloalkanes, respectively. No aromatics were identified [10]. These two examples already indicate how different products can be obtained, depending on the process conditions.

In general, it can be observed that raising the temperature and residence time can increase arenes creation and can also raise the alkane to alkene ratio in the product, for example, in [11]. Aromatics content can also be significantly raised by the use of certain catalysts, like zeolites. The type of catalyst also influences the alkane/alkene ratio [12]. It should be noted that catalysts can be rapidly deactivated, limiting their use in continuous processes [13]. An increase in the temperature can increase the yield of long-chain hydrocarbons due to reduced residence time. However, it can also favour increased gas and lower molecular weight product formation by increasing the number of secondary reactions if the residence time is long enough. The majority of cracking processes are conducted at atmospheric pressures. However, some investigations present that higher pressure can increase the gas formation at lower temperatures, but with the increase of the temperature, the effect was diminished. A decrease of double bonds formation with the pressure increase was also observed [14]. What is more, different polymers can have a synergistic effect on co-pyrolysis [15]. Polystyrene (PS) is also a valuable feedstock for pyrolysis. The product of PS cracking is almost fully aromatic, with styrene monomer as a major product [16]. It can also be co-pyrolysed with polyolefins. Poly(methyl methacrylate) (PMMA) is another polymer that can be pyrolysed [17]. Poly(vinyl chloride) (PVC) produces large quantities of corrosive hydrogen chloride and can contaminate all – gas, liquid, and residue. PET gives low yields of oil, and the thermal cracking of polyurethanes provides products reach in organic nitrogen components [18]. Pyrolysis of biomass converts waste into oil with high oxygen content and increase coke formation [19]. That is why most of the research and developed technologies are based on polyolefins, optionally with the addition of polystyrene, while other plastics and biomass are treated as impurities.

Pyrolysis of plastics is a complex process with many variables that produce hydrocarbons from polyolefin feedstock. It makes the process difficult but flexible at the same time. That is why many different solutions are used (other types of reactors), but also different product types for different applications are obtained.

3.3 Product applications

As described in the previous section, cracking of polyethylene and polypropylene can lead to many different products. The composition of the products – hydrocarbon type and chain length – will determine their application.

3.3.1 Plastic-to-intermediate

Pyrolysis oil obtained during thermal or thermocatalytic cracking of polyolefins is a complex mixture of hydrocarbons with different chain lengths (5 to 30 and more carbon atoms). Linear and branched paraffins and olefins, together with aromatics: mono and polyaromatics – with and without aliphatic side chains, are obtained. Such a complex mixture does not have a direct application without additional treatment. However, as a hydrocarbon product, it can be mixed with refinery and petrochemical streams and processed together with crude. The process is simple, consisting of only a cracking reactor, product cooling system, residue discharge system and gaseous product burning unit (for energy production).

However, the capacity of commercial chemical recycling plants is limited due to plastic waste availability and the process itself – polymers have a low thermal conductivity which makes the scale-up of the pyrolysis reactor challenging. The biggest pyrolysis plants have a capacity of about 100 000 tons per year which is very small compared to the standard refinery size of about 4–10 million tons per year. This means that the recycled stream is highly dissolved in the refinery. As a result, the product can be contaminated, so there is no need for expensive detailed sorting and washing of the plastic waste or purification of the pyrolysis oil. Even though this makes the process much cheaper, the solution is not economically feasible.

Cracking is an endothermic process that requires a lot of energy to melt the plastic and break polymer bonds, as plastics are excellent insulators. The residue obtained from cracking is usually a high-calorific by-product and contains a high level of contamination, limiting its use in incinerators, especially if the raw material used for pyrolysis was not properly separated and cleaned. For example, the presence of PVC significantly raises the chlorine content, which is limited in incinerators' feedstock specifications. Special treatment of this residue increases the overall cost of operations.

Pyrolysis of plastic waste into feedstock for refineries was very popular a couple of years ago. For example, in Poland until 2007, many commercial (Technology Readiness Level, TRL = 9) plants were operated, but their profitability was based on relief in excise tax. When regulation changed, they were all bankrupted. Low prices of crude oil caused the closure of other companies worldwide, or they changed their profile. For example, Agilyx from the US had to shut down its flagship plant in Tigard in 2016, later changing its activity profile to polystyrene recycling [20].

3.3.2 Plastic-to-fuel

The most popular application for products from the chemical recycling of polyethylene and polypropylene are fuels and fuel components. Hydrocarbon product can be separated into more narrow fractions like gasoline, diesel, light and heavy fuel oil. Hydrocarbons with the highest molecular weights (waxes) can be circulated back to the cracking reactor or cracked in an additional catalytic process. Proper process parameters can also limit the quantity of waxy hydrocarbons, but it usually causes high gas yields. Pyrolysis reactors are followed by distillation units. The ratio between different products depends on technology. An example of two companies' products slates are presented in **Table 1** [21].

From one side, fuels obtained from the pyrolysis of polyolefins are characterised by low concentrations of sulphur (less than 0,1%) and are easily burned as hydrocarbon fractions. On the other hand, high olefin content reduces oxidation stability. Furthermore, reactive alkenes relatively easily undergo polymerisation reaction creating gums with high molecular weights. That is why products from PE and PP cracking should not be stored for a longer time. This tendency to polymerisation can also cause issues in distillation units where resins deposit at surfaces of elevated temperatures, reducing heat transfer coefficient in heaters and heat exchangers, also plugging the distillation columns and reducing mas flow in these units. Foaming during distillation is also observed [22, 23].

Hydrotreatment (catalytic reactions with hydrogen) of the products from pyrolysis could be a solution – olefins can be saturated into paraffins, stabilising the product. But it would raise the total cost of the process as it usually carries out at elevated pressures and requires special, separate units. What is more, products reach in linear paraffins may have high pour point of diesel and light fuel oil fractions. Unsuitable gasoline fraction octane number and cold-temperature behaviour of heavier fractions limit their use in combustion. To keep proper fuel parameters, blends of hydrotreated fractions from chemical recycling of plastics and commercial fuels can be prepared. But to keep proper parameters, a maximum of 1% of gasoline fraction, 10% of diesel fraction and 20% of light fuel fraction from polyolefins' cracking can be used [24]. If the process is controlled to produce a highly aromatic product, then higher octane gasoline and lower cetane diesel could be obtained.

Fuels and fuel components obtained from plastic waste compete in price with fossil fuels, making the profitability of the process challenging. Also, this type of application in European regulation is seen as energy recovery, so it is not considered chemical recycling. Nevertheless, there are several companies that are focused on the production of fuels. For example, Bightmark Energy is building a 100 000 t/a commercial facility (TRL 9) in Ashley (US), which is planned to be commissioned in 2021. Braven Environmental is planning a 65 000 t/a plant in Virginia. Nexus Fuels commercial-scale plant's capacity is 50 t/d, similar to one module of Integrated Green Energy Solutions' plant constructed in Amsterdam.

Type of product	Cynar	PK Clean
Kerosene, %	10	—
Diesel, %	70	66
Light fuel oil/Naphtha, %	20	33
Wax, %	—	re-circulated

Table 1.
Product slate examples [21].

Polish company Handerek Technologies develops technology for fuels production by pyrolysis of plastics and further hydrotreatment at atmospheric pressure, using syngas (mixture of hydrogen and carbon oxide). The process at an early stage of commercialisation, presenting only a small scale pilot plant (TRL 4–5) with plans to build commercial plants with a capacity of 10 000 t/a.

3.3.3 Plastic-to-monomer

The production of monomers that would be firther used for polymerisation could be the only solution for closed-loop recycling of PE and PP waste. Depolymerisation of polyolefins is not easy as bonds between carbon atoms in the chain are relatively strong. As described before, thermal or thermocatalytic cracking leads to a mixture of hydrocarbons with different chain lengths. Gas fraction is produced, but alkenes content in it is usually low. A research was conducted to maximise the gas fraction and olefin content, but the highest ethylene concentration in the gas stream achieved was 25%. Still, a minimum of 40% of liquid product was obtained [7]. In other research, steam was used in a fluidised bed reactor for pyrolysis of plastic waste to maximise olefin yields. 20–31% ethene and 14–18% propene were obtained. Additionally, 19–23% gasoline was produced. These yields are similar to standard naphtha steam cracker's product, but the research was run at lab-scale, and no information about scale-up is available [25].

The most extensively investigated process currently is the production of feedstock for commercial steam crackers. Plastic waste is cracked thermally or catalytically into a liquid fraction with proper boiling range and further mixed with fossil feedstock. Some research presented that it would be possible to use liquid from plastic pyrolysis only, receiving results similar to use naphtha [26]. But to avoid coking of the colder section of the steam cracking reactor and contamination of created streams with heteroatoms present in pyrolysis oils, co-cracking with standard feeds is preferred. Purification of the pyrolysis oil by hydrotreatment upfront steam cracking could be a solution that increases the total cost of the process. Considering the capacities of chemical recycling plants and current steam crackers (millions of tons of ethylene), significant dissolution of pyrolysis oil in fossil-based feedstock might be a solution for this issue, as pyrolysis oil can be only a small part of the inlet stream to the steam cracker.

Although making monomers and then polymers from polyolefin waste look like a very promising route for closed-loop recycling of polyethylene and polypropylene, several concerns should be considered. First, the basic question mark is about yields of the process. If it is assumed that pyrolysis oil has similar properties to commercial naphtha, then yields of ethylene and propylene that can be used for further polymerisation is limited, as presented in **Table 2** [27]. In the case of higher boiling fractions (like gasoil or diesel fraction) or oils reach in branched or aromatic

Product	% wt.
Residual gas (methane, hydrogen)	16,0
Ethylene	35,0
Propylene	15,0
C4 fraction	8,5
Fraction >C4 (C5, pyrolysis gasoline, residual oil)	25,5

Table 2.
Products of steam cracking of naphtha [27].

hydrocarbons, yields of ethylene and propylene are lower, and the yield of liquid products rises [28]. These products can be used as feedstock for chemical processes but are currently used primarily as fuels, which is not considered as recycling according to European regulations.

As the plastic-based liquid has to be blended with naphtha or other steam cracking fossil-based feeds, it is challenging to trace the flow of materials along the supply chain. For that purpose, a mass balance accounting system is required. It is a set of rules for allocating the recycled content to different products in order to be able to claim the recycled content. Products can be accredited by the independent scheme, for example, the International Sustainability and Carbon Certification Plus (ISCC) scheme. NGOs challenge the currently used calculation method as requiring more clarification and a more strict approach as it can be misused, claiming incorrect recycled contents [29]. It is understandable as long as detailed and correct data is not shared. For example, in one of the published Life Cycle Assessments (LCA) for the process, the considered amounts of naphtha from chemical recycling that is needed to produce 1 kg of LDPE were 1,2–2,0 kg with the baseline of 1,5 kg of naphtha per 1 kg of LDPE produced. These numbers are not in line with ethylene yields from fossil naphtha (as presented in **Table 2**) and require a broader explanation [30].

The final consideration is about the overall environmental impact of the process. For evaluating the influence of the process or product, a systematic analysis of the environmental impacts, called Life Cycle Assessment (LCA), is used. Currently, only two executive summaries of LCA were published, which were also criticised by NGOs [31]. As not enough data is publicly available, it is difficult to evaluate these concerns. What is sure, the process chain is very long and complex, as presented in **Figure 3**, and requires the use of fossil-based feedstocks and only part of the plastic pyrolysis oil is converted back to a polymer. The yield of the fraction that can be processed in a steam cracker in the plastic pyrolysis process is unknown. In this case, LCA analysis should consider yields and processing of other products from plastic pyrolysis and steam cracking to present the whole impact. Lastly, the feasibility of these processes are a matter of concern, especially if hydrotreatment is used for pyrolysis oil purification.

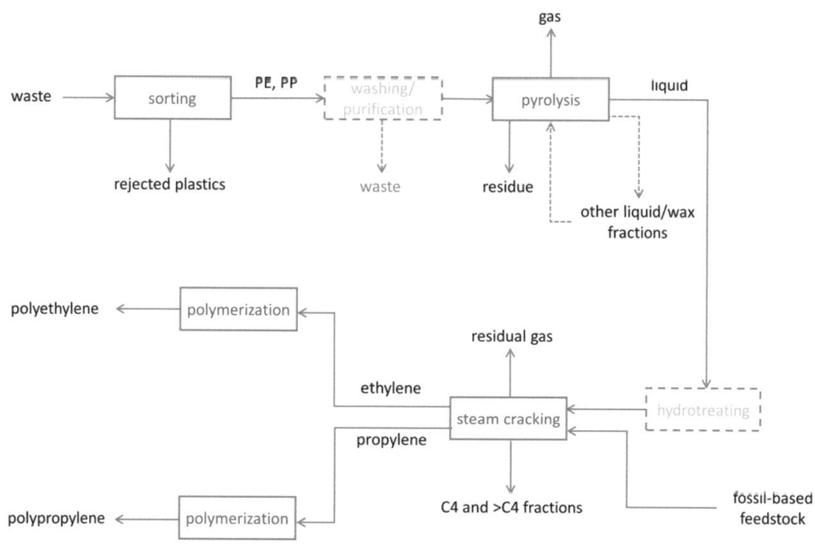

Figure 3.
Scheme of polyethylene and polypropylene from the plastic waste production process.

Currently, big polyolefin producers are involved in projects for the chemical recycling of plastics into monomers, like BASF, SABIC, Borealis or Chevron Philips Chemical, cooperating with companies experienced in pyrolysis, like Plastic Energy, Quantafuel or Nexus Fuels. In Geleen, the Netherlands, a plant for cracking of polyolefins is constructed and a hydrotreating system for purification of pyrolysis oil, which will later be fed to a steam cracker. This project is a joint investment of Plastic Energy and SABIC [32]. The plant is expected to have a capacity of 15–20 000 t/a and to become operational in 2022.

3.3.4 Plastic to chemicals (upcycling)

The production of valuable chemicals from waste, called upcycling, is an interesting alternative. The mixture of hydrocarbons obtained from polyolefins' pyrolysis can be upgraded or separated into different hydrocarbon types. What is more, the flexibility of the cracking process enables the maximisation of target fractions.

The major advantage of plastic-to-chemicals processes is that most of the proposed solutions offer final market products that do not require further processing in petrochemical plants. In this case, a mass balance approach is not required as products are based entirely on plastic waste. As products are not dedicated to be burned for energy production, these technologies can be classified as open-loop recycling also under European regulations. What is more, special, niche applications enable higher margin than compared to naphtha or fuel. On the other hand, these applications are limited when products are produced from waste, require high purity (virgin) polymers or complex pre- or post-treatment and purification, which may significantly influence the feasibility.

Benzene, toluene and xylenes (BTX) are important aromatics used by petrochemical industry to produce valuable chemicals like polystyrene, nylons, methacrylates, polyurethanes, plasticisers and many more. The pyrolysis process of polyethylene and polypropylene can be controlled to maximise aromatic hydrocarbons. A presence of polystyrene in the raw material could increase yields of BTX fraction. Nevertheless, it is possible to obtain 53% and 32% BTX from PP and PE, respectively [33–35]. These aromatics have to be further separated from the pyrolysis oil.

Encina from the US is an example of a company that provides a technology of catalytic cracking for BTX and propylene production but is currently not at a commercial scale. The planned unit will produce about 90 000 t/a of chemicals [36].

Polyolefin waxes can also be produced by the cracking of polymers. These kinds of waxes are widely used in PVC production, surface modifiers, additives to other waxes etc., and can be produced as a by-product of polyolefins production. Some companies, like Mitsui Chemicals America, Hana Corp., EPChem or Merlob, crack virgin polymers for the purpose of wax production. In this case, an issue of contamination by additives does not exist. If polyolefin wastes are considered, then a proper purification process should be implemented, or the application range would be significantly limited to those where colour and presence of inorganics is not an issue.

GreenMantra Technologies from Canada produces waxes by catalytic pyrolysis of plastic wastes at elevated pressure (4,5–25 bar). Wax products under the name of Ceranovus can be used for bitumen and asphalt modification, polymer processing or adhesives production. As an addition, fuel oil is produced. The current plant (TRL 9) has a capacity of 5 000 t/a [37]. Another company producing waxes ("EnviroWax") from plastic waste through pyrolysis is Trifol from Ireland. The by-products are fuels: diesel/kerosene and naphtha. The company has a pilot plant (TRL 6–7) with plans for scale-up to 37 000 t/a [38].

Figure 4.
Clariter products.

Clariter carries out the most complex process for plastic waste conversion into chemicals. Aliphatic solvents ("Solventra"), white oils ("Oilter") and paraffin waxes ("Clariwax") of high purity are produced from polyolefin waste via thermal cracking, hydrotreatment, and distillation and are alternatives to similar fossil-based products available in the market (**Figure 4**). To maximise on profit, target products are deeply purified from heteroatoms and hydrogenated so they can potentially be used in the cosmetic industry. Other applications are: paints, inks, degreasers, wax emulsions, paper and wood impregnation, lubrication, car or furniture polishes, silicone sealants and others. The company owns a pilot plant in Poland (TRL 5) and an Industrial-scale Plant in South Africa (TRL 7) with scale-up plans for 60 000 t/a facilities. Most interestingly, the company claims to achieve a net negative carbon footprint which is unique compared to other LCA's published [39].

4. Gasification

Gasification is a well-established process for the conversion of many non-renewable sources, like petroleum resids, petcoke or coal. In general, it can process any carbonaceous material into a valuable mixture of hydrogen, carbon monoxide and carbon dioxide – called syngas (synthetic gas). Syngas itself is a source of hydrogen and carbon monoxide used for various chemical processes or as fuel when separated into pure streams, or as a high calorific fuel or feedstock for chemical synthesis into other chemicals, like methanol, ethylene glycol, acetic anhydride and hydrocarbons (via Fischer-Tropsch synthesis) when not separated. Examples of syngas applications are presented in **Figure 5**. As polyolefins consist of carbon and hydrogen, these can be very good feedstock for gasification.

4.1 Gasification reaction

Gasification is a complex process with several reactions taking place between carbon-based material, oxygen and steam at high temperatures (700–1500°C). Examples of reactions that take place during gasification are presented in **Table 3**.

The hydrogen to carbon oxide ratio in obtained syngas varies significantly (from 0,7 to 6 for different fossil fuels) depending on the type of the raw material and technology. For different applications, different ratios are required. It is technically possible to obtain every required syngas composition from every feedstock, but it has economic limitations [28].

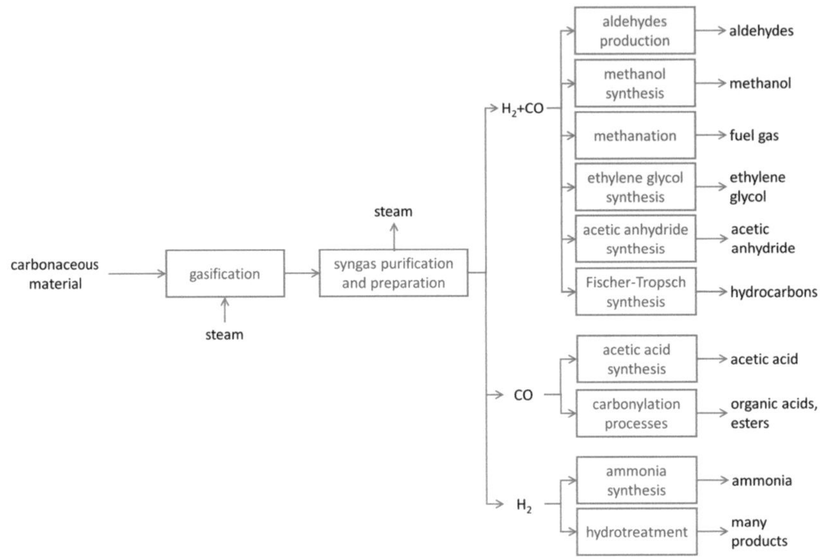

Figure 5.
Overview of potential syngas applications [28].

Reaction	ΔH, kJ/mol
Primary reactions	
$C + H_2O \rightleftarrows CO + H_2$	+118,9
$C + CO_2 \rightleftarrows 2CO$	+160,9
$C + 2H_2 \rightleftarrows CH_4$	−87,4
$2C + O_2 \rightleftarrows 2CO$	−246,3
Secondary reactions	
$C + H_2O \rightleftarrows H_2 + CO_2$	−42,3
$CO + 3H_2 \rightleftarrows CH_4 + H_2O$	−205,8

Table 3.
Examples of gasification process reactions [28].

4.2 Gasification of plastic waste

Polyolefin waste is potentially a very good source of carbon for gasification. What is more, as biomass and other plastic wastes are also reach in carbon, mixed wastes streams can be used. Gasification of waste has the following steps: drying, pyrolysis, cracking and reforming, char gasification. Drying of plastics has an insignificant role but might be important when plastics are processed with biomass, unlike the pyrolysis step, which is key in the gasification of plastics. The poor heat conductivity, sticky nature, high content of volatiles created during pyrolysis of polyolefins, and relative low char and high tar yields make the process different from coal or biomass gasification challenging but also promising. Air and steam gasification processes are proposed with different pros and cons. Finally, co-gasification with fossil-based feed-stock or other waste streams is possible, expanding the flexibility of the process. The variability of quality and composition of the feedstock is a significant problem [40].

Gasification seems to be a promising alternative to pyrolysis for chemical recycling of plastic waste, which can produce chemicals or fuels without detailed separation and washing of the feedstocks. It is attractive due to the versatility of potential products. On the other hand, it is related to high capital and operational costs [41]. What is more, currently, only fuels are produced, which might be related to instability of the waste streams, which may cause problems with keeping proper hydrogen to carbon oxide ratio.

Canadian company Enerkem provides the most advanced waste pyrolysis technology. The company currently operates a commercial (TRL 9) plant in Edmonton with a capacity of 38 000 m^3/a of methanol and ethanol used as a fuel. Another plant in Canada for fuels and chemicals production with a capacity of 125 000 m^3/a is under construction, and two more projects with capacities of 270 000 m^3/a of methanol each are developed currently in Europe [42]. Ebara Environmental Plant and Ube Industries from Japan provide gasification of plastic waste through partial oxidation by oxygen and steam. 70 000 tons of plastic waste is processed at Showa Denko's Kawasaki Plant since 2003. Currently, a feasibility study for the next plant in South Korea is being processed [43].

5. Conclusions

Chemical recycling of polyethylene and polypropylene gained much attention over the last years due to the rising plastic waste issue. Many projects of demonstration or commercial facilities are currently being developed. Although the definition of chemical recycling is inconsistent worldwide, four main types of products can be identified: intermediates for further chemical processing, fuels and fuel components, monomers and final chemical products. Pyrolysis and gasification are processes that can be used for polyolefins providing mainly open-loop recycling solutions. These technologies require deep feasibility and environmental impact analyses due to their complexity, different values of products, need for co-processing with fossil-based feedstock or high capital cost. Nevertheless, intensive growth of plastic waste volumes is an opportunity for these technologies to be further developed, optimised and commercially used.

Conflict of interest

The author declares no conflict of interest.

Thanks

This review is dedicated to Dr. Mieczysław Steininger's (1950–2021) memory, a well-respected academic, extraordinary scientist and wonderful mentor. On behalf of the Clariter Board, Management, all the staff and especially myself, I would like to thank you for all the research, involvement, impact and inspiration of your extraordinary mind and charming personality.

Author details

Daria Frączak
Clariter Poland Sp. z o.o., Warsaw, Poland

*Address all correspondence to: daria.fraczak@clariter.com

References

[1] European Parliament and the Council of the European Union. Directive (EU) 2019/904 of the European Parliament and of the Council of 5 June 2019 on the reduction of the impact of certain plastic products on the environment. 2019/904 2019.

[2] He P, Chen L, Shao L, Zhang H, Lü F. Municipal solid waste (MSW) landfill: A source of microplastics? Evidence of microplastics in landfill leachate. Water Research. 2019;159:38-45.

[3] European Parliament and the Council of the European Union. Directive 2008/98/EC of the European Parliament and of the Council of 19 November 2008 on waste and repealing certain Directives. 2008/98/EC 2008.

[4] ISO 15270 – Plastics — Guidelines for the recovery and recycling of plastics waste.

[5] Frequently Asked Questions [Internet]. [cited 2021 May 8]. Available from: https://plastics.americanchemistry.com/ARAP-Frequently-Asked-Questions/

[6] Meys R, Frick F, Westhues S, Sternberg A, Klankermayer J, Bardow A. Towards a circular economy for plastic packaging wastes the environmental potential of chemical recycling. Resour Conserv Recycl. 2020;162:105010.

[7] Diaz-Silvarrey LS, Zhang K, Phan AN. Monomer recovery through advanced pyrolysis of waste high density polyethylene (HDPE). Green Chem. 2018;20(8):1813-23.

[8] Gascoin N, Navarro-Rodriguez A, Gillard P, Mangeot A. Kinetic modelling of high density polyethylene pyrolysis: Part 1. Comparison of existing models. Polym Degrad Stab. 2012;97(8):1466-74.

[9] Zhang Y, Ji G, Chen C, Wang Y, Wang W, Li A. Liquid oils produced from pyrolysis of plastic wastes with heat carrier in rotary kiln. Fuel Process Technol. 2020;206:106455.

[10] Ahmad I, Khan MI, Khan H, Ishaq M, Tariq R, Gul K, et al. Pyrolysis study of polypropylene and polyethylene into premium oil products. Int J Green Energy. 2015;12(7):663-71.

[11] Onwudili JA, Insura N, Williams PT. Composition of products from the pyrolysis of polyethylene and polystyrene in a closed batch reactor: Effects of temperature and residence time. J Anal Appl Pyrolysis. 2009;86(2): 293-303.

[12] Seo Y-H, Lee K-H, Shin D-H. Investigation of catalytic degradation of high-density polyethylene by hydrocarbon group type analysis. J Anal Appl Pyrolysis. 2003;70(2):383-98.

[13] López A, de Marco I, Caballero BM, Adrados A, Laresgoiti MF. Deactivation and regeneration of ZSM-5 zeolite in catalytic pyrolysis of plastic wastes. Waste Manage. 2011;31(8):1852-8.

[14] Al-Salem SM, Antelava A, Constantinou A, Manos G, Dutta A. A review on thermal and catalytic pyrolysis of plastic solid waste (PSW). J Environ Manage. 2017;197:177-98.

[15] Dubdub I, Al-Yaari M. Pyrolysis of mixed plastic waste: I. kinetic study. Materials. 2020;13(21):4912.

[16] Achilias DS, Kanellopoulou I, Megalokonomos P, Antonakou E, Lappas AA. Chemical recycling of polystyrene by pyrolysis: potential use of the liquid product for the reproduction of polymer. Macromol Mater Eng. 2007;292(8):923-34.

[17] Esmizadeh E, Khalili S, Vahidifar A, Naderi G, Dubois C. Waste Polymethyl Methacrylate (PMMA): Recycling and

High-Yield Monomer Recovery. In: Martínez LMT, Kharissova OV, Kharisov BI, editors. Handbook of Ecomaterials. Cham: Springer International Publishing; 2018. p. 1-33.

[18] Czajczyńska D, Anguilano L, Ghazal H, Krzyżyńska R, Reynolds AJ, Spencer N, et al. Potential of pyrolysis processes in the waste management sector. Therm Sci Eng Prog. 2017;3:171-97.

[19] Zhang L, Bao Z, Xia S, Lu Q, Walters K. Catalytic pyrolysis of biomass and polymer wastes. Catalysts. 2018;8(12):659.

[20] Oregonian/OregonLive AB| T. Agilyx of Tigard retools to produce styrene [Internet]. oregonlive. 2016 [cited 2021 May 11]. Available from: https://www.oregonlive.com/business/2016/02/economy_took_steam_out_of_gas.html

[21] Ocean Recovery Alliance. Plastics-to-fuel Project Developer's Guide [Internet]. Hong Kong: The American Chemistry Council; 2015 Jun [cited 2021 May 11]. Available from: https://www.oceanrecov.org/about/plastic-to-fuel-report.html

[22] Walendziewski J. Continuous flow cracking of waste plastics. Fuel Process Technol. 2005;86(12):1265-78.

[23] Doliński B, Dolińska-Kobińska A, Dolińska M, Hrebieniuk W, Mucha J, Paćkowski Z, et al. Sposób otrzymywania komponentów paliw płynnych z produktów krakingu odpadów poliolefinowych. PL199407, 2008.

[24] Budzyńska-Józwiak A, Sosnowska-Maciukiewicz L, Szumacher S, Swat M. Zastosowanie hydrorafinowanego destruktu z krakingu odpadowych tworzyw sztucznych jako bazy komponentów paliw ciekłych. Probl Ekspl. 2009;(3): 19-38.

[25] Simon CM, Kaminsky W, Schlesselmann B. Pyrolysis of polyolefins with steam to yield olefins. J Anal Appl Pyrolysis. 1996;38(1-2): 75-87.

[26] Angyal A, Miskolczi N, Bartha L, Tungler A, Nagy L, Vida L, et al. Production of steam cracking feedstocks by mild cracking of plastic wastes. Fuel Process Technol. 2010;91(11):1717-24.

[27] Weissermel K, Arpe H-J. Industrial organic chemistry. 3rd ed. Weinheim: VCH; 1997. 464 p.

[28] Grzywa E, Molenda J. Technologia podstawowych syntez organicznych. IV. Vol. 1. Warszawa: Wydawnictwa Naukowo-Techniczne; 2008.

[29] Determining recycled content with the 'mass balance approach' – 10 recommendations for development of methods and standards [Internet]. Zero Waste Europe. [cited 2021 May 12]. Available from: https://zerowasteeurope.eu/library/determining-recycled-content-with-the-mass-balance-approach-10-recommendations-for-development-of-methods-and-standards/

[30] Quantis. Life Cycle Assesment of Plastic Energy technology for the chemical recycling of mixed plastic waste [Internet]. Plastic Energy; 2020 [cited 2021 May 12]. Available from: https://plasticenergy.com/sustainability/lca-report/

[31] Understanding the environmental impacts of chemical recycling – ten concerns with existing life cycle assessments [Internet]. Zero Waste Europe. [cited 2021 May 12]. Available from: https://zerowasteeurope.eu/library/understanding-the-environmental-impacts-of-chemical-recycling-ten-concerns-with-existing-life-cycle-assessments/

[32] World Economic Forum 2020: SABIC is collaborating with PLASTIC ENERGY on the chemical recycling of plastics, as part of their TRUCIRCLE portfolio of solutions [Internet]. Plastic Energy. 2020 [cited 2021 May 12]. Available from: https://plasticenergy.com/world-economic-forum-2020-sabic-is-collaborating-with-plastic-energy-on-the-chemical-recycling-of-plastics-as-part-of-their-trucircle-portfolio-of-solutions/

[33] Kaminsky W, Kim J-S. Pyrolysis of mixed plastics into aromatics. J Anal Appl Pyrolysis. 1999;51(1-2):127-34.

[34] Jung S-H, Cho M-H, Kang B-S, Kim J-S. Pyrolysis of a fraction of waste polypropylene and polyethylene for the recovery of BTX aromatics using a fluidized bed reactor. Fuel Process Technol. 2010;91(3):277-84.

[35] Cho M-H, Jung S-H, Kim J-S. Pyrolysis of mixed plastic wastes for the recovery of benzene, toluene, and xylene (BTX) aromatics in a fluidized bed and chlorine removal by applying various additives. Energy Fuels. 2010;24(2):1389-95.

[36] Encina to Invest Hundreds of Millions to Make Products From Waste Polypropylene [Internet]. Waste360. 2021 [cited 2021 May 13]. Available from: https://www.waste360.com/business/encina-invest-hundreds-millions-make-products-waste-polypropylene

[37] GreenMantra Technologies. Closed Loop Fund to invest up to $3 Million tn GreenMantra™ Technologies [Internet]. [cited 2021 May 13]. Available from: https://www.prnewswire.com/news-releases/closed-loop-fund-to-invest-up-to-3-million-in-greenmantra-technologies-300494427.html

[38] Taylor C. Irish company Trifol raises €300,000 in 24 hours from crowdfunding [Internet]. The Irish Times. [cited 2021 May 13]. Available from: https://www.irishtimes.com/business/energy-and-resources/irish-company-trifol-raises-300-000-in-24-hours-from-crowdfunding-1.3946915

[39] News – Clariter [Internet]. [cited 2021 May 13]. Available from: https://clariter.com/news/

[40] Lopez G, Artetxe M, Amutio M, Alvarez J, Bilbao J, Olazar M. Recent advances in the gasification of waste plastics. A critical overview. Renew Sustain Energy Rev. 2018;82:576-96.

[41] Waste Gasification & Pyrolysis: High Risk, Low Yield Processes for Waste Management [Internet]. Global Alliance for Incinerator Alternatives. 2017 [cited 2021 May 13]. Available from: https://www.no-burn.org/gasification-pyrolysis-risk-analysis/

[42] Facilities & Projects | Clean Technology Around the World [Internet]. Enerkem. [cited 2021 May 13]. Available from: https://enerkem.com/company/facilities-projects/

[43] JGC to conduct feasibility study for Korea's first plastic-waste gasification plant [Internet]. Chemical Engineering. 2021 [cited 2021 May 13]. Available from: https://www.chemengonline.com/jgc-to-conduct-feasibility-study-for-koreas-first-plastic-waste-gasification-plant/

Chapter 5

Polyethylenes: A Vital Recyclable Polymer

Macdenis Egbuhuzor, Chima Umunankwe and Peter Ogbobe

Abstract

Polyethylene (PE) is a synthetic polymer made from the polymerization of ethylene. It is the most widely used plastic in the world. Its production, processing, usage, applications, and disposal system had made the study of this plastic very significant. The role played by this plastic in the world has made the knowledge of its usage, disposal system, processing, recycling, and applications inevitable. The chapter discussed the general overview of this plastic product, the production, properties, and disposal systems. The processing of recycled polyethylene is vital in its end-use through collecting, sorting, cleaning, separation, and compounding, and this was extensively treated. We also discussed the opportunities, applications, and limitations of polyethylene recycling. At the end of this chapter, one will know the production, processing, recycling, and applications of polyethylene plastic and the dangers posed by this plastic if a proper disposal system is not followed.

Keywords: disposal, high-density polyethylene, low-density polyethylene, monomer, polymers, recycling, resin

1. Introduction

Polyethylene (PE) is a synthetic resin made from the polymerization of ethylene [1]. It is an ethylene polymer with the structural formula of $(\text{-CH2-CH2-})_n$. This is generated at high pressures and temperatures in the presence of a catalyst, based on the desired characteristics and properties for the final product. It is manufactured as branched low-density polyethylene (LDPE), linear high-density polyethylene (HDPE), and many other variants. This olefin plastic can be combined with other elements, compounds, and monomers to form other polyethylene brands and co-polymers. The primary production processes employed are the Ziegler-Natta, metallocene, and chromium/silica catalysts in their manufacture, and this production pathway affects the final products' mechanical and end-use properties. Water, acids, alkalis, most solvents, and chemicals do not affect polyethylene. Polyethylene offers superior low-temperature resistance, excellent chemical resistance, excellent power insulation, intense pressure, and high radiation resistance. Polyethylene is highly susceptible to environmental stress (both chemical and mechanical) and has low heat-aging resistance. Polyethylene characteristics and properties vary based on the molecular structure and density of the polymer [2]. PE is used in various items and packaging, including milk jugs, drinking straws, bottle caps, detergent bottles, cream tubs, waste bins, water pipes, children's toys, and films, as well as plastic bags. Most unwanted items are disposed of in landfills, burned, and recycled, while some are inappropriately disposed of and strewn on streets and highways, wreaking havoc on our environment and marine life. Improperly disposed polyethylene waste poses many public health concerns and harms flora and fauna and the environment, mainly when it is not collected and disposed of properly [3]. Efforts are made to convert or reuse polyethylene products through recycling. Polyethylene waste disposal, reduction, and recycling generate many benefits if handled well [4]. This chapter will discuss polyethylene production and processing, properties, application and usage, disposal systems, and recycling.

2. Production and processing

Polyethylene is composed of carbon (C) and hydrogen (H), and these elements can be combined in several ways to make different types of polymer [5]. It is produced by modifying natural gas (methane, ethane, and propane blend), fermentation, ethanol dehydration via acetaldehyde hydrogenation, pyrolysis, and catalytic cracking of crude petroleum product or its distillation into gasoline. Branched low-density polyethylene (LDPE) was discovered first through free radical polymerization under high pressure and temperature by ICI Laboratories UK in 1933 [6]. Linear High-Density Polyethylene (HDPE) is manufactured at low pressures using Ziegler-Natta catalysts in slurry or gas-phase processes. Metallocene linear low-density polyethylene (mLLDPE) is also produced using low-pressure polymerization technology, which copolymerizes ethylene with another monomer, such as butene-1 or hexene-1, with the help of metallocene catalyst. The metallocene catalyst leads in resins with very consistent and specific properties, such as superior toughness and stiffness balance. Low-pressure polymerization technology employs transition-metal catalysts to produce Medium Density Polyethylene (MDPE) and linear low-density polyethylene (LLDPE) products. However, comonomers are introduced into the reaction to form small short-chain branches on the linear molecule, causing the density of the polymer to decrease [7]. The production process of PE involves the polymerization of ethylene into polyethylene. This process takes five routes from ethylene to the preparation and production of polyethylene, namely high pressure, metallocene, Ziegler-Natta, Standard Oil, and the Phillips processes [8].

High-Pressure polymerization is usually done at high pressures (0.1–0.3KN/mm^2) and temperatures from 80 to 300°C, using free-radical initiators like azo-diisobutyronitrile. This is accomplished by reducing the exotherm using flowing or running water through a jacketed reactor by utilizing a high cooling surface volume ratio in the right portion of the continuous reactor. In this process, 10–30% of the monomer is converted to polymer, which is then extruded as granules [8–10].

PEs with short-chain branching are produced through metallocene manufacturing routes using metallocene catalysts. Metallocene catalysts are made by infusing zirconium or titanium, or other transition compounds into a cyclopentadiene-based structure. In this technique, a monomer in gaseous form and a metallocene catalyst are loaded into a fluidized bed reactor at pressures less than 24 MPa and temperatures just under 100°C. The process is very adaptable, and different kinds of PE can be produced by modifying the reaction conditions using the catalysts. Short branches of PE are formed by adding small amounts of propene, butene, hexene, or octene into the monomer feed. They are designed using existing polymerization processes, giving way for metallocene-PEs having different properties and grades [8].

Ziegler Processes are the result of Ziegler's, Natta's, and co-workers' efforts. This polymerization allows the generation of a coordination complex due to the reaction between the initiator and the catalysts. This complex controls how the monomer approaches the growing chain. In this process, ethylene is supplied under low pressure into a reactor containing a liquid hydrocarbon that serves as a diluent and the catalyst. The catalyst is composed of titanium tetrachloride and aluminum triethyl. The catalyst complex can be produced ahead of time and then fed into the vessel, or it can be prepared in situ by feeding the components directly into the main reactor. In the absence of oxygen and water, the reaction can reach temperatures of up to 70°C. The polymer precipitates from the solution, forming a slurry, and the reactants are emptied into a catalyst decomposition tank [9].

The Standard Oil Company Process utilizes a transition metal oxide in combination with a promoter. Temperatures and pressures in the reactant vary from 230 to 270°C and 4 MPa–8 MPa, respectively. As a catalyst, molybdenum oxide is utilized to fast track the reaction, while sodium or calcium as metals or hydrides are used as promoters. The reaction occurs within a reactor in a hydrocarbon solvents [8, 10].

The Phillips Process includes dissolving ethylene in a liquid hydrocarbon solvent and then polymerizing it at 130-160°C and 1.4–3.5 MPa pressure with a supported 5 percent chromium oxide catalyst on a finely split silica-alumina catalyst. The major role of the solvent is to dissolve the polymer as it develops while simultaneously serving as a heat transfer medium. The recommended catalyst is a finely split silica-alumina catalyst that has been activated by heating to about 250°C and includes 5% chromium oxides, principally CrO_3. The combination is then transported through a gas–liquid separator, where the ethylene is flashed off, the catalyst is removed from the separator's liquid product, and the polymer is removed from the solvent [11].

These processes of polymerization reaction involve three stages: pre-treatment stage, reaction stage, and separation stage, and the process is sensitive to a catalyst that initiates the reaction to produce free radicals. The production process takes place in a continuous operation that requires a source of pure ethylene, suitable compression equipment at high pressure, and a high-pressure reactor to perform quick and high exothermic polymerization control. In the pre-treatment stage, the ethylene stream is pressurized into a compressor and heat exchangers, and this is maintained at 2000 bar and 150°C with the initiator before entering the reaction zone. The reaction zone is modeled in high-pressure tubular reactors with a water vapor stream to ensure steady temperature conditions in the reactor. At this stage, the initiators generate free radicals that link up with different ethylene chains to form a vinyl polymer. Then, other radical and monomer transfers occur at propagation to increase the number of chains formed gradually. Finally, the polymerization process is completed at termination by temperature conditions that disrupt the chemical bond formed, producing low-density polyethylene. The final stage involves the removal of the polymer and its separation. Also, Flash equipment is used to evaporate ethylene that does not react and then reused. LDPE is polymerized, forming large chains and generating enough entropy in the order of their bonds between monomers, causing a reduction of their density.

Three processes are used in the manufacture of high-density polyethylene. The first is the solution process, where the catalyst, initiator, and monomer are dissolved in a solvent in a continuously stirred tank reactor [12]. The solvent is removed when the polymerization is completed while isolating the polymer. The second is the slurry process, where the catalyst and polymer are suspended in a liquid medium in a continuously stirred tank reactor or tubular reactor where polymerization takes place. The catalyst and polymer are not dissolved in the medium and are separated after polymerization. The final and third process is the gas phase process, where no solvent is involved. The monomer and the catalyst are blown into the fluidized bed reactor for polymerization to take place. The processes of PE production are cataloged in **Table 1**.

PE can also be produced from other routes rather than from high and low-pressure processes. Linear polyethylene with the repeat unit of ↭$(CH_2)n$↭ by

PE type	Year	Company	Process	Catalyst
LDPE	1933	ICI Laboratory, UK	High-pressure free-radical polymerization 2000–3000 bar, 80-300°C	Low quantity of oxygen
LDPE	1943	Union Carbide & Du Pont	Film extrusion and injection molding	Free radical catalysts using initiators (peroxides)
HDPE	1953	Philips Petroleum, Ziegler	Gas-phase processes in solution or slurry	Ziegler-Natta organo-metallic catalysts
HDPE	1955	Hoechst A G in Germany	Low-pressure polymerization	Organo-metallic catalysts
MDPE			Low-pressure polymerization	Transition-metal catalysts
LLDPE	1950s	Du-Pont	Low-pressure solution process	Transition-metal catalysts
HDPE	1973	Union Carbide	Their fluidized bed gas-phase process	Ziegler-Natta
LLDPE	1977	Union Carbide	Low-pressure gas-phase process	Free radical catalysts using initiators (peroxides)
LLDPE VLPE	1980s	Dow Chemicals	Solution process	Ziegler/Natta, Cr/Mo oxide
mLLPE	1990	EXXOM MOBIL	Gas-phase PE Process	Metallocene catalysts

Table 1.
Manufacturing & processing development of polyethylene.

condensation polymerization of an ethereal solution of diazomethane [13]. Also, Linear PE with a molar mass of up to 1.3 kg/mol is produced by the reaction of decamethylene dibromide with sodium in the Wutz reaction. Other processes include reducing carbon monoxide by modified Fischer Tropsch process, reducing polyvinyl chloride (PVC) with lithium aluminum hydride, and hydrogenation of polybutadiene [8].

3. Properties

Polyethylene as a polymeric material possesses excellent properties, which makes it useful as an engineering material. These properties depend on the density, molecular weight, molecular weight distribution, degree of long-chain branching, and short branching. It possesses low water absorption, moderate to low gas permeability, good toughness, flexibility at low temperatures, relatively low heat resistance, and many chemicals. The use of PE as an engineering material is due to these excellent properties greatly affected by density, molecular weight, and molecular weight distribution, as enumerated in **Table 2** [15].

Polyethylene has the following engineering properties in near-absolute terms. The crystallinity of LDPE is between 40 and 60%, depending on the degree of branching and thermal history, while that of HDPE is between 60 and 90%, depending on the cooling rate and thermal history. It melts in the absence of air at 300°C to form transparent liquid except for its cross-linked polymer, which does not melt. Its specific heat at 20°C is 1330–2400 $JKg^{-1} K^{-1}$ for HDPE, and 1900–2300 $JKg^{-1} K^{-1}$ for LDPE, coefficient of linear expansion is 0.00017–0.00022 K^{-1} for LDPE and 0.00013–0.00020 K^{-1} for HDPE, Thermal conductivity at 23°C is

Property	Increase in density	Increase in molecular weight	Increase in molecular weight distribution
Tensile strength	Increases	Increases	—
Impact strength	Decreases	Increases	Decreases
Stiffness	Increases	Increases	Decreases
Stress crack resistance	Decreases	Increases	Increases
Abrasion resistance	Increases	Increases	—
Chemical resistance	Increases	Increases	—
Melt strength	—	Increases	Increases
Softening point	Increases	—	Increases
Low-temperature brittleness	Increases	Decreases	Decreases

Table 2.
Effects of density, molecular weight and molecular weight distribution on some properties of polyethylene [14].

0.33Wm^{-1} K^{-1}, melting point (MP) 109–125°C for low-density polyethylene and 130–135°C for high-density polyethylene. The specific gravity of LDPE is between 0.915–0.94, while that of HDPE is 0.94–0.97. PE has a refractive index of 1.51–1.52. PE also has excellent electrical properties. Its electrical resistivity is 1017–1019 ohm-m, dielectric strength 20 kV–160 kV/mm, and dielectric constant of 2.28. Polyethylenes, when exposed to moisture for 1 year, increase their weight by 0.2%. This shows excellent water resistance. PE's average elastic modulus is between 0.565–1.500GPa for HDPE and 0.190–0.520GPa for LDPE, flexural modulus is between 0.28–1.86GPa for HDPE and 0.152–2.200GPa for LDPE. It also has Flexural yield strength of 13.8–75.8 MPa, compressive yield strength of 4 MPa-23 MPa for HDPE, and Tensile strength (TS) at yield is 7 MPa-16 MPa and that at the break at 23°C is 32 MPa-60 MPa for LDPE. LDPE has a hardness of 50–60 while its HDPE counterpart is 65–70 with an impact of 0.92 for LDPE and 0.96 for HDPE [16].

3.1 Mechanical properties

Some of the significant mechanical properties include Tensile strength, compressive strength, flexural strength, impact strength, Viscoelasticity, and abrasion resistance.

3.1.1 Tensile strength

Tensile strength is the ability of a material to resist a force that tends to pull it apart. It is a basic provision for classifying the properties of given polymer materials at a specified loading rate and temperature. The ultimate tensile strength of PE at 0-70°C is 11 MPa–25 MPa, while its yield strength at the same temperature range is 6 MPa-30 MPa. This polymer property depends on percentage crystallinity, the thermodynamic stability of the primary PE chain, chain orientation, and packing density of the crystalline chain [17]. Polyethylene's strength, rigidity, friction, and hardness are low but have high impact strength and ductility. Three main tensile strengths are essential in the study of the properties of a polymer. The yield strength is the stress the PE can withstand without permanent deformation. The ultimate tensile strength is the maximum stress the PE can withstand, while the breaking strength is the stress coordinate on the strain–stress curve at the point of rupture.

Polyethylene shows excellent creep under an applied force. Elastic strain is the strain in the region of stress/strain curve of material under deformation, which recovers its shape on the release of applied stress. Elastic strain is reversible. Young's modulus is the slope of the graph ab covering from a being the origin of the stress–strain graph to point b, representing the reversible region of the stress–strain in **Figure 1**. After point b, strain is no longer proportional to stress, and the slope of the stress–strain graph changes at an increasing rate, and the strain is irreversible. The materials continue to deform after point c until it breaks at point d [18]. The percentage elongation we get during a tensile experiment is significant because it provides information on the ductility of the polyethylene under investigation. Materials with a high degree of elongation will exhibit high ductility. This is because the force necessary to sustain sample elongation and finally break the specimen changes very little at the yield point, and this makes the value of the yield strength and the breaking strength quite close [14].

3.1.2 Compressive strength

Compressive strength is the ability of a polymer material to resist the direct pressure of applied compressive force [19]. It is the ability of the polyethylene material under test to resist loads applied, thereby shortening the length of the material under compression. Its force acts in the opposite direction to the tensile force applied to the load. Compression is a force that pushes the particles of material closer together, thereby reducing the size of the materials. The compressive modulus is equal to the elastic modulus at minor strains and gives a reliable compressive stress and strain ratio. When a compressive force is applied to a PE material, it yields slowly and hardly fails.

3.1.3 Flexural strength

Flexural strength is defined as a material's resistance to distortion on the application of load, while flexural modulus measures the capacity of the test sample to bend. Flexural strength represents the amount of force necessary to break a test sample with a specific diameter.

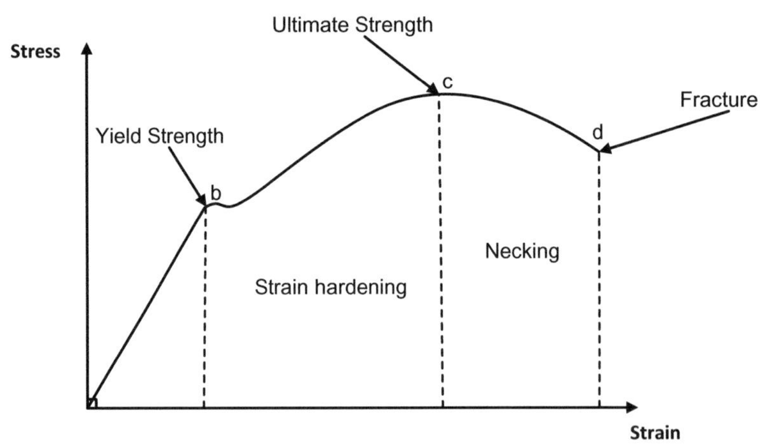

Figure 1.
Stress vs. strain curve of a typical PE.

3.1.4 Impact strength

Impact strength can be defined as the capacity of a test polymer sample to withstand fracture when a sudden force is applied to it. It is the energy absorbed by this sample without breaking. The most frequent tests for plastic materials are the drop-weight test, the Izod Impact Test, and the Charpy Impact Test. Both approaches assess a PE test sample's capacity to absorb energy upon failure. PE's Izod impact resistance values range from 0.534 to 0641kj/m at ordinary room temperature. It determines how brittle or rigid material will be when subjected to a suddenly applied load. It is affected by the volume of the specimen, the presence of a notch, cold working, and water absorption in a polymer [20].

3.1.5 Viscoelasticity

PE is a viscoelastic material. It exhibits elastic and viscous behavior when stress is applied. The application of this force results in an instantaneous elastic strain followed by a viscous, time-dependent strain [21]. This makes the material display and behave partly as a crystalline metal and partly a very high viscosity fluid. This viscoelastic behavior of polyethylene polymer makes the curve of the stress–strain of any test sample of polyethylene be divided into three essential segments.

The viscoelastic nature of PE provides for creep and stress relaxation, which are two unique engineering characteristics employed in the HDPE design [22]. This property of the polymer materials is fundamental because the functionality and applicability of the plastic can change after a while and may lead to defects and loss of functionality over sometime [23].

3.1.6 Abrasion resistance

The capacity of a polymer to withstand the wear caused by contact with another surface is referred to as abrasion resistance. PE has excellent abrasion resistance in a variety of end-use situations. Furthermore, the abrasion resistance of this material has led to its extensive usage in engineering and technical applications. Generally, the abrasive wear resistance of polymers correlates with the reciprocal product of their ultimate tensile stress and the elongation at break [24]. It increases with increasing molecular weight, the molecular weight distribution, and the degree of crystallinity of polymeric materials [25]. **Table 3** shows the common properties of some grades of polyethylene.

3.2 Electrical properties

PE is an excellent electrical insulator with good tracking-resistant properties. It is easily electrostatically charged. Its arc resistance is 200–250 seconds, with surface resistivity greater than 10^{13}ohm. PE has a dielectric strength between 450 and 1000 volts/mil and a dielectric constant of 2.25–2.35 @ 60 Hz. Also, PE's Volume resistivity is greater than 10^{16}ohm-m with a Dissipation factor greater than 0.0005 @ 60 Hz [26]. These properties make the polymer be used in any engineering application that requires high insulation requirements.

3.3 Optical properties

PE can vary between almost transparent as in LDPE, translucent as in LLDPE, and opaque as in HDPE. The ability of polyethylene films to scatter light is

Polyethylene type	Mol. weight (g/mol)	Density (g/cm³)	Degree of crystallinity (%)	Melting temp (°C)	Enthalpy of melting (j/g)	Yield strength (MPa)	Melt index rate (g/10 min)
HDPE	>250,000	0.940–0.970	70–90	120–130	203	20–40	0.1–150
LDPE	10,000–50,000	0.915–0.940	45–55	105–115	103	4–16	0.1–150
LLDPE	50,000–200,000	0.915–0.926	35–45	112–124	160	8–45	0.1–150
MDPE	100,000–200,000	0.930–0.940	30–60	115–130	141.9	15–40	0.1–150

Table 3.
Important PE grades and properties.

determined by the quenching conditions and sample's thermal history. Based on recent quantitative studies, the light scattered by a thin polyethylene sample quenched to 0°C from 125°C is much lower than that scattered by a heat-treated sample [27]. Because the refractive index along the tangent to the PE spherulites is lower than that along the radius of the spherulites, extruded polyethylene films have a slight positive birefringence. As a result, the clarity of polyethylene film is determined by the light scattered by it [28].

3.4 Thermal properties

Polyethylene exhibits a low thermal conductivity. It has a thermal expansion coefficient of 0.26 mm/°C and thermal conductivity of 0.4 W/m per °C. The specific heat of PE depends on temperature. Its low-density form has a specific heat capacity of 2.3 J/g at room temperature and 2.9 J/g at 120–140°C. The higher the molar mass and the more the branching, the lower the brittle point. Polyethylene is sensitive to surface imperfections. The standard commercial grade of high-density polyethylene's melting point is between 120 and 180°C, and LDPE is 105–115°C [29]. The zero-shear rate apparent viscosity of linear PE is related to the weight-average molar mass by Eq. (1) for polymers with a molar mass over 5 kg/mol.

$$(\eta a;0) = KMw^{3.4} \tag{1}$$

Polymers with long branches do not fit in the above equation, and Eq. (2) represents a relationship between polymers of different degrees of long branching. In many cases

$$\log(\eta a;0) = A + BMn^{1/2} \tag{2}$$

where $\log(\eta a;0)$ = zero shear rate apparent viscosity, K and A are constants, M_w = weight average molecular weight, and Mn = number average molecular weight [30].

Generally, as the branching and molecular weight distribution increases, the viscosity of the PE increases, and its shear rate decreases. The increase in the molecular weight of the polymer increases its viscosity, decreases its melt flow index, and decreases its critical shear rate. Finally, an increase in the molecular weight distribution decreases the PE flow behavior index [11, 13, 31, 32].

4. Applications and usage

Polyethylene has been used extensively in food and beverage packaging because of its excellent properties and cheapness. It is the most used engineering material finding application in food packaging, construction, industrial and chemical industries, automobiles, and other allied companies. It can be used in the film, container, and tubing forms produced by extrusion, blow molding, injection, thermoforming cast, and other-oriented processes. It can be combined with other polymers to form plastics with improved performance characteristics and properties such as polyvinyl chloride (PVC), polypropylene (PP), polyethylene terephthalate (PET), and ethylene vinyl alcohol (EVOH), low-density polyethylenes are mainly used for films production. In contrast, high-density polyethylenes are used for containers and pipings [4, 33].

Different production methods and processes are employed in the manufacture of polyethylene products. They include film-making methods, injection molding, blow molding, and extrusion methods to produce various products. HDPE can be used in packaging and film applications, consumer goods, fibers, textiles, pipes, and fittings. Applications of HDPE in packaging include jerry can, domestic and industrial containers, chemical containers, drums, crates, dustbins, detergent containers, garbage containers, housewares, iceboxes, toys, sports nets, ropes, fishing nets, different types of pipes for gas, water, sewage, drainage, sea outfalls, industrial application, cable protection, steel pipe coating, industrial and decorative fabrics, fuel tanks and sheeting for telecommunication and energy cables due to its excellent chemical resistance and good mechanical and physical properties [34].

LDPE's most popular application is in plastic bags. It is with LDPE that containers, various types of laboratory equipment are formed. Also, it finds application in dispensing bottles, tubings, wash bottles, plastic bags for computer parts and components. LDPE is used for packaging pharmaceutical and squeezes bottles, closures and caps, trash bags, liners, and films for packaging frozen and dry food items and laminations. Other applications include housewares, agricultural bags and films, hoses for water pipes, sub conductor insulators, and cable jackets due to their excellent plasticity, strength, and low water absorption properties. It is used to manufacture milk carton linings, bowls, buckets, squeezable bottles, and cling films [35].

Other applications of polyethylene, including LLPDE, are its application in general purpose films, garment packaging, agricultural films, stretch films, food boxes, and coating cables. It can be used for heat-sealed overwrapping film and container liners for bulk transport, etc.

5. Disposal systems for polyethylene

Polyethylene usage poses a significant threat to public hygiene and the aquatic environment. PE gets accumulated at dumping sites, waterways, gutters, drains, agricultural fields, residences, and roadsides, resulting in a refuse stockpile. The littering of the environment with polyethylene has made the disposal system very difficult. Governments, corporate organizations, and individuals are looking for a cleaner environment by developing different ways of disposing of the waste generated through polyethylene. Some of the proposed disposal systems include (1) recycling, (2) Composting, (3) Landfills, and (5) incineration [36].

5.1 Recycling

Recycling is one of the methods of reducing polyethylene waste littering in our environment. Polyethylene recycling is a process by which polyethylene waste is collected, recovered, and converted into valuable products. These materials can be found in agriculture, packaging, food, pipes, beverage bags, toys, electronics construction, electronic and allied sectors [37, 38]. The most straightforward plastic recycling processes involve collecting, sorting, washing, shredding, melting, and pelletizing. The actual process varies, and this is based on the type of recycled product to be formed. The waste materials collection involves picking all the plastic bags and bottles from dumpsites, waste bins, riversides, homes, businesses, and institutions and gathering them at collection points, recycling facilities for reuse. These collected plastics are sorted into each polyethylene group, separated manually or with machines, and washed to remove impurities. Washing removes the impurities such as adhesives, labels, oil that affect operation. Then they are shredded into fine particles and are fed into various component machines for use. Also, shredded plastic pieces can be used for other applications, such as a binder for paving stones and block making and additive within the asphalt. Recycling involves three processes, namely mechanical, chemical and energy recovery processes [39].

The mechanical recycling process involves using cleaned, sorted, and shredded plastic granules or pellets to form new products like trash cans, toys, bottles, bags, and other reusable products. The primary recycling process involves the conversion of the cleaned shredded plastic waste into similar products using injection molding and extrusion principles. The quality of the manufactured product can be improved by the introduction of virgin raw materials, reinforcers, stabilizers and master batches to the scrap or plastic waste. The difference between the primary recycled product and that gotten from the virgin raw materials is in the quality of the manufactured products. Severino et al. [40] studied the effect of nanofillers and compatibilizers on the mechanical properties of extruded low-density polyethylene waste and the results showed enhanced properties when compared to ordinary LDPE waste. The primary and secondary types of recycling involve the reuse of the products in their original form and structure. **Figures** 2–5 show the mechanical processing of used polyethylene films taking place at the University of Nigeria Nsukka, Enugu State, Nigeria. These films are used water sachets were used to make paving stones for the construction industry. Virgin polyethylene granules can be added here to improve the desired mechanical properties. This mechanical processing technique also involves the collection of plastic waste, sorting, cleaning, drying and reusing of plastic waste. The cleaned waste plastic can be cut into size, agglomerated and then extruded into pellets before being used to manufacture the desired products [41]. The pellets or granules can be fed into an injection molding machine, blow molding machine or an extruder to form different products. Also, the melt can be mixed with other materials to form varieties of composites.

It is in the secondary mechanical process that plastic blocks, paving stones and varieties of products are formed for construction and structural purposes. It can be used as partition walls in buildings for non-load bearing applications. Plastic waste has been used as aggregate in asphalt to improve the skid and crack resistance of pavements [42]. Kumi- Larbi and associates studied the effect of particle size of sand on some properties of water sachet/sand composites used as paving stones/bricks and the results showed that the LDPE (water sachet) bonded sand showed an improvement in the durability, compressive strength, specific heat and thermal diffusivity of the bricks and can be used to as a substitute for cement in some building applications [43].

The third method or tertiary recycling of polyethylene products involves a chemical recycling process in which polymers are chemically converted to monomers or depolymerized to monomers and oligomers through a chemical reaction. The chemical inertness of the polymer and its variable structural nature has limited most studies on polyethylene to pyrolysis. The chemical recycling process is a

Figure 2.
Typical dumpsite where waste collection is being carried.

Figure 3.
Sorting of the polyethylene film waste collected.

process that reduces a polymer to its original monomeric form. The technology uses chemical reactions, heat or both to break down used plastics into raw materials for other chemicals, fuel or new plastics. PE can be processed through thermal or catalytic pyrolysis to produce monomers and small organic molecules. Thermal pyrolysis uses heat under high pressure to break down the PE to different smaller monomeric molecules while catalytic pyrolysis uses a catalyst to reduce both the reaction time and temperature required to break down the PE into its component

Figure 4.
Drying of sorted waste.

Figure 5.
Waste conversion.

monomers [44]. Major research is being carried out on the chemical recycling of the world's biggest plastic waste from PE and the results have been impressive and when concluded will help to reduce the menace associated with PE disposal. Ha'jekova and his colleagues investigated the recycling of LDPE using co-pyrolysis with naphtha at the temperature range of 740°C -820°C and found a high yield of alkenes as its primary products and coke [45].

The energy recovery process is the fourth method of polyethylene recycling, also known as quaternary recycling. This is the recovery of the plastic's energy content on the application of external heat. Incineration is an energy recovery method that reduces a large volume of organic materials in an enclosed system. This method generates a high amount of energy from the polyethylene fed inside the incinerator but poses a higher health risk due to the high production of toxic substances that are carcinogenic to human health.

5.2 Plastic composting

This is the breakdown of waste plastics into small natural substances such as carbon dioxide, water, and methane by microorganisms. Composting can be done through aerobic or anaerobic methods. Aerobic composting is the breakdown of the polyethylene wastes by microorganisms in the presence of oxygen or air, while anaerobic composting is the decomposition or degradation of plastic waste substances by microorganisms in the absence of oxygen. Generally, microorganisms bind themselves to the plastics' surface, colonize the exposed surface, catalyze enzymatic degradation of the plastic into lower molecular weight monomers, dimers, and oligomers, and finally form carbon dioxide and water as its bye products [46]. Some of the microorganisms that biodegrade polyethylene include, Acinetobacter sp., Ideonella sakaiensis, Bacillus sp., Staphylococcus sp., Streptococcus sp., Diplococcus sp., Micrococcus sp., Pseudomonas sp., and Moraxella sp. ss bacteria and, *Aspergillus niger*, *A. ornatus*, *A. cremeus*, *A. flavus*, *A. candidus*, *A. ochraceus, A. nidulans*, and *A. Glaucus* as fungi [47].

5.3 Landfills

Plastic waste constitutes a more significant proportion of municipal solid waste. Most of these solid wastes end in landfills. Research has it that it takes more than 300 years for polyethylene to degrade entirely, and this means that plastic waste disposal is a problem that needs to be talked about globally. Landfills remain the safest way of polyethylene disposal, and the waste is subjected characteristically to mechanical stress, decomposition, leachate, and chemical reactions both by aerobic and anaerobic conditions. Landfills have become a breeding ground for dangerous reptiles and insects. It poses environmental pollution, facilitating the release of harmful pollutants in the air when the landfill is set on fire. Carcinogenic and estrogenic compounds are leased, which are dangerous to life. These compounds harm human and aquatic health, causing diseases like breast and ovarian cancer [48]. The policy on solid waste disposal hinges on the reduction, reuse, and recycling of plastic waste to reduce landfill waste [49].

5.4 Incineration

Incineration is a waste treatment process that involves burning waste materials through the application of heat. Incineration is a widely used disposal method because it takes little space but is not generally used because of concerns about producing toxic gases as the plastics are burnt. The process of incineration involves waste collection, sorting, storage, handling, waste combustion, pollution control, to residue collection and handling. The incinerator is designed to capture the component gases for industrial purposes and remove the feedstock produced. The main product of incineration is carbon dioxide, and this is captured through modern technology to avoid associated problems of CO_2 like ozone layer depletion, which causes global warming. However, only a tiny fraction of the petroleum supply is used to produce polymers, out of which less than 2% of the used products are incinerated. Carbon dioxide and carbon monoxide emissions from incinerated sources are not significant compared to carbon dioxide production arising from the burning of fossil fuels. Incineration and gasification are some of the technologies which convert waste to wealth. Incineration produces high-temperature heat, while combustible gasses are a product of gasification [50].

5.5 Bio- and photo-degradation

The biodegradation of PE is a process whereby microorganisms are used to modify and consume the polymer as a primary source of energy and, by so doing, change the physical and chemical properties leading to structural deterioration, weight loss, and eventual gas evolution. Polyethylene biodegrades naturally over a long period. It takes polyethylene films up to 300 years to biodegrade and polyethylene terephthalate more than 450 years to do the same too. Biodegradation of this polymer can be enhanced by introducing microbes, sunlight, moisture and increasing the hydrophilic properties through modification during production. The Physico-chemical method of PE degradation includes UV and thermal treatment, oxidation of PE with nitric acid followed by the microbial attack, and catalytic enzyme degradation. All these cause reductions of polymer chain size, modification of the structure of the polymer, and eventual weight loss.

6. Conclusion

Polyethylene achieved its dominant position because it has excellent mechanical and physical properties, the ease with which it can be manufactured and converted into various packaging forms, and its relatively low cost. These excellent characteristic properties have made the polymer the materials of the future. One of the problems associated with polyethylene is that it leaches additives. This leaching is increased by acids, how long the polyethylene stays in contact with food items, and the increase in heat applied during service conditions. However, meeting the challenges posed by plastics is not simple, and there exists a lack of awareness surrounding the plastic waste problem. Also, the world is putting efforts and investing in cheaper and more biodegradable polymers, better ways of disposal, and a cleaner environment. Polyethylene is a commodity plastic. Its widespread usage has made disposal of end-use products a problem. Knowledge of the production and processing pathways and their properties will significantly aid businesses and consumers in the proper application and best disposal systems. This will help get a world that is clean, safe, and habitable.

Conflict of interest

The authors declare no conflict of interest.

Author details

Macdenis Egbuhuzor[1*], Chima Umunankwe[2] and Peter Ogbobe[3]

1 Department of Materials and Metallurgical Engineering, University of Nigeria, Nsukka, Enugu, Nigeria

2 Department of Agric and Bioresources Engineering, University of Nigeria, Nsukka, Enugu, Nigeria

3 National Board for Technology Incubation, Abuja, Nigeria

*Address all correspondence to: onyekachi.egbuhuzor@unn.edu.ng

References

[1] Narici L, Casolino M, Fino LD, Larosa M, Picozza P, Rizzo A, et al. Performances of Kevlar and Polyethylene as radiation shielding on-board the International Space Station in high latitude radiation environment/639/166/984/639/766/25/128 article. Scientific Reports. 2017;**7**(1):1-11. DOI: 10.1038/s41598-017-01707-2

[2] Zhong X, Zhao X, Qian Y, Zou Y. Polyethylene plastic production process. Insight-Material Science. 2018;**1**(1):1. DOI: 10.18282/ims.v1i1.104

[3] Akinro AO, Ikumawoyi OB, Yahaya O, Ologunagba MM Environmental impacts of polyethylene generation and disposal in Akure City, Nigeria. Global Journal of Science Frontier Research Agriculture and Biology. 2012;**12**(3):1-8

[4] Andreoni V, Saveyn HGM, Eder P. Polyethylene recycling: Waste policy scenario analysis for the EU-27. Journal of Environmental Management. 2015;**158**:103-110. DOI: 10.1016/j.jenvman.2015.04.036

[5] Patel RM. Types and basics of polyethylene. In: Handbook of Industrial Polyethylene and Technology. 2017. pp. 105-138

[6] Tice P. Polyethylene for food packaging a pplications. In: Packaging Matrials. 1st ed. Vol. 4. Brussels, Belgium: ILSI Europe; 2003. pp. 1-28

[7] Jimenez-Varon CF, Soto RT. Design of a production plant of LDPE in Columbia. Vector Manizales (Columbia). Vol. 10. Enero-diciember 2015. p. 122. ISSN: 1909-7891

[8] Ronca S. Chapter 10 - Polyethylene. Brydson's Plastic Materials, Eighth Edition. Butterworth-Heinemann; 2017. pp. 247-278. ISBN: 9780323358248. DOI: 10.1016/B978-0-323-35824-8.00010-4

[9] Kennedy JP, Langer AW. Recent advances in cationic polymerization. Fortschritte der Hochpolym. 1964;**3**(3):508-580. DOI: 10.1007/BF02189446

[10] Peters EF, Zletz A, Evering BL. Solid catalysts in ethylene polymerization. Industrial and Engineering Chemistry. 1957;**49**(11):1879-1882. DOI: 10.1021/ie50575a035

[11] Ertl G, Knözinger H, Weitkamp J. Handbook of Heterogeneous Catalysis. 2nd ed. Weinheim: VCH Verlagsgesellschaft mbH; 1997

[12] Nowlin TE. Business and Technology of the Global Polyethylene Industry: An In-depth Look at the History, Technology, Catalysts, and Modern Commercial Manufacture of Polyethylene and Its Products. New Jersery: John Wiley & Sons; 2014

[13] Carothers WH, Hill JW, Kirby JE, Jacobson RA. Polyethene. Journal of the American Chemical Society. 1930;**52**:5279

[14] AWWA. Engineering properties of polyethylene. In: PE Pipe—Design and Installation. 1st ed. USA: American Water Works Association; 2006. pp. 1-19

[15] Kubik L, Zeman S. Tensile properties of polyethylene films and paper. In: 5th International Conference on Trends in Agricultural Engineering 2013. 2013. pp. 707-711

[16] Christensen M, Behner B. Engineering properties of American Water Works Association;**44**(5):1-31. DOI: 10.12999/AWWA.M55ed2

[17] Cheremisinoff NP. Condensed Encyclopedia of Polymer Engineering Terms. Boston: Elsevier Inc; 2001

[18] Porter RS, Southern JH, Weeks N. Polymer modulus and morphology: The tensile properties of polyethylene. Polymer Engineering and Science. 1875;**15**(3):213-218. DOI: 10.1002/pen.760150315

[19] Defonseka C. Polymeric Composites with Rice Hulls: An Introduction. Berlin, Germany: De Gruyter; 2019

[20] Mizera A, Chalupa P, Hudec I. Impact resistance of high-density polyethylene against falling penetrator with different potential energy. In: MATEC Web of Conferences, CSCC. Vol. 2017. 2017. pp. 1-4. DOI: 10.1051/matecconf/20171250 CSCC 2017 2050

[21] Choi HR, Jung K, Chuc NH, Jung M, Koo I, Koo J, et al. Effects of prestrain on behavior of dielectric elastomer actuator. In: Bar-Cohen Y, editor. Proc SPIE 5759, Smart Structures and Materials: Electroactive Polymer Actuators and Devices (EAPAD). San Diego, CA; 2005. pp. 283-291

[22] Ligia G, Deodato R. Viscoelastic behaviour of polymers. In: Physicochemical Behavior and Supramolecular Organization of Polymers. Dordrecht, Netherlands: Springer Netherlands; 2009. pp. 43-16

[23] Emri I, Zupančič B, Gergesova M, Saprunov I, Gonzalez-Gutierrez J, Bek M. Importance of viscoelastic characteristics in determining functionality of time-dependent materials. DNYA. 2012;**79**:97-104

[24] Tervoort TA, Visjager J, Smith P. Abrasive wear of polyethylene. Macromolecules. 2002;**35**:8467-8471. DOI: 10.1021/ma020579g

[25] Utchings IM. Tribology: Friction and Wear of Engineering Materials. 2nd ed. London: Arnold; 1995

[26] Cosmoiu A. Engineering properties of polyethylene. In: American Water Works Association Manual. M55 ed. Denver, CO: American Water Works Association; 2005. pp. 1-19

[27] Keane JJ, Norris FH, Stein RS. Light scattering from oriented polyethylene films. Journal of Polymer Science. 1956;**20**(94):209-211

[28] Aggarwal SL, Sweeting OJ. Polyethylene: Preparation, structure, and properties. Chemical Reviews. 1957;**57**(4):665-742. DOI: 10.1021/cr50016a004

[29] Gilbert M. Chapter 4 - Relation of Structure to Thermal and Mechanical Properties. Brydson's Plastics Materials, Eighth Edition. Butterworth-Heinemann; 2017. pp. 59-73. ISBN: 9780323358248. DOI: 10.1016/B978-0-323-35824-8.00004-9

[30] Brydson JA. Aliphatic polyolefins other than polyethylene, and diene rubbers. In: Plastics Materials 7. 7th ed. Oxford: Butterworth-Heinemann; 1999

[31] Allal A, Lavernhe A, Vergnes B, Marin G. Relationships between molecular structure and sharkskin defect for linear polymers. Journal of Non-Newtonian Fluid Mechanics. 2006;**134**(1-3):127-135. DOI: 10.1016/j.jnnfm.2005.12.010

[32] Combs RL, Slonaker DF, Coover HW. Effects of molecular weight distribution and branching on rheological properties of polyolefin melts. Journal of Applied Polymer Science. 1969;**13**(3):519-534. DOI: 10.1002/app.1969.070130312

[33] Lotha G. Polyethylene, In major industrial polymers: Polyethylene (PE). Encyclopedia Britannica; 2019. Retrieved from: https://www.britannica.com/science/polyethylene [August 23, 2021]

[34] Omnexus. Polyethylene (PE)-Complete Guide. In: Advanced Guide for A to Z Information on PE; 2021.

Retrieved from: https://omnexus.specialchem.com/selection-guide/polyethylene-plastic [August 22, 2021]

[35] R. Asit, Industrial Uses of Polyethylene, 2015. Available from: http://www.kplintl.com/BLOG/INDUSTRIAL-USES-OF-POLYE THYLENE [Accessed: August 23, 2021].

[36] Grover A, Gupta A, Chandra S, Kumari A, Khurana SP. Polythene and environment, International Journal of Environmental Sciences. 2015;**5**(6): 1091-1105. DOI: 10.6088/ijes.2014050100103

[37] Soto JM, Blázquez G, Calero M, Quesada L, Godoy V, Martín-Lara MA. A real case study of mechanical recycling as an alternative for managing of polyethylene plastic film presented in mixed municipal solid waste. Journal of Cleaner Production. 2018;**203**:777-787. DOI: 10.1016/j.jclepro.2018.08.302

[38] Al-Salem SM, Antelava A, Constantinou A, Manos G, Dutta A. A review on thermal and catalytic pyrolysis of plastic solid waste (PSW). Journal of Environmental Management. 2017;**197**:177-198

[39] Hopewell J, Dvorak R, Kosior E. Plastics recycling: Challenges and opportunities. Philosophical Transactions: Biological Sciences. 2009;**364**:2115-2126 Available from: http://www.jstor.org/stable Accessed: August 25, 2021

[40] Severino PRP, Montanheiro TLDA, Ferro O, Passador FR, Montagna LS. Protective low-density polyethylene residues from prepreg for the development of new nanocomposites with montmorillonite: Recycling and characterization. Recycling. 2019;**4**(4):45. DOI: 10.3390/recycling4040045

[41] Grigore M. Methods of recycling; Properties; and applications of recycled thermoplastic polymers. Recycling. 2017;**2**(4):24. DOI: 10.3390/recycling2040024

[42] Awoyera PO, Adesina A. Plastic wastes to construction products: Status, limitations and future perspective. Case Studies in Construction Materials. 2020;**12**:e00330. DOI: 10.1016/j.cscm.2020.e00330

[43] Jnr AK-L, Silva M, Galpin R, Lenkiewicz Z, Cheeseman C. Recycling plastics in developing countries: properties of novel polymer-sand composites produced from waste plastics e. In: Thessaloniki 2021 8th International Conference of Sustainable Solid Waste Management. 2021. pp. 1-16

[44] Thiounn T, Smith RC. Advances and approaches for chemical recycling of plastic waste. Journal of Polymer Science. 2020;**58**(10):1347-1364. DOI: 10.1002/pol.20190261

[45] Hájeková E, Bajus M. Recycling of low-density polyethylene and polypropylene via copyrolysis of polyalkene oil/waxes with naphtha: Product distribution and coke formation. Journal of Analytical and Applied Pyrolysis. 2005;**74**(1-2):270-281. DOI: 10.1016/j.jaap.2004.11.016

[46] Tokiwa Y, Calabia BP, Ugwu CU, Aiba S. Biodegradability of plastics. International Journal of Molecular Sciences. 2009;**10**(9):3722-3742. DOI: 10.3390/ijms10093722

[47] Kathiresan K. Polythene and plastics-degrading microbes from the mangrove soil. Revista de Biología Tropical. 2003;**51**:3-4 Available from: https://www.scielo.sa.cr/scielo.php

[48] Canopoli L, Coulon F, Wagland ST. Degradation of excavated polyethylene and polypropylene waste from landfill. Science of the Total Environment. 2019;**698**:134125. DOI: sci-hub.se/10.1016/j.scitotenv.2019.134125

[49] Usapein P, Chavalparit O. Options for sustainable industrial waste management toward zero landfill waste in a high-density polyethylene (HDPE) factory in Thailand. Journal of Material Cycles and Waste Management. 2014;**16**:373-383. DOI: 10.1007/s10163-013-0198-6

[50] Council NR. Waste incineration overview. In: Waste Incineration and Public Health. 2000th ed. Washington, DC: The National Academies Press; 2000. pp. 1-334

Chapter 6

An Evaluation of Recycled Polymeric Materials Usage in Denim with Lifecycle Assesment Methodology

Sedef Uncu Aki, Cevza Candan, Banu Nergis and Neslihan Sebla Önder

Abstract

Today, World economy is only 8.6% circular, which creates a huge potential in materials reuse. To close the Emission Gap by 2032, this percentage needs to be doubled. The circular economy ensures that with less virgin material input and fewer emissions. With the help of effective recycling technologies, virgin material use can be decreased and especially petroleum based materials impact can fall within planetary boundaries. This book chapter analyzes different chemical and biological recycling technologies, their advantages and challenges in denim production. Moreover, Life Cycle Assessment (LCA) analysis will be used to evaluate the environmental impact of recycled polymeric materials usage in denim fabrics. Finally, it concludes by challenges and the future of chemically recycled materials in denim production and opportunities to evaluate waste as a raw material to design circular systems.

Keywords: denim, recycled polymeric materials, life cycle assessment, design for sustainability, circularity

1. Introduction

Recycling has been placed on the political agenda in the early 1990s with the rapid increase in the production and consumption of plastics and the increasing concerns over limited resources and plastics waste. As a result, a new industry, the plastics recycling industry, emerged. The main concerns behind the emergence of this industry were the environmental impact of plastics production, its waste management and the fact that almost all synthetic polymers are non-degradable. In many countries, used plastics end up in landfills and the landfill space, as well known, is limited. Two approaches, as a solution to these challenges were discussed to achieve the goal of sustainable development for material consumption [1]: dematerialisation and transmaterialisation.

Dematerilization, increasing the rate of recycling, is highligted as one of the three solutions in a recent report published by Ellen Mac Arthur Foundation [2]. Global Commitment Report stated these three solutions as: to eliminate the plastic items we do not need; innovate so all plastics we do need are designed to be safely reused, recycled, or composted; and circulate everything we use to keep it in the economy and out of the environment. Transmaterilization, a shift to bio-based raw materials, on the other hand, have attracted much attention due to the public concerns over limited fossil fuels and climate change recently [3, 4]. Global bioplastics production capacity is set to increase from around 2.1 million tonnes in 2020 to 2.8 million tonnes in 2025 [5].

Denim industry has started to vastly use recycled polymeric materials in the last 5 years. Recycled PET and recycled elastane as an example of synthetic polmers and recycled modified cellulose as a bio-based polymer are the main materials in use. Many novel bio-based plastics, such as PLA and stratch plastics are still at their early stage in use for denim fabrics.

A research gap exists on the environmental impact of denim fabrics from recycled polymeric fibers using the LCA methodology in the context of Turkey. For this reason, this chapter was conducted such that it firstly addresses sources of polymeric recycled materials as well as their chemical and biological recycling process. Finally, Life Cycle Assessment (LCA) methodology is adopted to quantify and compare the environmental impact of denim fabrics from recycled synthetic/man-made fibers.

2. Sources of polymeric recycled materials

Being part of the European Green Deal, the new Circular Economy Action Plan points out textiles and plastics as two of the key value chains that will be addressed as a matter of priority [6]. The textiles' system is characterized by significant greenhouse gas emissions and a high use of resources: water, land and a variety of chemicals [7, 8]. Textile industry is the third consumer of plastics after Packaging and Building & Construction Industries, representing a quarter of the world carbon dioxide budget [9–11]. Every year, 80 billion new garments are produced for fast-growing fashion industry [12], and apparel business utilizes over 97% natural and synthetic based (principally plastic) virgin materials, only 12% of which is recycled into another item after disposal [13–15]. Moreover, apparel industry represents up to 2% of worldwide oil request, and in this manner a portion of the 300 million tons of plastic is created consistently. The production of fibers, their finishing processes and the chemicals for the required functional properties add up to about 20% of worldwide modern water contamination credited to the sector [13–15]. There is no need to say that effective recycling of such wastes will bring ample environmental and economic benefits such that about 7.5 million cubic yards of landfill space, 17 million tons of CO2, and 4.2 trillion gallons of water can, for example, be saved [12, 16, 17]. In addition to these, it should also be noted that the world population of 7.6 billion people is anticipated to reach 8.6 billion in 2030, 9.7 billion in 2050 and almost 11 billion in 2100, thus resulting in a further increased consumption of textiles and apparel, and in turn that of synthetic (polymeric) fibers [18].

The cause of the plastic pollution is mostly due to the fact that plastic, being generally landfilled or incinerated, does not decompose in the environment, continually accumulating in the waterways, agriculture soils, rivers and oceans, massively contributing to global warming [10, 11], causing damages to biodiversity and ecosystem services, and leading to social and economic drawbacks [19, 20]. Plastics have tended to replace traditional materials such as wood, glass, etc. because of their lower weight, flexibility, and simple processing. They can be made from a single polymer, or multiple layers of different polymers, or other materials. Generally, plastics can be classified into petrochemical-based or bio-based, depending on the material from which they are made. Petrochemical-based plastics can be separated into thermoplastics and thermosets.

Thermoset plastics are permanently cross-linked together and therefore difficult to be reformed whereas thermoplastics can be remelted and reformed, and thus are the most commonly used plastics in the economy [21]. Bio-based plastics are, however, derived from biomass (e.g., starch, sugar, and vegetable oils) excluding materials from geological formations or fossilized, as defined by the European Standard EN 16575 [22–24]. Bio-based plastics can be classified into two groups, namely polymers made entirely from biomass and polymers made partly from biomass [23]. Furthermore, all plastics, regardless of whether they are petrochemical or bio-based, can be designed to behave in two distinct ways: biodegradable and non-biodegradable. Biodegradable plastics can be decomposed in the environment by the activity of microorganisms (bacteria or fungi) into water, carbon dioxide (CO2), methane (CH4), and biomass (e.g., growth of the microbial population), though their biodegradability can vary based on the plastic's inherent and designed properties [25, 26], climatic and process specific conditions, and degradation speed [21, 27, 28]. Biodegradable plastics may also be compostable, which are capable of undergoing biological decomposition in a compost site as part of an available program, and the resultant breakdown fragments are completely used by the microorganisms under the certain conditions certified by the international standards ISO 17088, EN 13432 (Europe), ASTM D400, and D6868 (United States) [28, 29]. It should, though, be noted that not all biodegradable plastics are compostable, but all compostable plastics are biodegradable [30].

A slower development within the field of recycling of plastics in terms of the methods employed, created added value, properties of recycled polymers, etc., causes some problems regarding the inclusion of such polymeric materials in the economic cycle [30, 31]. Among typical examples of waste stream products, are there short-life packaging materials (bags, bottles, etc.), used goods (computers, cell phones, etc.), demolition materials from buildings, and disposables. About 1 million plastic bottles, for example, are wasted every minute and are estimated to double in the next 20 years [32].

With about 13% of the market share, textiles are an important source of man-made polymers. Nearly 63% of the textile fibers are made from petrochemical materials such as nylon, acrylic, polyester, and polypropylene, and these fibers' production and fate give rise to significant carbon dioxide (CO2) emissions [33]. From the perspective of plastics, the scope of this section is, therefore, limited to textiles made of synthetic fibers (**Figure 1**). Polyester is one of the most popular fiber in the textile industry [34], which is followed by polyamide (nylon).

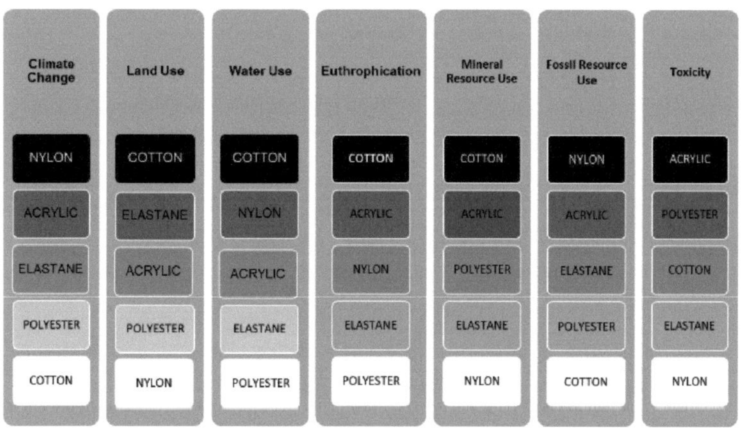

Figure 1.
Comparison of the environmental impacts of the manufacturing of 1 kilogram of dyed, woven fabric (black = worst, white = best) (adapted from [35]).

Polyester alone had a market share of around 52% of total global fiber production, and approximately 58 million mt of polyester was produced in 2019. Polyamide, on the other hand, accounted for 5.6 million mt and approximately 5 percent of the global fiber market in 2019 [36]. In the last years, recycled PET (rPET) production has enhanced dramatically, but only 30% of PET bottles were recycled [37, 38]. In 2019, the estimated rPET share of polyester staple fiber was as high as around 30 percent whereas that of polyester filament was at around 6 to 7% [36]. Recycled PET fibers have the potential to replace virgin PET (v-PET) fibers, and these fibers can be blended with other polymers to create the required properties for each relevant application. But more research appears to be needed to uncover the further potential of rPET fiber based applications [31, 39–42].

Polyamides (PAs), also known as nylons, are other polymeric materials which are widely used in many engineering applications including textile fibers. This is due to their excellent mechanical properties, chemical resistance, wear resistance, dimensional stability, low friction, etc. and ease of processing. Unfortunately, these useful properties are also the ones causing significant environmental consequences. In fact, nylon accounts for about 10% of the debris, mostly in the form of fishing nets in oceans. Contamination is also another issue as far as nylon is concerned. This is mainly because of the fact that nylon is melted at lower temperatures, meaning some contaminants, i.e. non-recyclable materials and microbes or bacteria, can survive. Therefore, all nylon waste must be cleaned thoroughly before a recycling process [41].

As a final note, there is a number of textile companies that have successfully applied the various recycling technologies to produce commercially available raw polymeric materials as is presented in **Table 1** [43, 44]. Sportswear brands are in particular increasingly using recycled synthetic fibers. Most use rPET made from PET bottles [45, 46], but some brands work with recovered ocean plastics, recycled nylon made from discarded fishing nets, and/or with recycled elastane. Also, there are several brands looking for plant-based fibers such as lyocell, Tencel to replace polyester [47, 48].

2.1 Sources of biologically recycled materials

Due to the public interest in the environment, climate change and the limited resources of fossil fuel, bio-based plastics which could be divided into three principal groups, have been investigated for some time [49]:

Manufacturer	Product	Material	Source
Unifi	Repreve®	Recycled polyester	Plastic bottles, industrial waste, fabric scraps
Tenjin	Ecocircle®	Recycled polyester	Consumer and industrial waste
Aquafil	Econyl®	Recycled nylon	Post-consumer waste, fishing nets, old carpets, etc.
Speedo	Powerflex Eco®	Recycled nylon	Fabric scraps, offcuts, fishing nets, old carpets
Jeplan	BRING Material™	Recycled polyester	post- and pre-consumer textiles
Polygenta	perPETual	Recycled polyester	recycled plastic bottles and pre-consumer textiles

Table 1.
Samples of recycled polymer based commercial products [43, 36].

1. bio-plastics that are biodegradable, like starch plastic, cellulose polymers, proteins, lignin and chitosan plastics, polylactic acid (PLA), polyhydroxy alkanoates (PHAs), but also polyhydroxybutyrates (PHBs), polyhydroxyvalerate (PHV) and their copolymers in different percentages (PHBV). This class includes polymers such as PVC, PE, PP, PET, nylon and polyamides (PA), named as bio-plastics because the starting monomers could be obtained from biological resources;

2. bio-plastics based on petroleum resources, which are 100% biodegradable, like polycaprolactone (PCL), polybutylene succinate (PBS), polybutylene adipate (PBA) and its copolymers with synthetic polyesters like polybutylene adipate-terephthalate (PBAT) and polyvinyl alcohol (PVOH);

3. bio-plastics obtained by using monomers from both biological and petroleum resources like polyesters obtained with petroleum-derived terephthalic acid and biologically derived ethanol, 1,4-butanediol and 1,3-propanediol, such as polybutylene terephthalate (PBT), polytrimethylene terephthalate (PTT), polyethylene-co-isorbite terephthalate (PEIT), polyurethane (PUR) and epoxy resins (thermoset plastic).

The actual development stage of emerging bio-based polymers is given in **Figure 2**. While the main application of bio-based plastics is in packaging, around 11% of global production (240.000 tones), are used in textiles, mainly polytrimethylene terephthalate (PTT) and PLA [20, 35, 50, 51].

Bio-based synthetic fibers are often mentioned as environmentally friendly alternatives to traditional, virgin fossil-based ones. Mostly bio-based synthetic fibers, such as bio-PET, are developed to have the same properties and therefore chemical composition as their fossil-based counterparts [35]. It is important to note

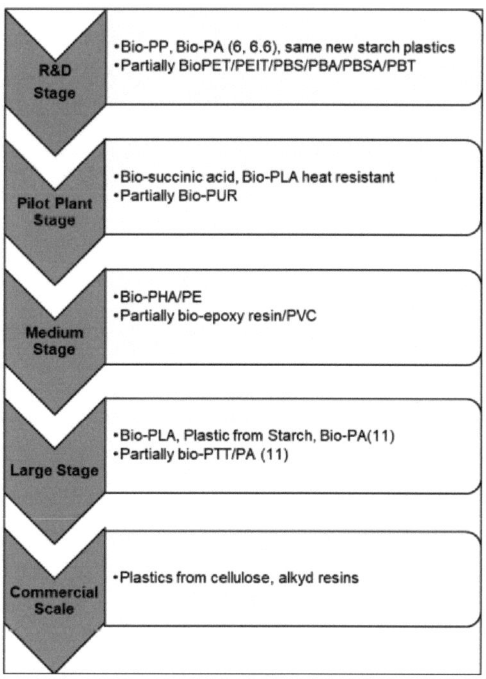

Figure 2.
Development stage of emerging bio-based polymers (adapted from [49]).

that bio-based origin does not imply that the fibers are bio-degradable [13–15]. The key to bio-based synthetics lies in bio-based feedstocks, and the production of these feedstocks has some sustainability issues such as the use of land. This is due to the fact that the cultivation of biomass for bioplastic production can compete with food production for arable lands [35, 52].

As is well known, there are differences in terms of environmental impact between crop based-and waste-based feedstocks. Crop-based corresponds generally to commodity crops such as corn or sugar cane whereas biomass or waste-based raw materials employ agricultural residues and organic waste. There are commercially available bio-based fibers and yarns and some of them are given in **Table 2** [36]. Obviously, feedstocks from waste will most likely be more environmentally preferred and more efficient, as they generally do not require new production of crops and they reutilize residues that would otherwise end up discarded [52]. Additionally, although there is growing interest in textile and apparel industry for bio-based fibers, some barriers such as production cost and low process efficiencies, hinder their further commercialization. Also, some new bio-based fibers, having different structures and properties than conventional ones, cannot yet be handled in current textile manufacturing processes.

Addressing climate change is one of the most urgent action areas for the textile industry. If apparel industry continues on its current path, by 2050, it could use

Bio-based polyester
Far Eastern: TopGreen®, bPET filament made with 30% bio-based feedstock from sugarcane.
Far Eastern: bio-based PTT and bio-based PLA made with NatureWorks Ingeo™, which is made from corn.
INVISTA: LYCRA® T400® EcoMade fiber. More than 65% of the overall fiber content comes from a combination of chemically recycled plastics (PET bottles) and renewable plant-based resources (corn).
Palmetto Synthetics: bio-based PLA staple fiber made with NatureWorks Ingeo®, which is derived from corn.
Radici: CornLeaf filament yarn based on Ingeo™ PLA biopolymer, which is made from corn. And a 30% bio-based polyester filament yarn produced from bio-PET resins made with plant-based bio-MEG.
Toray: Ecodear® PET, a 30% plant-based polyester fiber derived from sugarcane and Ecodear® PTT, a 30% plant-based and a 100% bio-based PLA filament.
Trevira: bio-based PLA fibers and filaments made with Nature Works LLC Ingeo™ which is made from grain (corn).
Bio-based polyamide
Cathay: TERRYL®, a bio-based polyamide line offering PA56, PA510, PA512, PA514 and co-polymers chips and filament with 31 to 100% renewable shares.
Fulgar: EVO®, a 100% bio-based polyamide yarn made from castor oil.
RadiciGroup: Biofeel®, a 64 to 100% bio-based polyamide filament yarn derived from castor oil and agricultural waste and Dorix® 6.10 is 64% bio-based polyamide staple fiber/polyamide yarn
Toray: ECODEAR® PA 6.10, a bio-based polyamide filament derived from the castor bean.
Recycled/bio-based elastane
Asahi Kasei: Roica™ EF, The Global Recycled Standard (GRS) certified recycled elastane, polyurethane filament
Invista: LYCRA® EcoMade fiber, elastane made with recycled content (20% pre-consumer content, diverting waste, and keeping materials in use) and LYCRA® 162 R fiber, an elastane fiber with 70% bio-based content derived from corn.

Table 2.
Bio-based fibers and yarns (adapted from [36]).

more than 26% of the carbon budget associated with a 2 °C pathway. Therefore, in addition to academic institutions several companies have been exploring innovative approaches to recycle carbon and directly use it as feedstock for textiles. Covestro is working with university partners and various textile manufacturers to develop the production process on an industrial scale and aim to make the innovative fibers ready for the market. The company announced in 2019 that they have succeeded in making elastic textile fibers based on CO2 and so partly replacing crude oil as a raw material [53]. Fairbrics, being another example, has developed a novel process to create the components of polyester from waste CO_2, and with Airwear, became a Global Change Award winner in 2020 [36, 54]. The significant amount of textile waste generated is also a potential feedstock for bio-based products [55]. A detailed discussion for this very resource and its conversion to feedstock is given in Section 4.

3. Chemical recycling methods

Heaps of polymer wastes are generated due to the extensive use of polymers in many applications and generally life time of the product is exceeded by that of the polymers that the product is made of. The recycling strategies allow production of new polymeric materials from waste. There are existing and emerging approaches for recycling polymer waste which may contain thermoplastics or crosslinked polymers. The categories of recycling technologies that include primary, secondary (mechanical), tertiary (chemical), and quaternary approaches may be summarized as follows [17, 31, 44, 56–66]:

Primary recycling, also termed as closed-loop recycling, refers to reprocessing of industrial byproducts and pre-consumer scrap materials to give a product that will be used for the same purpose as the original one without loss of properties. Primary textile waste may be single or complex polymers that are usually easy to recycle.

Secondary recycling involves mechanical applications such as grinding, melting, and reforming, for processing post-consumer products into new ones with different physical/chemical properties. In contrast to primary recycling, extraction/dissolution and purification of materials is needed for secondary recycling since materials with unknown composition and purity are treated.

Tertiary recycling approach covers chemical processes such as pyrolysis and hydrolysis, in which chemical structure of waste is broken and converted to basic chemical constituents, monomers, or fuels. It is also known as feedstock recycling.

Quaternary recycling is waste-to-energy conversion process where the energy of fuel value of waste is recovered via incineration or pyrolysis.

Polymers, either orinating from natural or synthetic sources, can be recycled commercially by using one of the recycling processes aforementioned. **Figure 3** presents the polymer recycling approaches.

Other than the recycling approaches presented, biodegradable recycling is an emerging approach especially performed for natural based textile materials. Special microorganisms, enzymes, diverse bacteria and fungi are utilized for degradation of biological polymers (cellulose, chitin, wood, hemp) and organic compounds (PET, polylactic acid-PLA, 1,4butanediol, etc.) [59].

To recycle polymeric waste, chemical recycling technologies offer complementary solutions to mechanical recycling. Chemical recycling, in other words feedstock recycling, breaks down the synthetic fibers for repolymerization and yields the monomers of the polymers (or partially depolymerises to oligomers) by performing processes such as hydrolysis, pyrolysis, gasification, condensation, glycolysis, hydrocracking, dissolution etc [57, 59].

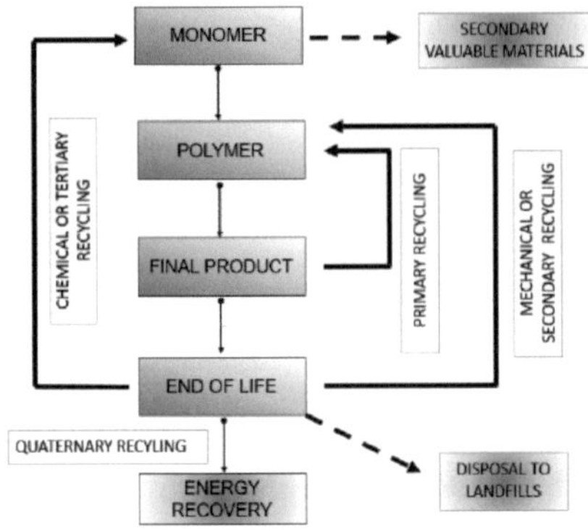

Figure 3.
Polymer recycling approaches (adapted from [56]).

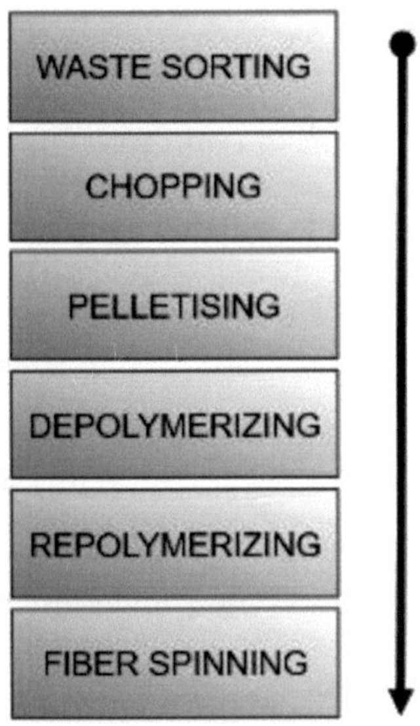

Figure 4.
Simplified process diagram of chemical recycling (adapted from [60, 67].

For chemical recycling the synthetic fibers are chopped, pelletized, depolymerised by chemicals and repolymerised for fiber formation. An overview of the process is shown in **Figure 4**.

Before chemical recycling, the feedstock has to be carefully sorted and all non-textile elements and contaminants (buttons, zips, etc.) need to be removed. Missing or washed-out care labels in garments make manual sorting inaccurate and automated near-infrared identification techniques are required for successful sorting. After sorting, the feedstock is shredded into small fragments to facilitate the dissolving process. The recovery of synthetic fibers involves depolymerisation followed by the production of polymer pellets. A cellulosic pulp is obtained from the cotton. Solvents used are typically recovered to minimize waste and reduce processing costs [35].

Pyrolysis, sometimes referred to as thermolysis, is a thermo-chemical process that polymers are subjected to various temperatures and pressure levels in the presence of catalysts or reactive gases and, decomposed. Pyrolysis processes are economically viable for polyamide 6 polymers. For depolymerisation of polyester, frequently studied approaches are glycolysis, hydrolysis and metanolysis that use glycol, water and methanol, respectively [57].

Solvolysis is depolymerization of cross-linked polymers by using solvents to break covalent bonds in the tridimensional network. The process can be conducted in a wide range of temperature and pressure by employing different solvents. The need for high temperature, pressure and harsh chemicals limit the industrial adoption of solvolysis [17].

Chemical processes allow successful fiber-to-fiber recycling since undesired non polimeric constituents such as colorants, catalysts, surface treatments, backing materials, and other auxiliary chemicals used in textile production are removed. On the other hand, contamination rates as high as 20–30% by weight is economically feasible for chemical recycling. As a result of degradation or contamination of the physical quality during mechanical recycling, chemically recycled polymers offer better inherent quality properties. Another benefit for chemical recycling is that form of the polymer -bottle, jacket, industrial scrap, automobile component, etc.-to be recycled does not matter [58].

Despite its advantages, there are also limitations for large scale applications of chemical recycling processes due to the requirement of considerable amounts of energy inputs for the present methods and due to many uncertainties about their environmental impacts. Even though processes for chemical recycling are technically viable, the ecological and economical impacts need to be questioned. Technologies for chemically recycling of polyester and some other polymers have been existing for some time however, building and operating costs of chemical recycling facilities are higher than those of mechanical recycling facilities. Presence of additives and chemicals used during polymerization might also complicate the processing and affect the purity and quality of the monomers obtained after chemical recycling [35, 58, 59, 65].

Existing chemical processes for recycling polyester are expected to increase and new chemical recycling approaches for polyester in laboratory scale are expected to be developed until 2030. Chemical recycling of cotton and other cellulose based fibers are expected to be developed in full scale by 2030 [68].

3.1 Chemical recycling approaches for commonly consumed polymers

3.1.1 Chemical recycling of polyester (PET)

Commercially available chemical recycling options for polyester waste are [31, 56, 66, 69]:

- Glycolysis in which Bis(2-Hydroxyethyl) terephthalate (BHET) and oligomers are derived using glycol. Virgin PET cannot be produced by glycolysis since waste is partially depolymerized to BHET, and colorants or dyes are not removed. It is a simple, economical and flexible process that can be applied to conventional PET production plants.

- Methanolysis in which polymer is depolymerized to dimethyl terephthalate (DMT) and ethylene glycol (EG) by using methanol. DMT is then purified by crystallization and distillation. PET feedstocks with lower quality are acceptable in the methanolysis process as compared with the glycolysis because purification of DMT is easier compared to BHET obtained by glycolysis.

- Hydrolysis in which polymer is depolymerized to Terephthalic acid (TPA) and ethylene glycol (EG) by using water. High-pressure (1.4–2 MPa), high-temperature (200–250°C) and a long depolymerization time are needed for the process. The hydrolysis of PET can be performed as acid, alkaline, and neutral hydrolysis.

Condition of the input material is important for chemical recycling of PET. When the input material is dyed, for example, decolorization and the removal of dyes require an additional step in the process. Plastic bottles from PET with high molecular weights can be successfully recycled by chemical processes. Even though recycling reduces the chain lengths, recycled materials have an average molecular weight high enough for fiber production [64].

3.1.2 Chemical recycling of nylon

The most widely used nylons are Nylon 6,6 (derived by the polymerization of adipic acid and hexamethylenediamine) and Nylon 6 (obtained from caprolactam). Chemical recycling of nylon includes a depolymerization process followed by distillation to obtain and recover their monomeric constituents: caprolactam (for Nylon-6), and HMDA and adipic acid (for Nylon-6,6). Nylon can be efficiently depolymerized to monomer using chemical and thermal approaches [56, 61, 64, 66].

While chemical recycling, the molecular structure of the polymer is broken down using chemical reactions. After the reaction the products obtained can be purified and used to produce either the same or a related polymer.

During thermal recycling, breaking down of the polymer chemical structure relies on a reaction triggered by heat.

Various chemical processes demonstrated and developed for recycling Nylon 6 and Nylon 6,6 are:

- Hydrolysis of Nylon 6, patented by AlliedSignal, Inc., in which the scrap is dissolved using high pressure steam (963–997 kPa) at 175–180°C for 0.5 hour in a batch process. Then it is continuously hydrolyzed using super-heated steam (350°C) and 790 kPa to form caprolactam. This process needs no additional purification for the recovered monomer.

- Hydrolysis of Nylon 6,6, in which polymer is depolymerized to adipic acid and hexamethylene diamine (HMDA) by the hydrolysis of the polymer in concentrated sulfuric acid. The adipic acid is purified by recrystallization and the HMDA is recovered by distillation after neutralizing the acid.

- Ammonolysis is a process preferred by the DuPont Company for the depolymerization of nylon 6,6 carpet waste. The reaction of nylon 6,6 and nylon 6,6/nylon 6 mixtures with ammonia under 300 and 350°C temperature and 68 atmospheres in the presence of an ammonium phosphate catalyst to gives a mixture of monomers from both nylon 6 and nylon 6,6 polymers.

High costs, challenging materials issues, multiple processing steps requiring high operational knowledge are among the barriers to widespread adoption of chemical recyling methods for nylons.

3.1.3 Chemical recycling of cotton

For fiber to fiber recycling in textiles, the chemical recycling of synthetic materials is not new but studies on natural fiber chemical recycling is relatively new. Although energy products, bio-based products, bacterial cellulose, glucose, aerogels etc. are among the potential applications of chemically recycled cotton, chemical recycling of cotton waste is widely used to produce new fibers that can be used in a variety of products.

The two main routes for chemical recycling of cotton is based on the dissolution of cellulose [61, 70, 71]. Accordingly, either glucose monomers are depolymerized for use in other applications or a polymer dissolution route is followed where cellulosic fibers separate and regenerate by use of solvents. Via the latter process, chemically modified or pure cellulosic fiber products, which can be used as feedstock for regenerated man-made cellulosic fibers (MMCF), may be recovered. A simplified process diagram for chemical recycling of cotton is given in **Figure 5**.

In the Lyocell method, cotton fibers are dissolved (dissoving pulp) using N-methylmorpholine N-oxide (NNMO). The regenerated MMCF are produced by processing the dissolved pulp and and blending with other plant-derived pulp products such as wood, flax, hemp, etc.

Ionic Liquids (ILs) are organic salts in liquid state with low melting point, have chemical and thermal stability, and are non-flammable. They are less aggressive to the environment and considered to substitute organic solvents. The ionic liquid process, yet not commercially available, can be applied to all cotton and also to blends of polyester-cotton.

Another checical recycling approach for cotton is the Ioncell-F cellulose spinning process which uses ionic liquid DBNHOAc without co-solvents or stabilizers [62, 71].

3.1.4 Chemical recycling of blends

The shortcoming of the depolymerization of waste from blended fibers depend on the differences in the fiber regeneration processes. There are ongoing efforts for successfully chemical recycling of blended materials.

In the products from cotton and polyester, chemical recycling has proven to be successful as selective degradation method is used by which the fibers can be artificially isolated and transformed into new ones. When a process using n-methylmorpholine-N-Oxide is applied, cellulose is dissolved. The dissolved cellulose and polyester are then separated by filtration and the captured polyester is respun into a fiber, filament, or yarn while the broken down cellulose can be utilized for the formation of MMCF [44, 60].

Financially viability of chemical fiber-to-fiber recycling of polycotton blends depends on the price and availability of sufficient volumes of well- textile waste and the market value of the resulting cellulosic pulp and polyester pellets [35].

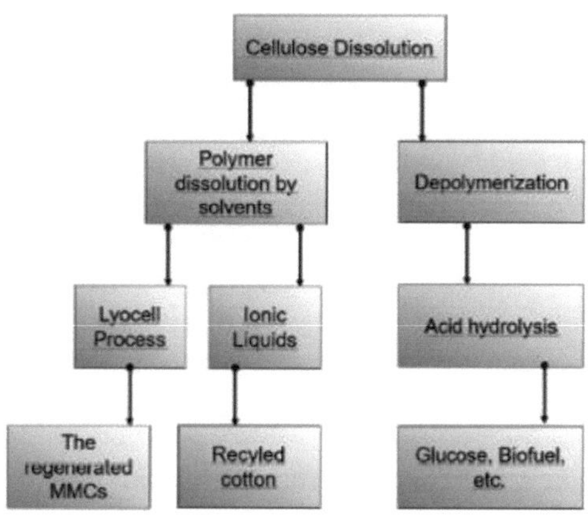

Figure 5.
A simplified process diagram for chemical recycling of cotton (adapted from [61]).

The Japanese Teijin process, for example, is able to handle mixes of 80 percent polyester and 20 percent cotton. However, the process is very sensitive, and only uses Teijin products as input [68].

Nylon and spandex are also two fibers that are commonly used together especially in sportswear and usually, the percentage of nylon is considerably higher than that of spandex. Spandex can be expelled from the blend by dissolving it in solvents, for example, N,N-dimethyl formamide. However, this is a costly process with adverse environmental issues. There has been another successful approach for removing spandex by first treating the blend with heat to degrade the spandex and then washing by ethanol. At the end of the process spandex is successfully evacuated leaving just the nylon [44, 60].

4. Biological recycling methods

Polymers can be degraded by a synergetic combination of different degradation mechanisms present in nature. Enzymes or by products produced by microorganisms such as bacteria, yeasts, fungi, enzyemes, are responsible for the occurance of microbiological degradation. Also, mechanical, chemical or enzymic aging of polymers can be caused by macro-organisms that eat and, sometimes, digest polymers. The main steps of biological degradation are depolymerisation, down to oligomeric or monomeric fragments, and mineralization. Enzymes are the biological catalysts, which can cause huge increases in reaction rates in an environment that is not favorable for chemical reactions [72].

Biological recycling, considered as an improved form of chemical recycling by some researchers, is an emerging approach and expected to be relatively cleaner than conventional textile recycling approaches [31, 73]. During biological recycling, certain polymers can be converted into compost or other substances under specific process parameters (pressure, pH, etc) within the presence of microorganisms [74]. There are several bio-recycling methods developed to recover cotton and PET fiber from textile waste. Research on pretreatment and hydrolysis of cellulose (biodegradable part in textiles) has been carried out to convert it into fermentable glucose [75]. However, polyester in the cotton-based textiles restricts bioconversion of textile waste, as it obstructs the enzymes from reaching the cotton during hydrolysis. Hence, current researches mainly focus on optimizing the conditions for pretreatment and enzymatic hydrolysis, for effective and efficient bioconversion of cotton–polyester blends [76]. According to one such biological recycling method PET fibers could be recovered from polyester-cotton blended wastes by hydrolyzing cotton via enzymatic methods to obtain glucose. The remaining non-biodegradable component-polyester can be re-spun into fibers. The carbon components of most man-made polymers cannot be broken down by the enzymes of microorganisms and that is why such polymers are resistant to biological degradation. Despite this, a commercial approach has been succesfully developed by CARBIOS for enzymatic recycling of PET in various plastics or textiles [58, 77]. Accordingly, the process biologically recycles PET by using an enzyme capable of specifically depolymerizing the polymer to its monomers.

A few organizations, in collaboration with several clothing brands, are also developing sustainable solutions to effectively manage textile waste for obtaining bio-based raw materials. The Hong Kong Research Institute of Textiles and Apparel (HKRITA), for instance, has joined hand with the Hong Kong Polytechnic University to develop new bio-based textiles that combine the properties of Polyactic (PLA) and polyhydroxybutyrate-co-hydroxyvalerate (PHBV) to function as a green alternative to non-biodegradable synthetic polymers in the market. PLA is derived from renewable sources such as cassava roots, corn, and sugar cane, whereas PHBV is naturally produced by fermentation process of bacteria [78, 79]. Worn Again Technologies has, however, focused on converting polyester and poly-cotton blended textiles, and PET plastic back into circular raw materials using their special recycling technology which is stated to be able to separate, decontaminate and extract polyester and cellulose from textile waste, polyester bottles and packaging to produce dual PET and cellulose outputs [80].

Clearly, textile waste with high cellulose content, mainly from cotton, can also be used as an alternative feedstock for man-made cellulosic fibers' (MMCF) production by recycling of post-production, pre-consumer (e.g., samples or stock that cannot be sold) or post-consumer textiles. Promising natural fibers with high α-cellulose content and low hemi-cellulose content are also found in the fibers of banana, pineapple, and abaca leaves. Man-made cellulosic fibers (MMCF) producers are, therefore, intensifying their research and development (R&D) activities to focus on alternative feedstock for cellulose production [81]. Regarding that, Circulose® is one of the few commercially available products. It is "dissolving pulp" from 100% textile waste, such as worn-out jeans and production scraps, used to manufacture viscose, lyocell, modal, acetate other types of regenerated fibers [82]. SaXcell, an abbreviation of Saxion cellulose, is another regenerated virgin textile fiber made from chemical recycled domestic cotton waste. The end product is SaXcell, a regenerated virgin cellulose fiber that can be spun into yarns and turned into fabrics [83]. As may be seen from these few examples, the use of alternative fibers could become an interesting option for the production of man-made cellulosic fibers (MMCF), but they still face some economic, technological and social barriers to scalability.

5. Usage of recycled polymeric materials in denim

Denim itself is very well-known for its durability and long-lasting properties. Although there is no certain lifetime for a jean, the common belief is that it lasts longer than any other garment. This belief relies on the fact that Jeans used to be the uniform of the miners during Gold Rush in the United States.

The fabric parameter that indicates the durability is the strength value of the fabric. Strength value is derived from both yarn strength and the construction. Since the construction remains the same for denim fabric, yarn strength should be analyzed to better understand the contribution of recycled fiber to the strength of the yarn. Fiber fineness (dtex), staple length (mm) and tenacity of the fiber (cN/Tex) are the most important parameters that contributes to the yarn strength. Fiber fineness directly determines the number of fibers in the cross-section of the yarn. The finer the fiber, the number of fibers contributing to the strength will increase and the yarn strength will be higher. Staple length refers to the average length of individual fibers. The shorter the staple length, the more difficult it can be to spin. More twist is needed if the staple length is short. Therefore, this parameter plays an important role on spinnability and strength of the yarn. Tenacity of the fiber directly relates to the yarn strength. The stronger the fiber, the stronger the yarn will be.

Mechanical recycling of natural fibers, mainly cotton, shortens the fiber length drastically. Therefore, a perfect mix of pre & post consumer recycled materials and alternative natural materials is essential. Chemical recycling overcomes this challenge since the required fiber length can be set in the process and the tenacity of recycled fibers are almost the same with the virgin versions in most of the cases. Hence, this ability creates a longer lifetime for a recycled denim with recycled polymeric materials. As seen in **Table 3**, PET and Recycled PET have the highest tenacity values in comparison to the Cotton, Tencel, Recycled Tencel and PLA. This means that using PET in denim increases its durability hence lifetime. On the

Physical Properties	Unit	Cotton*	Recycled Cotton*	Lenzing Tencel	Lenzing Tencel Refibra	Pla-Trevira	Recycled Pet-Unifi Repreve	Regular Pet
Fiber Fineness	dtex	1,77	1,65	1,3	1,7	1,7	1,56	1,39
Staple Length	mm	29	24	38	38	38	38	38
Tenacity	cN/Tex	30	< 30	36	33	33	56	57,5

**Cotton & Recycled Cotton-Company Specific Data.*

Table 3.
A comparison of physical properties of selected fibers [84–87].

other hand, both Tencel/Refibra and PLA have higher tenacity values than Cotton. In theory, using chemically recycled polymeric material and biologically produced PLA in denim production increases the lifetime of the jean.

However, the environmental impact of petrolueum based fibers are fairly high. Synthetic polymers, PET in this study, are produced from fossil fuels. The environmental concerns regarding the use of synthetic polymers bring the main discussion in this study: The environmental impact of recycled polymeric materials and bio-based materials in denim production.

In the previous study of the authors, the effect of using regular, organic and recycled cotton in a denim fabric on environment was analyzed [88]. As a continuum, this chapter analyzes the environmental impact of using polymeric and recycled polymeric fibers in denim fabrics. The past 5 years witnessed the commercialisation of recycled versions of elastane, Tencel®, T400® and many novel bio-based plastics, such as PLA, starch plastics. Today, main recycled polymeric materials that are used in denim production are as follows:

- Recycled Polyester
- Recycled Elastane
- Recycled Tencel- Tencel® x Refibra®
- Recycled Bio-based Polyester- LYCRA® T400® EcoMade fiber

In order to eliminate the effect of finishing processes on the lifecycle assesments of the fabrics, Recycled Elastane and Recycled Bio-based Polyester are excluded in this study. Instead, recycled bio-based polymer, Poly Lactic Acid (PLA) fiber is included. In this study, six fiber compositions for the same denim construction (**Table 4**) are determined to calculate and analyze the effect of different recycled polymeric materials in denim.

Article Code	Composition
Article 01	100% Cotton
Article 02	80% Cotton- 20% PET
Article 03	80% Cotton-20% Recycled PET
Article 04	80% Cotton 20% PLA
Article 05	80% Cotton-20% Tencel
Article 06	80% Cotton- 20% Refibra

Table 4.
Article specification.

Fiber	Source	Global Warming Potential (kg CO^2e/ kg fiber)	Water Use (l/kg fiber)	Land Use (m^2a / kg fiber)	Eutrophication Potential (kg PO_4^{3-} e /kg fiber)	Abiotic Depletion Potential (kg Sb e / kg fiber)
Cotton	Ecoinvent 3 Company specific blend	3,07	1861	7,91	0,013	0,018
PET	Ecoinvent 3	5,26	24,0	0,13	0,012	0,054
Recycled PET	Ecoinvent 3	1,84	1,78	0,07	0,004	0,012
PLA	Confidential Source	3,1	15	0,085	0,0052	0,040
Tencel	[89, 90]	2,32	20	2,4	0,0019	0,007
Refibra	Confidential Source	2,00	18,4	1,68	0,0019	0,007

Table 5.
Summary of raw material data used in this Study.

PET, recycled PET and PLA fibers are supplied from Reliance, Unifi and Trevira, respectively. Tencel & Refibra are lyocell fibers produced by Lenzing AG. More than 99% of wood and dissolving wood pulp used by the Lenzing Group is either certified by FSC® and PEFC™ or inspected in line with these standards. The trees grow quickly without the use of pesticides, fertilizers, irrigation or GMO. Tencel with Refibra Technology contains 30% recycled material as pre-consumer cotton scraps and post-consumer garments from the textile value chain as raw materials and produced with Tencel technology [89].

Table 5 below presents the raw material data set used in the calculation of lifecycle assesment values of the articles listed above.

6. Lifecycle assesment methodology and selected environmental impact categories

The life cycle assessment methodology takes into account all the impacts originated from the inputs and outputs of a system starting from the fiber cultivation/production till the end of life of the jean, cradle to grave. An experimental study was conducted to represent the environmental impact of using recycled polymeric materials in a denim fabric in comparison to using virgin materials. In doing so, the methodology given in the authors' previous study was employed [88]. All of the calculations were performed from cradle to denim factory gate. Furthermore, the inventory was based on the 2020 denim production figures of a denim company in Turkey. As the assessment tool, SimaPro software, developed by the Pré Sustainability, was used. SimaPro is one of the leading Life cycle assessment (LCA) softwares that has been used for more than 25 years by the industry and academics in more than 80 countries. SimaPro uses two types of data: primary and secondary [91]. Primary data involves the basic specifications of a denim fabric, for example Article X, 150 cm in width and 14.89 oz./yd^2 in weight.

This data is exclusive to the fabric production practices of the factory. Secondary data, however, comes from the database and it includes the impacts originated from producing that much raw material and all other inputs such as chemicals at each stage. For the secondary data, Ecoinvent database that is embedded into the software and is the most common Life Cycle Inventory (LCI) database worldwide, was utilized [92]. To be able to perform life cycle assessment of a specific good or service, one needs to have inventory data for the complete supply chain. Due to the huge amount of data needed for this purpose, it is practically impossible to collect and organize the data of the complete background system. In that respect, the Ecoinvent database provides this very system fulfilling the data required for the assessment. The Ecoinvent v3 database contains Life Cycle Inventory (LCI) data taken from various sectors such as energy production, transport, building materials, production of chemicals, metal production and fruit and vegetables. The entire database consists of over 10,000 interlinked datasets, each of which describes a life cycle inventory on a process level [93]. SimaPro software provides six libraries that each contain all the processes that are found in the Ecoinvent database, but use different system models and contain either unit or system processes [94]. The three Ecoinvent system models are as follows: "allocations at point of substitution", "cut-off by classification" and "consequential". For life cycle assessment (LCA) of a product, the production of an item (e.g. denim fabric) is simulated, using both consumption/production (primary) data of a factory and the corresponding secondary data from the Ecoinvent database. The next step is to choose the environmental impacts to be calculated. The whole process is given in **Figure 6**.

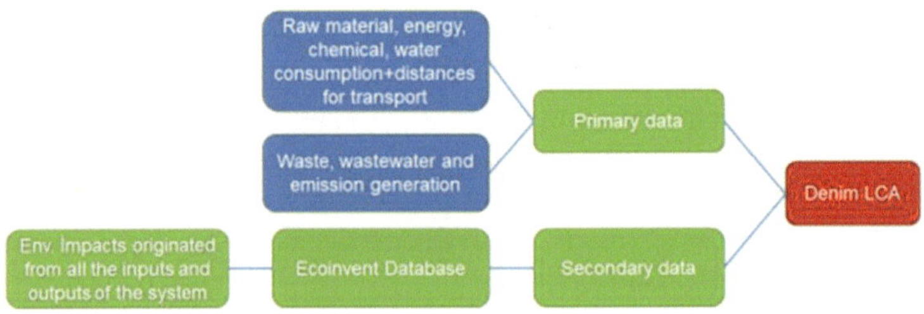

Figure 6.
LCA calculation process.

Indicator	Unit	Description	Example impact	Methodology [92–101]
Global warning potential	kg CO_2 eq (kilogram carbon dioxide equivalent)	Emission of greenhouse gases (GHGs)	Climate change	IPCC 2013 GWP 100a [95]
Freshwater use	lt (liters)	Excessive freshwater taken from the environment	Water scarcity	Life cycle inventory
Land use	m^2 a (meter square per annum)	The amount of agricultural area occupied	Deforestation	ReCiPe 2016 Midpoint (H) [96–97]
Eutrophication potential (EP)	kg PO_4^{3-} eq (kilogram phosphate equivalent)	Emission of substances to water contributing to oxygen depletion	Nutrient loading to water stream-water pollution	CML 2 baseline 2000 [98, 99]
Abiotic resource depletion	kg Sb eq (kilogram antimony equivalent)	Measure of mineral, metal, and fossil fuel resources used to produce a product	Mineral scarcity	CML 2 baseline 2000 [98, 99]

Table 6.
Selected environmental impact categories.

6.1 Selected impact categories

One of the most important parts of life cycle assessment (LCA) is the outputs, in other words the environmental impacts of the product. With SimaPro software, it is possible to calculate over 100 environmental impact categories. For this study, five impact categories were selected [88]. These impact categories, their definitions and calculation methodologies within the SimaPro software are presented in **Table 6**.

7. A comparison of recycled polymeric materials using LCA methodology

The results of the life cycle assessment (LCA) calculations are presented in **Figure 7**. The process flow in the calculations is as follows: Fiber production, transportation, spinning, warp preparation, dyeing and sizing, weaving, finishing, quality control & packaging. It should also be noted that the main differentiator in the calculation of impact categories is fiber mix.

7.1 Global warming potential

The highest Global Warming Potential was calculated for Article 02–80% cotton & 20% PET. The particularly high result for this article is primarily related to its fossil fuel source and high energy requirement of fiber production stage. Once the polyester is recycled as in Article 03, the global warming potential decreases to the lowest value since depletion of fossil fuels is eliminated. On the other hand, the impact value decreases by 13% when the bio-based polymer, PLA is used (Article 04). The fiber production stage is still an important contributor to the global warming potential value when Article 04 is considered.

Although production of pulp and wood-based fibers is also energy needed to process Tencel® and Refibra® does not only rely on fossil fuels. Lenzing site in Austria uses around 80% renewable energy whereas the pulp plant in Paskov employs 100% renewable fuels. Dissolving wood pulp production in Lenzing's biorefineries is not only self-sufficient in terms of meeting its own energy needs, it actually produces surplus energy. This surplus energy (steam and electricity) is used on-site for purposes such as fiber production or export to the local grid, which in turn lowers Lenzing products carbon footprint [84, 89].

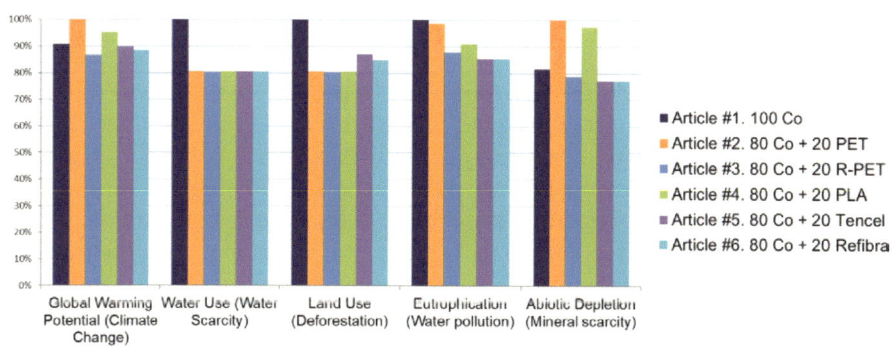

Figure 7.
The results of comparative LCA for different polymeric materials in denim fabric.

7.2 Water use

Water use is dominated by Article 1 with 100% cotton fiber content. This effect originates from the water use in cotton cultivation stage of the overall lifecycle of the denim fabric. Once 20% of cotton is replaced with polymeric materials, the level decreases nearly the same amount, by 20% for the rest of the articles.

Water usage of man-made polymeric fibers has a minor effect in this impact category. Man-made cellulosic fibers, on the other hand, are made from trees such as eucalyptus and beech that can grow with rain water and do not require irrigation which in turns reduces the freshwater use for fiber production hence the fabric [90]. Besides, TENCEL technology; manufactured in closed loop production which recycles water and reuses the solvents at >99% recovery rate. This avoids waste, ensures high resource utilization and results in less water consumption as well as fewer emissions.

7.3 Land use

When the land use for the production of bio-based fibers, cotton, Tencel®, Refibra® and PLA is taken into account, cotton requires a drastically high amount of land compared to the rest of the fibers. PLA land use, here has the lowest value. Tencel and Refibra are slightly higher than PLA. Data shows that cotton requires 65% more land than Tencel and 95% more land than PLA.

The other fibers are syntethised from petroleum, and therefore, Article 02 an 03 have the lowest impact in the impact category under discussion.

7.4 Eutrophication

Eutrophication is the indicator of water emission of substances, mainly nutrients contributing to oxygen depletion and is an indicator of water pollution. The eutrophication potential is mainly originated by the emission of phosphate and nitrate from fertilizers that are used in cotton cultivation.

Although 20% of cotton is replaced by PET in Article 02, the eutrophication level almost stays the same since fiber production requires high energy. Article 04 with PLA and Article 03 with recycled PET score third and fourth with regard to eutrophication impact. Articles with Tencel ® and Refibra® have the lowest impact because of the energy resources that used for their production. Besides, these fibers are compostable and biodegradable in freshwater, marine and soil conditions.

7.5 Abiotic depletion

Abiotic depletion reflects the environmental impacts from using non-renewable energy and material resources. As regular PET fiber is made from crude oil, it has higher abiotic depletion impact compared to other fibers (**Table 4**). This fact increases this impact for Article 2 compared to Article 1–100% cotton version. PLA follows PET for the abiotic depletion impact due to high energy consumption in the lactic acid and lactide production [102]. The differences between manufacturers must be identified and the same calculation should be conducted when renewable energy sources and heat recovery systems are used as in Lenzing® manufacturing system. Therefore, abiotic depletion values for Article 05 and 06 are recorded as the lowest in the study.

8. Discussion

This study mapped and discussed the environmental impact of recycled and bio-based polymeric fibers in a denim fabric. LCA was used as a framework since it is mostly recognized as a tool to quantify the overall impact from a systems perspective.

The data suggested that based on generic data, categorizing fibers as good or bad for the environment is incomplete. The production methodologies used for fibers and their influence on the subsequent product lifecycle are as equally important as fibers' direct environmental impact. A doubled life span of a garment decreases the average garment's climate impact by 52% considering the consumer will buy less garments in a given duration [103, 104]. Therefore, PET, and Recycled PET fibers in this study provided a longer lifetime to a product considering their fiber properties as stated in **Table 3**. On the other hand, fossil fuel and energy use related environmental impact categories such as Global Warming Potential and Abiotic Depletion were calculated as the highest for Article 02 where virgin PET was used. It should be noted that the energy system with which the fibers are produced here may change this data in either positive or negative direction. Once virgin PET is replaced with recycled PET, as in Article 03, these impacts reduce since fossil fuel usage is eliminated. The addition of Recycled PET content reflected the least amount of global warming potential impact. Both PET and Recycled PET presented the best impact for water use and land use categories. Since non renewable resources are already consumed for the virgin material production and post consumer PET remains as a waste for thousands of years in the system if not recycled, one should consider post consumer PET recycled materials (e.g. plastics from oceans) as a solution. On the other hand, the issue with microplastics shedding from synthetic/man-made based textiles which appears to be an important environmental problem, is not discussed in the LCA methodology.

Bio-based man-made polymers, such as Tencel, Refibra and PLA can be stated as the second group to increase the life span of a garment. Especially, Tencel and Refibra scored the lowest in every impact category analyzed in the study, except for the land use. In land use, the value is still lower than cotton. PLA, however, appears to have better values in every environmental impact category, when compared to PET. Although, the study shows that for Global Warming Potential, Eutrophication and Abiotic Depletion impacts, recycled PET is recorded better than PLA, it should be stated here that this result may change with the energy system used for the production of fibers. Besides, PLA is compostable and biodegradable in freshwater, marine and soil conditions.

To conclude, it is unlikely to state that one fiber can be the leader of sustainability alone as there is not one metric that is sufficiently broad enough to incorporate all the fields of sustainability. However, the key point for decreasing footprint of a denim fabric is to use both renewable resources and energy. Therefore, recycled and bio-based materials appear to be an effective solution for a lower environmental footprint in the life of a denim garment. Nevertheless, by bearing the fact that today's waste can be tomorrow's raw material in mind, sources of recycled PET (post or pre consumer) as well as recyclability of garments should be discussed in more detail, which are the topics of further studies.

Conflict of interest

"The authors declare no conflict of interest."

Author details

Sedef Uncu Aki[1], Cevza Candan[2*], Banu Nergis[2] and Neslihan Sebla Önder[1]

1 Orta Anadolu TAŞ, Istanbul, Turkey

2 Istanbul Technical University, Istanbul, Turkey

*Address all correspondence to: candance@itu.edu.tr

References

[1] Clark J H, Deswarte F E I. The biorefinery concept - an integrated approach. In: Clark JH, Destwarte FEI. Editors. Introduction to chemicals from biomass. Renewable Resources. John Wiley & Sons; 2008. p. 1-20.

[2] Ellen McArthur Foundation. Global Commitment 2020 Progress Report. [Internet]. 2020. Available from: https://www.ellenmacarthurfoundation.org/assets/downloads/Global-Commitment-2020-Progress-Report.pdf [Accessed 2021-5-31].

[3] Shen L, Worrell E, Patel MK. Present and future development in plastics from biomass. Biofuels Bioproducts & Biorefining. 2010; 4:25-40.

[4] Shen L, Worrell E, Patel MK. Environmental impact assessment of man-made cellulose fibres. Resources conservation and Recycling. 2010;55:260-274.

[5] Market Update 2020: Global bioplastic market set to grow by 36 percent over the next 5 years. [Internet]. Available from: https://www.bioplasticsmagazine.com/en/news/meldungen/20201202-Market-update-2020-global-bioplastics-market-set-to-grow-by-36-percent-over-the-next-5-years.php#:~:text=Global%20bioplastics%20production%20capacity%20is,continue%20to%20drive%20this%20growth [Accessed 2021-06-01].

[6] European Commission, 2020a, A new Circular Economy Action Plan for a cleaner and more competitive Europe. COM (2020) 98 final. [Internet]. 2020. Available from: https://eur-lex.europa.eu/resource.html?uri=cellar:9903b325-6388-11ea-b735-01aa75ed71a1.0017.02/DOC_1&format=PDF. [Accessed 2021-5-15].

[7] Textiles in Europe's circular economy. Resource efficiency and waste Briefing No 10/2019. EEA. [Internet]. 2019. Available from: https://www.eea.europa.eu/publications/textiles-in-europes-circular-economy. [Accessed 2021-5-15].

[8] ETC/WMGE, 2019b, Textiles and the environment in a circular economy. Eionet Report ETC/WMGE 2019/6. ETC/WMGE. [Internet]. 2019. Available from: https://www.eionet.europa.eu/etcs/etc-wmge/products/etc-reports/textiles-and-the-environment-in-a-circular-economy. [Accessed 2021-5-15].

[9] Utebay B, Celik P, Cay *A. Textile* waste: status and Perspectives. In: Korlu A. Editor. Waste in Textile and Leather Sectors. 2020. IntechOpen. ISBN: 978-1-83880-497-8 (online). DOI:10.5772/intechopen.90014.

[10] Plastic & Climate. The hidden costs of a plastic planet. In: Kistler A, Muffet C. Editors. CIEL Report. [Internet]. 2019. Available from: https://www.ciel.org/wp-content/uploads/2019/05/Plastic-and-Climate-FINAL-2019.pdf. [Accessed 2021-5-15].

[11] Ellen McArthur Foundation. A new textile economy: redesigning fashion's future. Ellen McArthur Foundation Report. [Internet]. 2017. Available from: https://www.ellenmacarthurfoundation.org/assets/downloads/publications/A-New-Textiles-Economy_Full-Report_Updated_1-12-17.pdf. [Accessed 2021-5-15].

[12] Navone L, Moffitt K, Hansen KA, Blinco J, Payne A, Speight R. Closing the textile loop: enzymatic fibre separation and recycling of wool/polyester fabric blends. Waste Manag. 2020;102:149-160.https://doi.org/10.1016/j.wasman.2019.10.026.

[13] Pandey R, Pandit P, Pandey S, Mishra S. Solutions for Sustainable Fashion and Textile Industry.

In: Recycling from Waste in Fashion and Textiles: A Sustainable and Circular Economic Approach. Pandit P, Ahmed S, Singha K, Shrivastava S. Editors. 2020. Scrivener Publishing LLC. USA. ISBN 978-1-119-62049-5. p 33-70.

[14] Teli MD, Pandit P, Singha K Pandit P. Economical and Sustainable Price Sensitive Fashion and Apparels Marketplace. In: Recycling from Waste in Fashion and Textiles: A Sustainable and Circular Economic Approach. Pandit P, Ahmed S, Singha K, Shrivastava S. Editors. 2020. Scrivener Publishing LLC. USA. ISBN 978-1-119-62049-5. p 278-296.

[15] Kumar V, Singha K, Pandit P, Kumar J, Maity S. Sustainability Innovations Coupled in Textile and Fashion. In: Recycling from Waste in Fashion and Textiles: A Sustainable and Circular Economic Approach. Pandit P, Ahmed S, Singha K, Shrivastava S. Editors. 2020. Scrivener Publishing LLC. USA. ISBN 978-1-119-62049-5. pp 299-307.

[16] Quartinello F, Vecchiato S, Weinberger S, Kremenser K, Skopek L, Pellis A, Guebitz GM. Highly selective enzymatic recovery of building blocks from wool-cotton-polyester textile waste blends. Polymers. 2018;10:1107. https://doi.org/10.3390/polym10101107.

[17] Valerio O, Muthuraj R, Codou A. Strategies for polymer to polymer recycling from waste: Current trends and opportunities for improving the circular economy of polymers in South America. Current Opinion in Green and Sustainable Chemistry. Special Issue on South America. 2020;25:100381. https://doi.org/10.1016/j.cogsc.2020.100381

[18] UN World Population Prospect. 2019. Available from: https://population.un.org/wpp/Download/Standard/Population/. [Accessed 2021-5-15].

[19] Boucher J, Friot D. Primary microplastics in the oceans: a global evaluation of sources. International Union for Conservation of Nature and Natural Resources (IUCN) Gland. Switzerland. [Internet]. 2017Available from: https://portals.iucn.org/library/sites/library/files/documents/2017-002-En.pdf. [Accessed 2021-5-15].

[20] Morganti P, Morganti G., Textile and Waste: Will Be Covid-19 A Catalyst for Changing?, J Textile Eng & Fash Tech. 2020;2 (3):26-31. Available from: https://www.opastonline.com/wp-content/uploads/2020/11/textile-and-waste-will-be-covid-19-a-catalyst-for-changing-jteft-20.pdf. [Accessed 2021-5-15].

[21] Hahladakis JN, Iacovidou E. Closing the loop on plastic packaging materials: what is quality and how does it affect their circularity? Sci Total Environ 2018;630:1394-1400. DOI: 10.1016/j.scitotenv.2018.02.330.

[22] EN 16575, 2014. Available from: https://standards.cen.eu/dyn/www/f?p=204:110:0::::FSP_PROJECT:38482&cs=17DD1C2CF17E88404A2E71F51A9D1E6D7. [Accessed 2021-5-15]

[23] Ashok A, Rejeesh C, Renjith R. Biodegradable polymers for sustainable packaging applications: a review. Int J Bionics and Biomater. 2016;1:11. Available from: https://www.researchgate.net/publication/318531449_Biodegradable Polymers for_Sustainable_Packaging_Applications_A_Review. [Accessed 2021-5-15]

[24] Van den Oever M, Molenveld K, van der Zee M, Bos H. Bio-based and biodegradable plastics: facts and figures: focus on food packaging in the Netherlands. Wageningen Food & Biobased Research. 2017. Available from: http://dx.doi.org/10.18174/408350. [Accessed 2021-5-15]

[25] Luyt AS, Malik SS. Can biodegradable plastics solve plastic solid waste accumulation?. In: Plastics to energy. Elsevier. 2019:403-423. DOI:10.1016/B978-0-12-813140-4.00016-9.

[26] Iacovidou E, Velenturf AP, Purnell P. Quality of resources: a typology for supporting transitions towards resource efficiency using the single-use plastic bottle as an example. Sci Total Environ. 2019;647:441-448. https://doi.org/10.1016/j.scitotenv.2018.07.344. [Accessed 2021-5-15]

[27] Iñiguez ME, Conesa JA, Fullana A. Recyclability of four types of plastics exposed to UV irradiation in a marine environment. Waste Manag. 2018;79:339-345. https://doi.org/10.1016/j.wasman.2018.08.006.

[28] Song JH, Murphy RJ, Narayan R, Davies GBH. Biodegradable and compostable alternatives to conventional plastics. Philos Trans R Soc Lond Ser B Biol Sci 2009;364(1526):2127-2139. doi: 10.1098/rstb.2008.0289.

[29] European Environment Agency. The circular economy and the bioeconomy: partners in sustainability. 2018. ISBN 978-92-9213-974-2. ISSN 1977-8449. doi: 10.2800/02937

[30] Hahladakis JN, Iacovidou E, Gerassimidou S. Plastic waste in a circular economy. In: Plastic Waste and Recycling. Letcher MT. Editor. 2020. Elsevier Inc. https://doi.org/10.1016/B978-0-12-817880-5.00019-0. ISBN 978-0-12-817880-5. P 481-510

[31] Ignatyev IA, Thielemans W, Vander BB, 2014. Recycling of Polymers: A Review. ChemSusChem. DOI: 10.1002/cssc.201300898. Available from: https://www.researchgate.net/publication/262151364. [Accessed 15 May 2021]

[32] Magnier L, Mugge R, Schoormans J. Turning ocean garbage into products–consumers' evaluations of products made of recycled ocean plastic. J Clean Prod. 2019;215:84-98. DOI: 10.1016/j.jclepro.2018.12.246

[33] Echeverria CA, Handoko W, Pahlevani F, Sahajwalla V. Cascading use of textile waste for the advancement of fibre reinforced composites for building applications. J Clean Prod. 2019; 208:1524-1536. https://doi.org/10.1016/j.jclepro.2018.10.227

[34] Majumdar A, Shukla S, Singh AA, Arora S. Circular fashion: Properties of fabrics made from mechanically recycled polyethylene terephthalate (PET) bottles. Resour Conserv Recycl. 2020;161:104915. DOI:10.1016/j.resconrec.2020.104915

[35] Plastic in textiles: potentials for circularity and reduced environmental and climate impacts. Eionet Report - ETC/WMGE 2021/1. European Topic Centre Waste and Materials in a Green Economy. [Internet].2021. Available from: https://wmge.eionet.europa.eu. [Accessed 2021-5-15].

[36] Preferred Fiber & Materials Market Report 2020. Textile Exchange. 2021. [Internet]. Available from: https://textileexchange.org/wp-content/uploads/2020/06/Textile-Exchange_Preferred-Fiber-Material-Market-Report_2020.pdf. [Accessed 2021-5-16]

[37] Abbasi M, Mohades Mojtahedi MR, and Kotek R, Experimental study on texturability of filament yarns produced from recycled PET. Text Res J. 2020; 90(23-24):2703-2713. https://doi.org/10.1177%2F0040517520925859

[38] Albini G, Brunella V, Placenza B, Martorana B, Guido Lambertini V. Comparative study of mechanical characteristics of recycled PET fibres for automobile seat cover application. J Ind Text. 2018;48(6):992-1008.

[39] Foolmaun RK, Ramjeeawon T. Comparative life cycle assessment and social life cycle assessment of used polyethylene terephthalate (PET) bottles in Mauritius. Int J Life Cycle Assess. 2013;18(1):155-171

[40] Zhang R, Ma X, Shen X, Zhai Y, Zhang T, Ji C, Hong J. PET bottles recycling in China: an LCA coupled with LCC case study of blanket production made of waste PET bottles. J Environ Manag. 2020;260:110062. DOI:10.1016/j.jenvman.2019.110062

[41] Sadeghi B, Marfavi Y, AliAkbari R, Kowsari E, Ajdari FB, Ramakrishna S. Recent Studies on Recycled PET Fibers: Production and Applications: a Review. Materials Circular Economy. 2021;3:4. https://doi.org/10.1007/s42824-020-00014-y.

[42] Turnbull J, Stevens K, Mzikian P, Bertele M, Cavadini F, Crippa M, Bodin M. Achieving a Circular Textiles Economy Facilitated by Innovative Technology. In: Sustainable Textile and Fashion Value Chains: Drivers, Concepts, Theories and Solutions. Matthes A, Beyer K, Cebulla H, Arnold MG, Schumann A. Editors. 2021. Springer Nature Switzerland AG. https://doi.org/10.1007/978-3-030-22018-1. ISBN 978-3-030-22018-1 (online). p 205-233.

[43] Ahmed MU. Consumer Engagement and Roles for Sustainable Textiles. In: Sustainable Textile and Fashion Value Chains: Drivers, Concepts, Theories and Solutions. Matthes A, Beyer K, Cebulla H, Arnold MG, Schumann A. Editors. 2021. Springer Nature Switzerland AG. https://doi.org/10.1007/978-3-030-22018-1. ISBN 978-3-030-22018-1 (online). P 248

[44] Leonas KK. The use of recycled fibers in fashion and home products. In: Muthu SS, editor. Textiles and Clothing Sustainability: Recycled and Upcycled Textiles and Fashion. 1st ed. Singapore: Springer; 2017. pp. 55-75. DOI: 10.1007/978-981-10-2146-6

[45] Ambiletics. 2021. AMBILETICS | Eco sportswear made from recycled polyester. Available from:https://www.mochni.com/ambiletics-the-new-eco-sportswear-label-from-berlin/. [Accessed 2021-5-15].

[46] Fair and sustainable sportswear - Pure Sportswear. [Internet]. 2021. Available from: https://puresportswear.nl/en/. [Accessed 2021-5-15].

[47] Bleed. 2021. Available from: https://www.bleed-clothing.com/english/sustainable-economy. [Accessed 2021-5-15].

[48] Recycled and Upcycled Clothing Brands You Need to Know About. [Internet] 2021. Available from: https://indiegetup.com/recycled-clothing-brands/. [Accessed 2021-5-15].

[49] Siracusa V, Blanco I. Bio-Polyethylene (Bio-PE), Bio-Polypropylene (Bio-PP) and Bio-Poly (ethylene terephthalate) (Bio-PET): Recent Developments in Bio-Based Polymers Analogous to Petroleum-Derived Ones for Packaging and Engineering Applications. Polymers. 2020;12:1641. DOI:10.3390/polym12081641.

[50] Yusuf M. Agro-Industrial Waste Materials and their Recycled Value-Added Applications: Review. In: Handbook of Ecomaterials. Martinex LMT, Kharissova OV, Kharisov BI. Editor. Springer International. 2019: 48-59.

[51] Smith M, Love DC, Rochman CM, Neff RA. Microplastics in Seafood and the Implications for Human Health. Curr Environ Health Rep. 2018;5: 375-386.

[52] Quick Guide to Biosynthetics. 2018. Available from: https://textileexchange.

org/wp-content/uploads/2018/01/Textile-Exchange-Quick-Guide-To-Biosynthetics-2018.pdf, [Accessed 2021-5-15].

[53] Covestro Solution Center. Article. Elastic textile fibers containing carbon dioxide. 2019. Available from: https://solutions.covestro.com/en/highlights/articles/stories/2019/elastic-textiles-made-from-co2. [Accessed 2021-5-15]

[54] Global Change Award. 2020. Available from: https://globalchangeaward.com/winners/airwear/. [Accessed 2021-5-15]

[55] Shen F, Xiao W, Lin L, Yang G, Zhang Y, Deng S. Enzymatic saccharification coupling with polyester recovery from cotton-based waste textiles by phosphoric acid pretreatment. Bioresour. Technol. 2013;130: 248-255. https://doi.org/10.1016/j.biortech.2012.12.025

[56] Andriotis L, Koutsidis IA, Louka DA, Nianias NP, Siafaka P, Tsagkalias I and Tsintzou G. Recent Advances in the Chemical Recycling of Polymers (PP, PS, LDPE, HDPE, PVC, PC, Nylon, PMMA).In: Achilias D, editor. Material Recycling - Trends and Perspectives. www.intechopen.com;2012. eBook (PDF) ISBN: 978-953-51-4100-6

[57] Björquist S. Chemical recycling of cotton and polyester textiles [thesis]. The Swedish School of Textiles: University of Boras; 2017.

[58] Greenblue. Chemical Recycling-Making Fiber-to-Fiber Recycling a Reality for Polyester Textiles. Report. [Internet]. 2017. Available from: https://greenblue.org/work/chemical-recycling/. [Accessed 2021-5-22]

[59] Hussain A, Kamboj N, Podgurski V, Antonov M, Goliandin D. Circular economy approach to recycling technologies of postconsumer textile waste in Estonia: a review. Proceedings of the Estonian Academy of Sciences. 2021;70(1):82-93. https://doi.org/10.3176/proc.2021.1.07

[60] Kumar PS, Yaashikaa PR. Recycled Fibres. In: Muthu SS. Editor. Sustainable Innovations in Recycled Textiles, Springer Nature Singapore Pte Ltd; 2018. pp.1-19. ISSN 2197-9871

[61] Le K. Textile Recycling Technologies, Coloring and Finishing Methods [report]. UBC Sustainability Scholar. 2018. Available from: https://sustain.ubc.ca/about/resources/textile-recycling-technologies-colouring-and-finishing-methods [Accessed 2021-5-19]

[62] Mait S, Mondal MK, Pandit P, Singha K. Future Mobilizations and Paths of Waste—Towards Best Solution. In: Pandit P, Ahmed S, Singha K and Shrivastava S. Editors. Recycling from Waste in Fashion and Textiles. Scrivener Publishing; 2020. pp.321-340. ISBN 978-1-119-62049-5

[63] Payne A. Open and closed loop recycling of textile and apparel products. In: Muthu SS, editor. Handbook of Life Cycle Assessment (LCA) of Textiles and Clothing. United Kingdom: Woodhead Publishing Series in Textiles, Elsevier Ltd; 2015. pp. 103-121. ISBN: 9780081001875

[64] Rahimi AR, García JM. Chemical recycling of waste plastics for new materials production. Nature Reviews-Chemistry. 2017;1(0046): 1-11. doi:10.1038/s41570-017-0046

[65] Sandin G, Peters GM. Environmental impact of textile reuse and recycling-A review. Journal of Cleaner Production. 2018;184:353-365. https://doi.org/10.1016/j.jclepro.2018.02.266

[66] Sherwood J. Closed-Loop Recycling of Polymers Using Solvents Remaking plastics for a circular economy. Johnson

Matthey Technol. Rev. 2020; 64(1):1-14 https://doi.org/10.1595/205651319X15574756736831

[67] Thompson P, Willis P, Morley N. A review of commercial textile fibre recycling Technologies. [Internet]. 2012. Available from: https://refashion.fr/ecodesign/sites/default/files/fichiers/A%20review%20of%20commercial%20textile%20fibre%20recycling%20technologies.pdf. [Accessed 2021-5-22]

[68] Ljungkvist H, Elander M. Report. Critical aspects in design for fiber-to-fiber recycling of textiles. [Internet]. 2018. Available from: https://www.researchgate.net/publication/305309055 [Accessed 2021-5-14]

[69] Zamani B, Svanstrom M, Peters G, Rydberg T. A Carbon Footprint of Textile Recycling A Case Study in Sweden. Journal of Industrial Ecology.2014;9(4):676-687. DOI: 10.1111/jiec.12208

[70] Johnson S, Echeverria D, Venditti R, Jameel H, Yao Y. Supply Chain of Waste Cotton Recycling and Reuse: A Review. Sustainability. 2020;7(1):19-31. DOI: 10.14504/ajr.7.S1.3

[71] Schuch AB. The Chemical Recycle of Cotton. Revista Produção e Desenvolvimento. 2016; 2(2):64-76. http://revistas.cefet-rj.br/index.php/producaoedesenvolvimento.

[72] Zee M. Biodegradability of Polymers – Mechanisms and Evaluation Methods. In: Bastioli, C. Editor. Handbook of Biodegradable Polymers Rapra Technology Limited; 2005, ISBN: 1-85957-389-4

[73] Hopewell J, Dvorak R, Kosior E. Plastics recycling: challenges and opportunities, review. Phil. Trans. R. Soc. B. 2009; 364: 2115-2126. doi:10.1098/rstb.2008.0311

[74] Subramanian K, Chopra SS, Cakin E, Li X, Lin CSK. Environmental life cycle assessment of textile bio-recycling – valorizing cotton-polyester textile waste to pet fiber and glucose syrup, Resources, Conservation and Recycling. 2020;161(104989). https://doi.org/10.1016/j.resconrec.2020.104989

[75] Huang YB, Fu Y. Hydrolysis of cellulose to glucose by solid acid catalysts. Green Chem. 2013;15: 1095-1111. https://doi.org/10.1039/c3gc40136g.

[76] Gholamzad E, Karimi K, Masoomi M. Effective conversion of waste polyester-cotton textile to ethanol and recovery of polyester by alkaline pretreatment. Chem. Eng. J. 2014;253: 40-45. https://doi.org/10.1016/j.cej.2014.04.109

[77] Enzymatic recycling: removing the constraints of current processes [Internet]. Available from: https://www.carbios.com/recyclage-enzymatique/ [Accessed 2021-5-27]

[78] HKRITA Report. Innovation for Sustainability. [Internet]. 2018. Available from: https://www.hkrita.com/old-files/Centre_Report/2018/00_HKRITA_Report_2018_Full.pdf. [Accessed 2021-5-15]

[79] Subramaniam K, Shauhrat SC, Cakin E, Xiaotong L, Carol Sze KL. Life cycle Approaches for Evaluating Textile Biovalorisation Processes: Sustainable Decision-making in a Circular Economy. In Waste Valorisation: Waste Streams in a Circular Economy. Ki Lin CS, Kaur G, Li C, Yang X. Editors. 2021. First Edition. John Wiley & Sons Ltd.

[80] Worn again technologies. 2021. Available from: https://wornagain.co.uk/. [Accessed 2021-5-15].

[81] Seisl S, Hengstmann R. Manmade Cellulosic Fibers (MMCF)—A Historical Introduction and Existing

Solutions to a More Sustainable Production. In: Sustainable Textile and Fashion Value Chains: Drivers, Concepts, Theories and Solutions. Matthes A, Beyer K, Cebulla H, Arnold MG, Schumann A. Editors. 2021. Springer Nature Switzerland AG. 2021. https://doi.org/10.1007/978-3-030-22018-1. ISBN 978-3-030-22018-1 (online), p 3-21.

[82] Circulose. Available from: https://www.renewcell.com/en/circulose/. [Accessed 2021-5-25].

[83] Regenerated 'virgin' cellulose fiber from cotton waste. [Internet]. 2021. Available from: https://saxcell.com/newfibre. [Accessed 2021-5-15].

[84] https://www.lenzing.com/. [Accessed 2021-5-27]

[85] https://www.trevira.de/en/trevira-brands-the-seal-of-quality-for-functional-textiles/trevira-sinfineco. [Accessed 2021-5-27]

[86] https://unifi.com/products/recycled-materials. [Accessed 2021-5-27]

[87] https://www.ril.com/ourbusinesses/petrochemicals/polyesters.aspx. [Accessed 2021-5-27]

[88] Uncu Akı S, Candan C, Uygun Nergis B, Onder N S. In: Understanding Denim Recycling: A Quantitative Study with Lifecycle Assessment Methodology. Korlu A. Editor. Waste in Textile and Leather Sectors. 2020. IntechOpen. ISBN: 978-1-83880-497-8 (online) DOI: 10.5772/intechopen.92793

[89] Lenzing Study. Turning Climate Commitment into Action. [Private Report]. 2020

[90] Shen L, Patel M. Life Cycle Assessment of man-made cellulose fibres. Lenzinger Berichte. 2010; 88: 1-59.

[91] Heerink M, Koenen A. Experiences and Bottlenecks with Recycling of Jeans [Student Research Project]. Netherlands: Sustainable International Business: Saxion University of Applied Sciences; 2013.

[92] DIN EN ISO 14040:2006-10, Environmental management - Life cycle assessment - Principles and framework (ISO 14040:2006) and Requirements and guidelines.

[93] Ecoinvent Centre. 2020. Available from: https://www.Eco-invent.org. [Accessed:2021-05-16]

[94] SimaPro LCA Package. Pré Sustainability [licensed software]. 2020. Available from: https:// www.pre-sustainability.com. [Accessed:2021-05-16]

[95] IPCC. GWP 100a: methodology developed by the Intergovernmental Panel on Climate Change. [Internet]. 2013. Available from: http://www.ipcc.ch/ [Accessed 2021-6-01]

[96] ReCiPe 2016 v1.1. A harmonized life cycle impact assessment method at midpoint and endpoint level Report I: Characterization. RIVM Report 2016-0104a [Internet]. 2017. Available from: http://www.rivm.nl/en/Topics/L/Life_Cycle_Assessment_LCA/Downloads/Documents_ReCiPe2017/Report_ReCiPe_Update_2017 [Accessed: 2021-06-01]

[97] Huijbregts MAJ, Steinmann ZJN, Elshout PMF, Stam G, Verones F, Vieira M, et al. ReCiPe2016: A harmonised life cycle impact assessment method at midpoint and endpoint level. International Journal of Life Cycle Assessment. 2017;22:138-147. DOI: 10.1007/s11367-016-1246-y

[98] Guinée, J.B.; Gorrée, M.; Heijungs, R.; Huppes, G.; Kleijn, R.; Koning, A. de; Oers, L. van; Wegener Sleeswijk, A.; Suh, S.; Udo de Haes,H.A.; Bruijn, H. de;

Duin, R. van; Huijbregts, M.A.J. 2002. Handbook on life cycle assessment. Operational guide to the ISO standards. Part III: Scientific background. Kluwer Academic Publishers, ISBN 1-4020-0228-9, Dordrecht, 692 pp.

[99] Huijbregts, M.A.J.; Breedveld L.; Huppes, G.; De Koning, A.; Van Oers, L.; Suh, S. Normalisation figures for environmental lifecycle assessment: The Netherlands (1997/1998), Western Europe (1995) and the World (1990 and 1995). Journal of Cleaner Production. 2003;11 (7): 737-748.

[100] Climate Change 2013: The Physical Science Basis. Working Group I contribution to the Fifth Assessment Report of the IPCC. [Internet]. 2006. Available from: http://www.climatechange2013.org [Accessed: 13-04-2021]

[101] Stevens E S. Green Plastics - An Introduction to the New Science of Biodegradable Plastics. Princeton, NJ: Princeton University Press; 2002.

[102] Vink E T H, Davies S. Lifecycle Inventory and Impact Assessment Data for 2014 Ingeo® Polylactide Production. Industrial Biotechnology. 2015;11 (3):167-180

[103] Roos S, Sandin G., Zamani B, Peters G. Environmental Assessment of Swedish Fashion Consumption. Five Garments-Sustainable Futures. Mistra Future Fashion, Stockholm, Sweden. Available from: http://mistrafuturefashion.com/wp-content/uploads/2015/06/Environmental-assessment-of-Swedish-fashion-consumption-LCA.pdf [Accessed 2021-6-01].

[104] Sandin G, Roos S, Johansson M. Environmental impact of textile fibers-What we know and what we don't know (part 2). Mistra Future Fashion.2019:03. ISBN:978-91-88695-91-8

Chapter 7

Retreatment of Polymer Wastes by Disintegrator Milling

Priit Kulu and Dmitri Goljandin

Abstract

Global introduction of waste utilization techniques to the polymer market is currently not fully developed but has enormous potential. Before reintegration of used material into a new product, it normally requires grinding, that is shredding, crushing, or milling. In traditional grinders, the generated stresses in the material to be ground are equal to or less than the strength of the material. If by traditional methods, the stresses generated are compressive + shift, so by milling based on collision are tension + shift. Due to the high stress-material strength ratio at collision, it is possible to crush not only brittle materials but also ductile materials. This process allows easily combining the grinding of composite materials with their separation into individual constituents. In the current study, the mechanical recycling of the following groups of polymer materials was studied: pure brittle and soft polymers (PMMA, HDPE and IER), blends of plastics (ABS+PMMA, PC + ABS), reinforced plastics (PMMA+GFP); elastomers (rubber and tyres), and printed circuit boards (PCB).

Keywords: mechanical recycling, disintegrator mill, selective milling, separation, plastic scrap, composite plastic waste, polymer wastes, plastic powder, electronic wastes, crumb rubber, tyres recycling, electronic wastes, printed circuit boards

1. Introduction

Plastics are inexpensive, easy to mold, and lightweight. These and many other advantages make them very popular candidates for commercial applications. In many areas, they have suppressed traditional materials. However, the problem on recycling still is a major challenge [1, 2]. There are both technological and economic issues that restrain progress in this field. Different types on recycling, primary, secondary, chemical, and biological recycling are discussed with related issues [3, 4].

Global introduction of waste utilization techniques to polymer market is currently not fully developed, but has an enormous potential [5]. Before reintegration of used material into a new product, it normally requires grinding, that is shredding, crushing, or milling. These processes make the material more homogeneous and easier to blend with additives. In traditional grinders like jaw crushers, roller mills, vibration and ball mills the generated stresses in material to be grinded are equal to or less than strength of material. If by traditional methods, the stresses generated are compressive + shift, so by milling based on collision are tension + shift.

Due to the high stress-material strength ratio at collision, it is possible to crush not only brittle materials but also ductile materials.

In current study the mechanical recycling of following groups of polymer materials were studied:

- pure plastics as examples of brittle (ABS and PMMA) and soft materials (HDPE);
- blends of ABS + PMMA, and PC + ABS plastics;
- glassfibre reinforced plastics, (PMMA + GFP);
- compounded plastics - circuit boards, (PCB);
- pure rubber and tyres.

2. Bases of disintegrator milling

2.1 Theoretical aspects of impact milling and separation in disintegrator

In traditional grinding using traditional jaw crushers, ball mills etc., the grindable material remains between the two grinding bodies (**Figure 1a**) and will be crush by pressure or shear stresses [6]. The generated stresses in the particle are equal to or less than the strength of the material.

At traditional grinding methods, will take place uniform size reduction. The selectivity of process is low. It means that by bring weak and strong particles in the active grinding zone both, the strong and weak pieces will be broken up.

In the disintegrator, grinding occurs when piece collides with the moving working blades. This collision is an unlimited hit with some speed. The speeds used in the disintegrator range from 30 to 200 m/s [7].

At the moment of impact of the moving piece on the grinding body, an intensive wave of compression stress begins to propagate in the piece from the area of contact with the working body. The piece remains as a whole, intact during the propagation of the compression wave, until it reaches the opposite side of the piece, where it is reflected as a stretching wave of the same intensity from which the destruction of the piece to particles occurs [8]. The stresses arising in the material in this case (**Figure 1b**) exceed its strength by an order of magnitude.

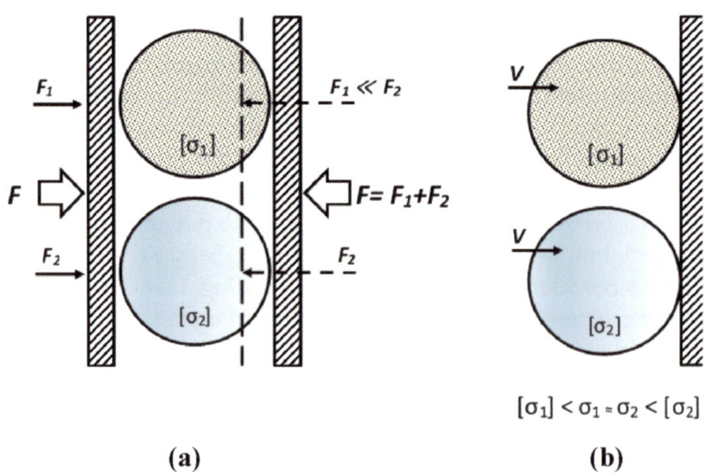

Figure 1.
Stresses in material: (a) Traditional milling; (b) Impact milling (image based on [8]).

Only by impact milling in the disintegrator, i.e. at a certain speed of impact, allows to implement a mode in which a less robust piece will collapse, and a more robust (or defect - free) one will remain intact, and selective size reduction occurs. The degree of selectivity of grinding depends on the materials properties (density, strength) and on the defectivity of the material crystallographic structure. It enables to grind selectively with disintegrator mills and is suitable for processing multicomponent materials and components, like polymer composites, blends and PCBs.

2.2 Developed disintegrators, separators and disintegrator milling systems

The DS series of multi-functional disintegrators were developed at Tallinn University of Technology (TalTech) for the processing of various materials, including polymer materials and plastic scrap [9]. The disintegrators of this series include:

- Special multipurpose disintegrator milling system – the laboratory disintegrator DSL-175 with a combined inertial-centrifugal classifier;
- Semi-industrial multi-functional disintegrator milling system DSL-115 of direct, separative and selective modes of treatment;
- DSA-series industrial disintegrators (DSA-2, DSA-158);
- Device SD-15 – an experimental module for studying the processes of the high-speed grinding of complexly deformed rubber composite;
- Device SD-25 – a module for simulating the processing of solid car tyres into rubber chips, with simultaneous separation of textile and metal cord at room temperature.

These systems are designed for processing various types of materials and are characterized as follows (**Table 1**):

- treats materials by collision at high rate velocities,
- operates on direct, separative or selective systems in portional or steady conditions,
- is flexible in transfer from one system to another,
- increases a material's chemical activity.

Disintegrators	Purpose	Productivity t/h
DSA-158	Precrushing	up to 1.5
DSA-2	Multistage direct milling	up to 5.8
DSL-115	Direct / Separative / Selective milling	up to 1.3 / 0.25 / 1.3
	Separators/classifiers	
Inertial (IC)	For separation of ground product	0.25[a]
Centrifugal (CC)	The same, higher sensitivity to compare with IC	0.15[a]

[a]*Set for separative milling by disintegrator DSL-115.*

Table 1.
Characteristics of used disintegrators and classifiers.

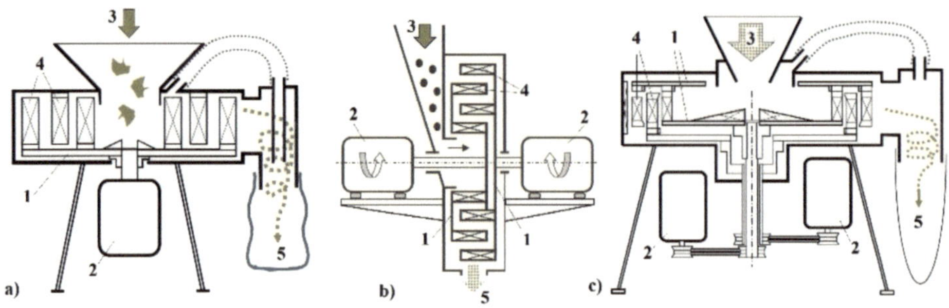

Figure 2.
Schematic representation of disintegrators DSA-2 (a), DSL-115 (b) and DSA-158 (c): Disintegrator equipment: 1 – Rotors; 2 – Electric drives; 3 – Material supply; 4 – Grinding elements; 5 – Output (authors image).

The developed disintegrators can work in the system of direct, separative (closed) or selective milling (**Figure 2**).

In the direct milling system, all the material passes through the disintegrator, undergoes a series of high-intensity impacts and flies out into the collector of the final product. Direct grinding is intended for the: testing the properties of materials, production of materials with a wide granularity, processing dry, wet or liquid materials, to combine the impact effect with other technological operations, such as mixing, homogenizing, cleaning and drying.

To obtain a product with a very fine grain, it is necessary to use the separative milling mode of disintegration, in which the crushed mass repeatedly passes the disintegrator and a classifier separates the product with the required grain size. Switching from one grinding mode to another requires only one simple operation - changing the classifier. An example of separative milling is the final operation of recycling acrylic waste – pre-crushed scrap is milled into particles of an adjustable size from 2 mm to 50 microns. The resulting powder can be effectively used as a starting material for creating new powder mixtures and for electrostatic spraying as a new coating.

Selective milling is a variant of separation in process, in which small and large particles are sent to different collectors. It allows to separate (enrich) materials with different density, strength, or get rid of microdefects in the structure of the material (**Figure 3**).

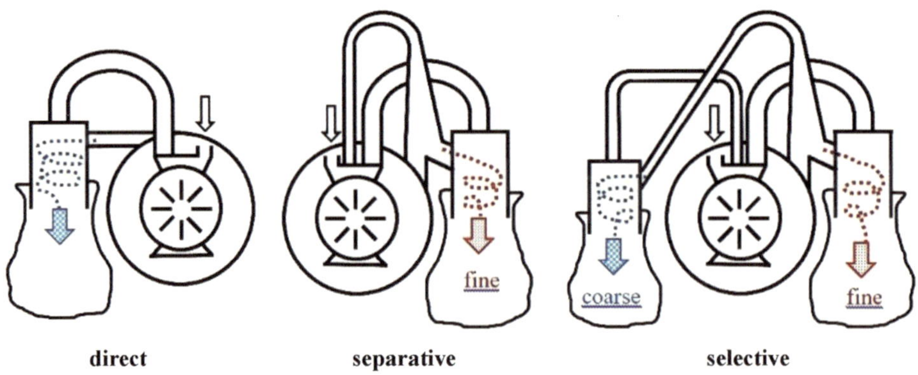

Figure 3.
Modes of disintegrator milling (authors image).

The purpose of this option is:

- treating multicomponent materials by refining one component mainly while leaving the other components as intact as possible,
- industrial wastes in order to extract valuable components.

3. Disintegrator treatment of different polymer-based materials

3.1 Treatment of brittle plastics

The plates of polymethyl methacrylate (PMMA) plastic preliminarily cut into pieces with the dimensions of length 100 mm, width 100 mm and thickness 3–5 mm were reprocessed by the disintegrator milling.

The reprocessing technology of the PMMA plastic in the disintegrators consisted of three stages:

- preliminary milling by the DSA-158 disintegrator in the conditions of direct milling;
- intermediate milling for the size reduction in the DSA-2 disintegrator in the conditions of multi-stage milling;
- final milling by the DSL-115 disintegrator system applying the direct or/and separative milling conditions.

The main kinetic parameter when processing the material in disintegrators is the specific treatment energy of E_S in kWh/t, which is important both from the point of view of the milling effect (grindability) and from the economic side of the process [10]. Grindability of PMMA plastic as a function of particle size of the specific energy of treatment was analyzed.

The powder particles size after preliminary, intermediate multi-step and final millings of the PMMA plastic are briefly given in **Figure 4**.

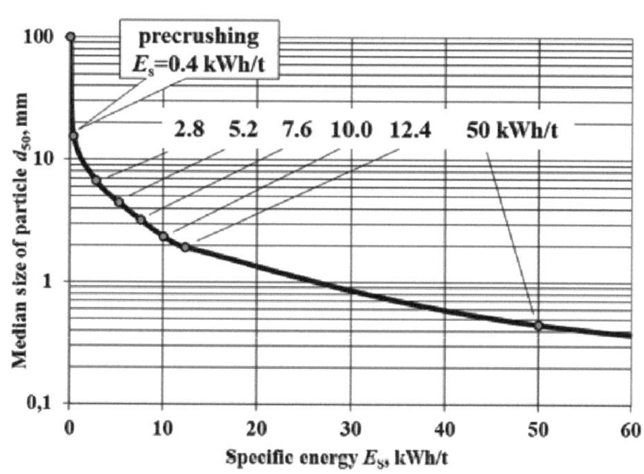

Figure 4.
Dependence of the median particle size d_{50} of the PMMA plastic powder on the specific energy of treatment E_S [9].

3.2 Treatment of ion exchange resins

Ion-exchange resins (ion exchangers – ionites, IER) are polymers of spatial three-dimensional structure and are divided into cationites and anionites.

By chemical nature, ionites are high-molecular "weight" "crosslinked materials" (polymers such as phenol-formaldehyde resins or "copolymers" of styrene and divinylbenzene), having functional groups-carriers of ion-exchange catalytic properties.

Resins that exchange positive ions are called cation - exchanging (cationites), and those that exchange negatively charged ions are called anion-exchanging (anionites).

Due to the ability to reversibly exchange their ions for an equivalent amount of other ions in solution, ionites have found wide application in various fields of technology (in hydrometallurgy for the separation and purification of rare elements, in the processing of radioactive waste, in the chemical and pharmaceutical industries, to eliminate water hardness; as catalysts for organic reactions of various types: alkylation, esterification, condensation, cyanethylation, hydrolysis, etc.)

In particular, cationides with a particle size of 1.5 mm ÷ 300 μm (**Figure 5**) are widely used as a component of water purification filters, and the remains of this material can be used as a catalyst in petrochemistry, provided that at least 90% of the material has a particle size of less than 125 μm.

The study of the possibility of obtaining a cationide powder with a dimension less than 125 μm by disintegrator grinding is the topic and task of this part. To obtain such a material, it is proposed to use a disintegrator mill with a separation mode of operation. The results of separative milling is presented in **Figure 6a, b**.

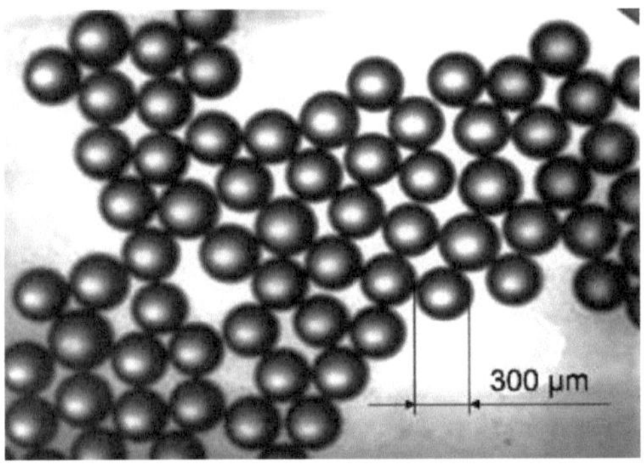

Figure 5.
Ion exchange resin (IER): Initial granules (authors image).

3.3 Treatment of soft polyethylene

High-density polyethylene (HDPE) is an example of thermoplastics.

Representing widespread polyethylene applications, HDPE offers excellent impact resistance, lightweight, low moisture absorption, and high tensile strength. HDPE is non-toxic with a high strength-to-density ratio. HDPE is used in the production of plastic bottles, corrosion-resistant piping, geomembranes, and plastic lumber.

Milling experiments to assess the grindability of HDPE (grade 277–73) were conducted in a semi-industrial disintegrator DSL-115. The parameter of grinding – the specific treatment energy E_S – was used to estimate grindability.

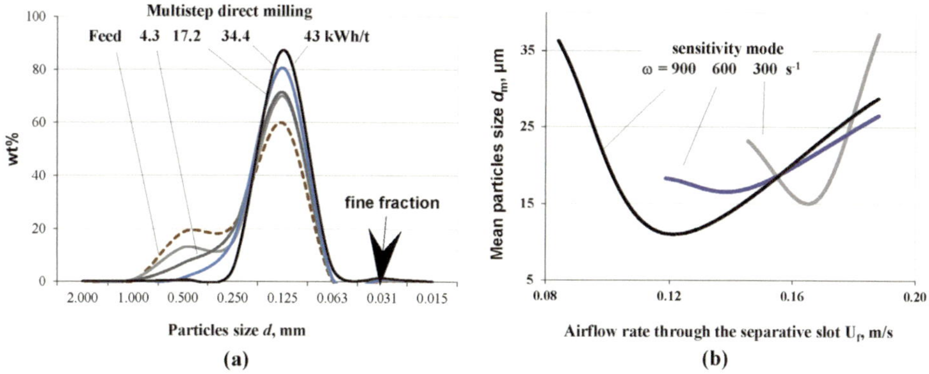

Figure 6.
Dependence of the particles size of the IER on the specific energy of treatment E_S and the grinding scheme: (a) Direct multistep milling, (b) results of separative milling with a centrifugal separator (authors image).

Initial HDPE granules with sizes from 1 to 10 mm were used as initial material for the following millings (**Figure 7** and **Table 2**).

The studied HDPE is difficulty to process with traditional disintegrator mills due to its plasticity, because most of the kinetic energy will be consumed in the deformation and heating of the material. However, still, there is a grinding method that allows obtaining a powder of the desired size and will turn out to be much less energy-consuming than melting.

Small particles of material are formed through low cyclic fatigue fracture, in which they peel off the surface in places of repeated plastic deformations formed during impact, therefore, this process requires a large number of impact cycles. Despite this, a fine fraction <355 μm of polyethylene can be obtained using separation grinding. It has been found that the size of the product can be adjusted within the required limits.

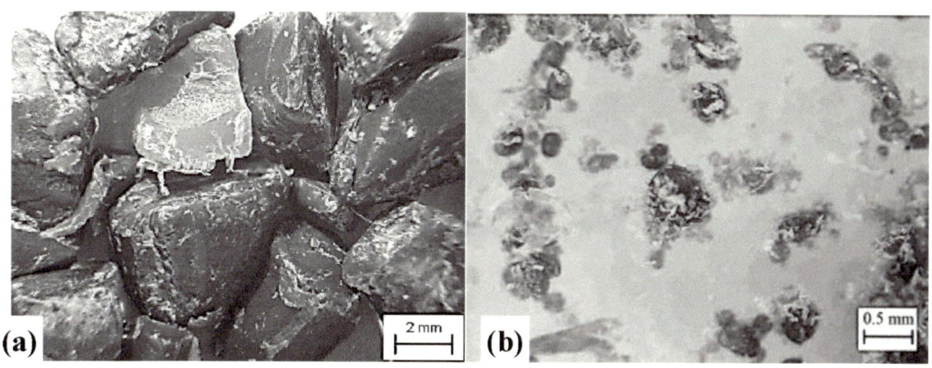

Figure 7.
HDPE particles: (a) initial; (b) disintegrator milled.

Sieves, mm	11.2	5.6	2.8	1.4	0.710	0.355	< 0.355
wt %	0	15.3	75.6	6.3	1.9	0.7	0.2

Table 2.
Granule size distribution of initial HDPE [9].

The reprocessing technology of the HDPE in disintegrators consisted of two stages:

- intermediate direct multi-step milling for the size reduction in the semi-industrial disintegrator DSL-115,
- final separative milling by the DSL-115 disintegrator with an inertial classifier.

Considering that the reason for the formation of small particles of the material is low-cycle fatigue failure, and the fact that small particles peel off from large ones in the places of repeated plastic deformations formed upon impact, it was assumed

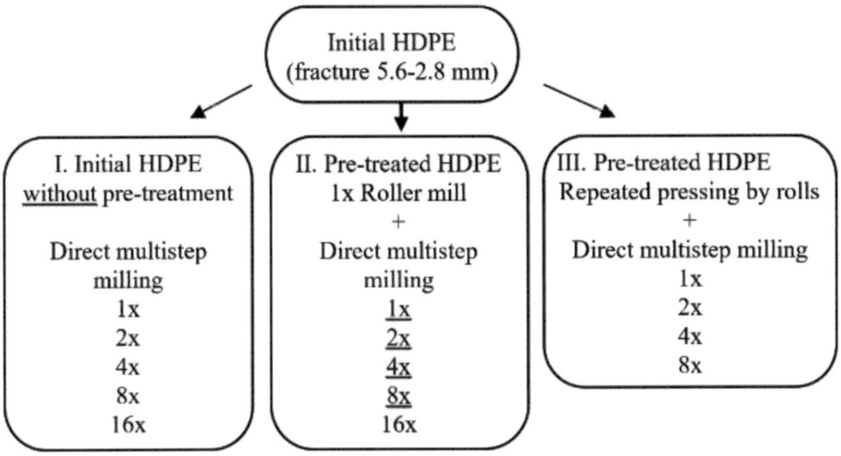

Figure 8.
Modes for treatment of HDPE by disintegrator DSL-115: I – Direct multiple grinding; II – Direct multiple grinding of single-deformed material; III – Direct multiple grinding of repeatedly deformed material (authors image).

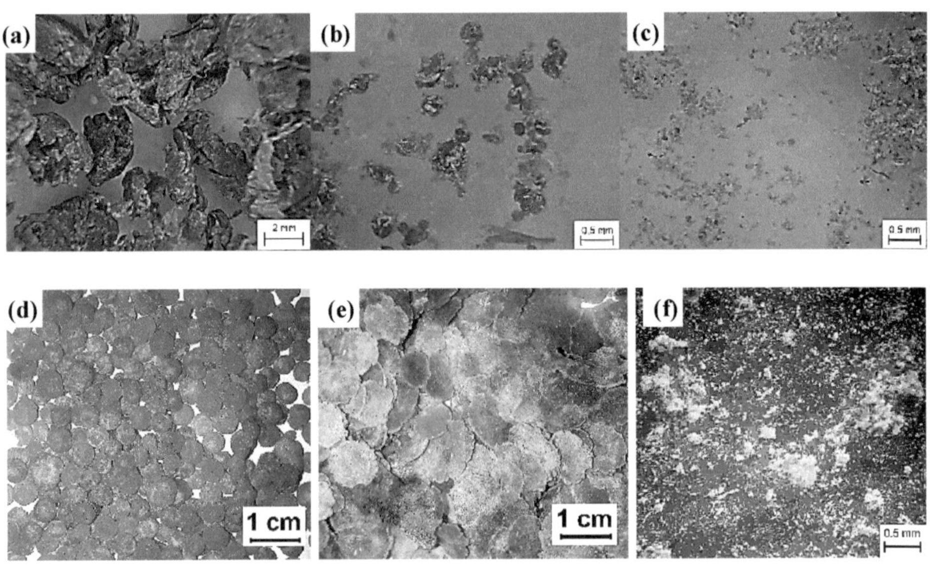

Figure 9.
Particles size and shape of HDPE after milling in DSL-115: Mode II (a – c) – Direct multiple milling at a single plastic deformation by rolls (a – 6.7 kWh/t, b – 33.5 kWh/t and c – 107.2 kWh / t; mode III (d – f) – Direct multiple milling after a multiple plastic deformation by rolls (e – Initial granules, f – Deformed by roll mill granules, g – Deformed and milled powder – 33.5 kWh/t) (authors image).

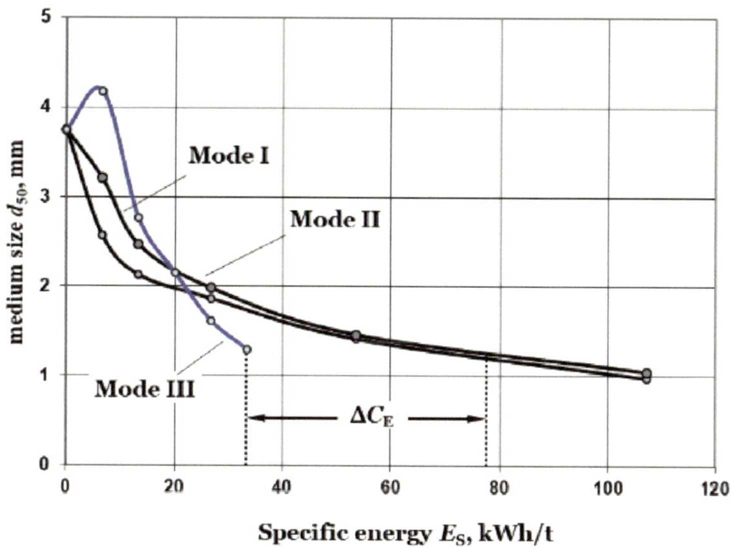

Figure 10.
*Dependence of the average size of the HDPE particles on the specific energy of treatment E_S and the grinding scheme: I, II, III – Modes of treatment (see **Figure 8**); ΔC_E – The difference between the energy costs of different treatment modes (authors image).*

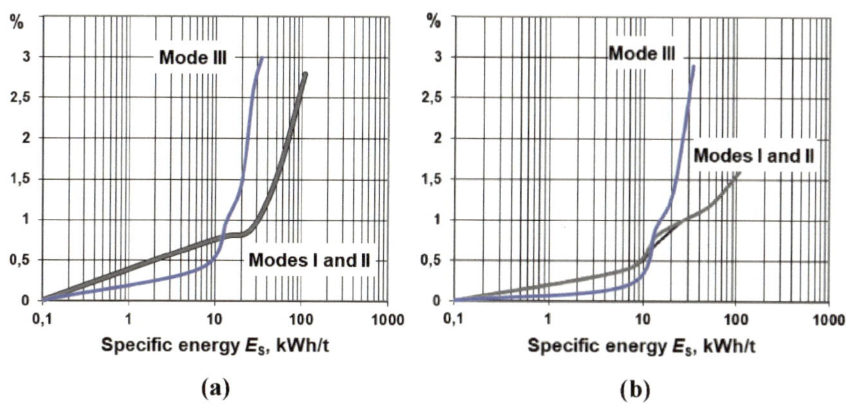

Figure 11.
Dependence of fine fraction of crushed HDPE at different milling modes on the specific energy of treatment E_S: (a) fraction −355 + 180 μm; (b) fraction <180 μm (authors image).

that a preliminarily strongly plastically deformed material could exhibit the property of easier formation of small particles. It is proposed to preliminarily pass the raw material through the rollers. The testing scheme will look like this in **Figure 8**.

Particle size and shape are given in **Figure 9**. The dependence of the particle size of multiple direct milling on the specific energy of treatment E_S is shown in **Figures 10** and **11**. Dependence of particle size at separative milling on the specific energy of treatment E_S is given in **Table 3**.

The production of powders of a given fine fraction is a process with an energy consumption of the order of E_S = 450 ... 1350 kWh/t of specific energy of treatment and total energy E = 700 ... 2200 kWh/t, depending on the separation mode.

A specific problem, arising at milling is the heating of the material during processing (on average 0.25 C° / kWh/t), which significantly reduces productivity. Possible options for solution of problem and increasing productivity may be:

Classifier type and specific energy of treatment E_S, kWh/t		Sieves, mm						d_m, mm
		1.4	0.710	0.355	0.180	0.090	< 0.090	
IC mode1	450	0	18.1	24	44	9.2	4.7	0.26
IC mode 2	690	0	5.5	12.5	30.7	31.9	19.4	0.13
CC mode 1	1350	—	0	6	32	39	23	0.11

Table 3.
HDPE powder particles distribution at disintegrator milling DSL-115 with an inertial (IC) and centrifugal (CC) classifiers.

1. Cooling of material to be treated

    a. air cooling (open-loop system of inertial or centrifugal separation, when air after separation of fine material does not return back to the working chamber) or

    b. adding water or dry ice to the material or

    c. pre-cooling of polyethylene in liquid nitrogen (perhaps the most effective option, which can notably increase the productivity of milling by an order of magnitude, since it will lead to embrittlement of a viscous particles).

2. Combination of disintegration with forming, for example, rolling between rolls (in this work the thickness reduction ratio R was 6:1). Increase in the yield of fraction −355 + 180 μm was up to 200%.

3. Effective non-impact (shear) crushing of material using compressive and friction loads. For this purposes was designed an experimental setup SD-25. The treatment is highly efficient and productive (50–75 kg/h with an energy consumption of 2–5 kWh/t); is technically easier to realize, but requires mandatory cooling (for example, with water). Results of shear crushing of HDPE are given at **Figure 12**.

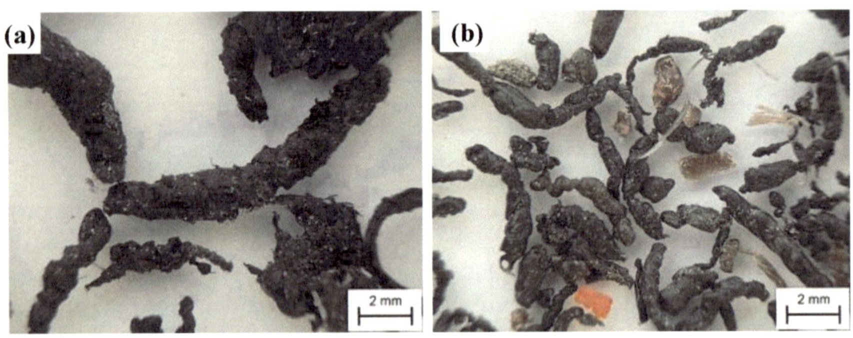

Figure 12.
Shear crushing of HDPE by the SD-25 with the gap of the cones: (a) 1 mm; (b) 0.5 mm (authors image).

3.4 Treatment of blends of plastics

Compounded plastics of the waste from electrical and electronic equipment (WEEE) subjected to recycling from dismantled personal computers (**Figure 13a**)

Figure 13.
Compounded plastics from dismantled personal computers: (a) initial material; (b) precrushed by disintegrator DSL-158 with the specific energy of treatment E_S = 1.4 kWh/t; (c) multistage milled by disintegrator DSA-2 with the specific energy of treatment E_S = 20.6 kWh/t [9].

Plastic	Tensile strength Rm, MPa	Modulus of elasticity E, GPa	Impact strength kJ/m^2	Density p, kg/dm^3	Elongation after fracture A, %
HDPE	25	1.03	20	0.96	500
PMMA	78	3.33	12	1.2	4
ABS	56	3.40	17.5	1.17	30
PC + ABS	88	3.40	34	1.20	150

Table 4.
Physical and mechanical properties of the studied plastics of the WEEE at RT [9].

are forming from processors (4 wt%), monitors (21 wt%) and keyboards (79 wt%). Physical and mechanical properties of the compounded plastics are given in **Table 4**.

Compounded plastics polycarbonate-acrylonitrile butadiene styrene (PC + ABS) and polyphenylene ether – polystyrene (PPE + PS) from monitor housings, forming 21 wt% of the monitors, were used as the compounded plastic waste.

The compounded plastics were cut into pieces and then preliminarily milled for the feed of DSA-158 (**Figure 13b**). The multistage milling (up to 16x) of the compounded plastics was performed in the DSA-2 disintegrator (**Figure 13c**). The final milling was performed in the semi-industrial disintegrator system DSL-115 with an inertial separator. The results of the preliminary and final milling of the compounded plastics from different parts of computers are given in **Table 5**. The photos of different precrushed and milled compounded plastics are given in **Figure 13b** and **c**.

Regarding the physical and mechanical properties of the plastics to be milled, the tensile strength and impact strength of the ABS plastics are lower than those of the compounded plastics PC + ABS (**Table 5**). After the fracture, the elongation of the PC + ABS plastics exceeds that of the ABS plastics for five times. These PC + ABS plastics were easy to cut with the guillotine shears for feeding the DSA-158 (100–150 mm). After the precrushing in the DSA-158, it was obvious that the material has a ductile fracture mechanism. The crushed material was appropriate for the feed of the DSA-2.

The multistage preliminary milling of the compounded material was performed by the DSA-2 disintegrator in the direct milling condition. The results of the preliminary crushing and milling of the compounded plastic PC + ABS in the disintegrator are shown in **Figure 14**.

As follows from **Table 5** and **Figure 14a**, the best results of size reduction were achieved after the two-stage preliminary milling in the DSA-158 when the mean

Milling step and type of disintegrator	Precrushing	Preliminary multi-stage milling						Final separative milling
Multiplicity of milling	2	1	2	4	8	12	16	1
	Median size d_{50}, mm							
PC + ABS monitors	12.0	7.5	6.2	5.3	4.1	—	3.6	2.2
ABS processors	9.9	—	—	—	4.1	3.9	3.8	2.3
ABS+ HIPS[a] keyboards	15.0	—	—	—	6.0	4.3	3.6	1.7
PS + ABS+PPE monitors	9.7	—	—	—	4.0	3.8	3.6	2.4
Specific energy of treatment E_S, kWh/t	1.4	3.8	6.2	11	20.6	30.2	39.8	67.0

[a]*High Impact Polystyrene (HIPS).*

Table 5.
Dependence of the milled compounded plastic powder particle medium size d_{50} on the disintegrator type and the specific energy of treatment E_S.

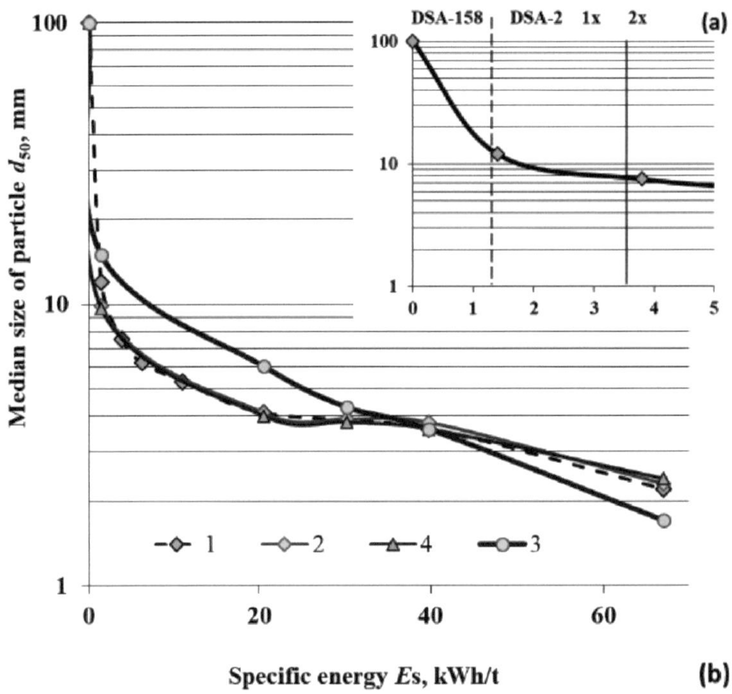

Figure 14.
Dependence of the median particle size d_{50} of compound plastic PC + ABS on the specific energy of treatment E_S after precrushing (a) and final multistage milling (b): 1 – PC + ABS monitor housings; 2 – ABS processors; 3 – ABS+HIPS keyboards after precrushing (a) and final multistage milling (b); 4 – PS + ABS+PPE monitor housings [9].

particle size was about 11 mm, and after milling in the DSA-2 – particles size was about 5 mm. During the next stages of milling in DSA-2 (from stage 4 to 8), the size reduction is less effective (from 5 mm to 4 mm only). Thus, for fracturing the particles of the ductile material, the number of collisions must be more than ten.

After eight stages of preliminary milling in the DSA-2, the particle size was in the range of 4–6 mm (**Figure 13c**). The final direct milling in the DSL-115 disintegrator reduced the particle size by 40–50% (from 3.6–3.8 to 2.4–1.7 mm).

Parameter		Fraction, mm			
		+ 1.25	+ 0.63	+ 0.315	< 0.315
Median size d_{50}, mm		1.88	0.94	0.47	0.16
Mean size	d_m, mm	2.63	0.37	0.69	2.63
	d_m^v, mm	3.60	0.46	0.74	3.60
Aspect, *AS*		1.52	1.55	1.62	1.67
Roundness, *RN*		1.32	1.43	1.53	1.57

Table 6.
The size, aspect and roundness of the milled PC + ABS powder fractions [9].

As follows from **Table 6**, the shape parameters (roundness and aspect) of powder particles milled PC + ABC compounded plastics are slightly increasing when the size of particles is decreasing.

Different types of plastics have their own dynamics of size reduction with the same applied crushing energy. If this dynamics are considerably different, it is possible to separate different types of plastics during grinding from each other in one-step.

Different particle size reduction rates allow to separate soft/ductile and hard/brittle plastics from each other in a single operation. When the viscous plastic pieces is crushed into relatively large pieces, at same time the brittle plastic pieces has time to reduce considerably in size and can be separated by a separation system (**Figures 15** and **16**).

Figure 15.
Difference in particle size of the different types of plastics at the same crushing energy E_S = 4.8 kWh/t: (a) sheets of material; (b) ductile (ABS); (c) brittle (PMMA) plastics (authors image).

3.5 Treatment of glass fiber reinforced plastic composite

Glass fiber reinforced plastic (GFP) scrap consisted of acrylic plastic (PMMA) with glass fiber reinforcement in polyester resin matrix.

Polymethylmethacrylate (PMMA) sheets were formed in vacuum (shower trays and bathtubs) and reinforced with fiberglass in a polyester resin matrix. At the end of the life cycle, the product was disassembled (cut into pieces). A scrap of this composite was used for experiments (**Figure 17**).

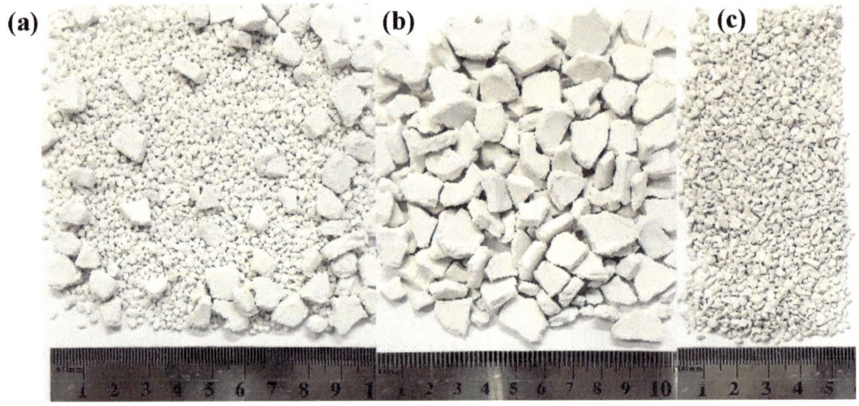

Figure 16.
The result of simultaneous grinding-separation of a mixture of ductile (ABS) and brittle (PMMA) plastics: (a) ground mixture of ABS+PMMA; (b) separated ABS; (c) separated PMMA (authors image).

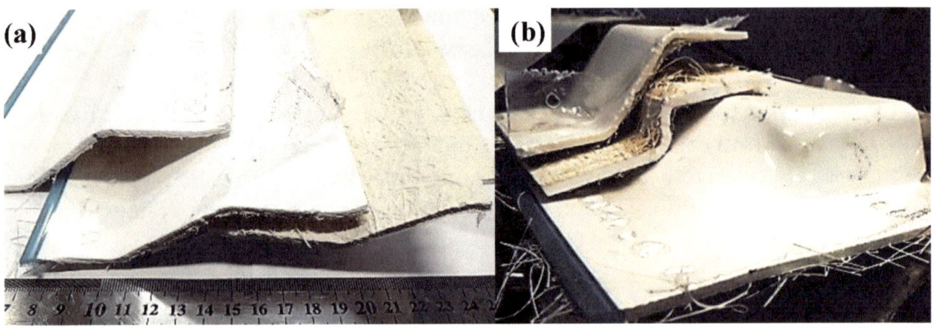

Figure 17.
Initial plastic composite scrap of PMMA+GFP (authors image).

For the milling of composite scrap, different disintegrator mills were used. To treat composite plastic scrap, the focus was on the size reduction of the acrylic plastic constituent and on the separation of the glass fiber constituent.

Disintegrator milling enables size reduction with simultaneous separation of components of low toughness. Composite plastic strips (PMMA+GFP) with dimensions of 100 × 100 × 5 mm were retreated.

The reprocessing technology of composite plastic scrap in disintegrators consisted of two steps:

1. Preliminary direct milling of reinforced acrylic strips with the disintegrator DSL-158 with following separation of glass fiber from the milled material by sieving or multi-stage milling with the semi-industrial disintegrator DSA-2. Sieve analysis were taken and the percentage of the separated glass fiber was determined;

2. Direct or separative final milling with the DSL-115 disintegrator milling system to remove glass fiber from the milled material.

3. The results of separation glass fiber and acrylic plastic are presented in **Figures 18–20**.

The results obtained during the preliminary grinding of PMMA + GFP composite plastic in disintegrator mills are shown in **Figures 18** and **20 a, b**. The particle

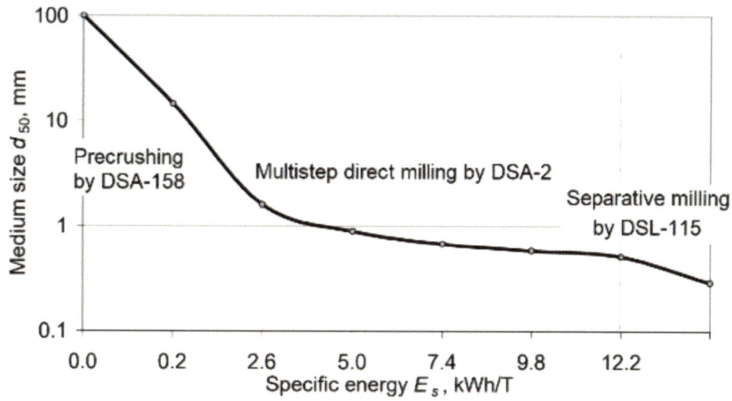

Figure 18.
Dependence of the particle size d_{50} of the milled composite plastic PMMA+GFP on the specific energy of treatment E_S [11].

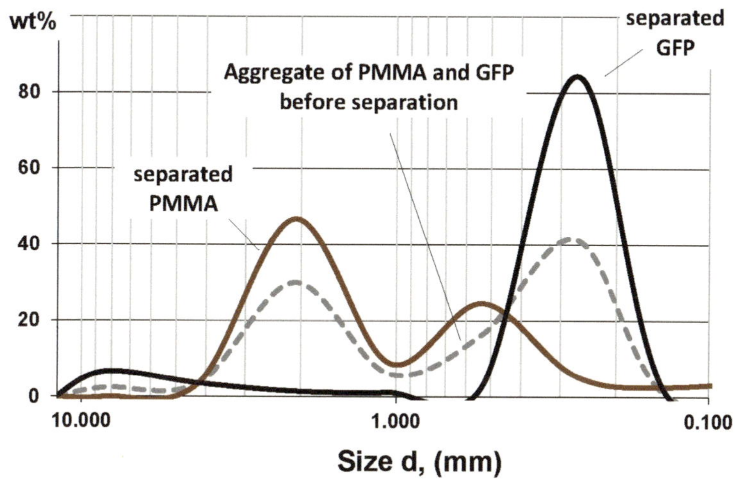

Figure 19.
Separation of glass fiber from acrylic plastic [12].

size at the outlet of the DSA-158 disintegrator was approximately 13–25 mm. The precrushed material is suitable for direct grinding in the DSA-2 disintegrator.

Before the separation, the curve of mixed plastics has two modes (**Figure 19**). After separation the acrylic plastic has two main fractions: 46% of 1.4 mm and 25% of 0.355 mm and glass fiber has one main fraction 0.180 mm, which is more than 85%. The results of separating glass fiber from scrap of composite plastic are shown in **Table 7**.

As follows from **Table 7**, the total amount of separated GFP was 45 wt%. As a result, 55% of acrylic plastic from composite plastic scrap can be reused. GFP can be reused as reinforcement in the production of polymer concrete products.

Plastic powder with a particle size of about 1–2 mm can be obtained by two-stage grinding, and 95% by weight of the glass fiber content can be separated by final selective grinding.

The recovered material can be reused in the same production process in which it was obtained. The crushed PMMA powder is applicable as a filler in the casting technology.

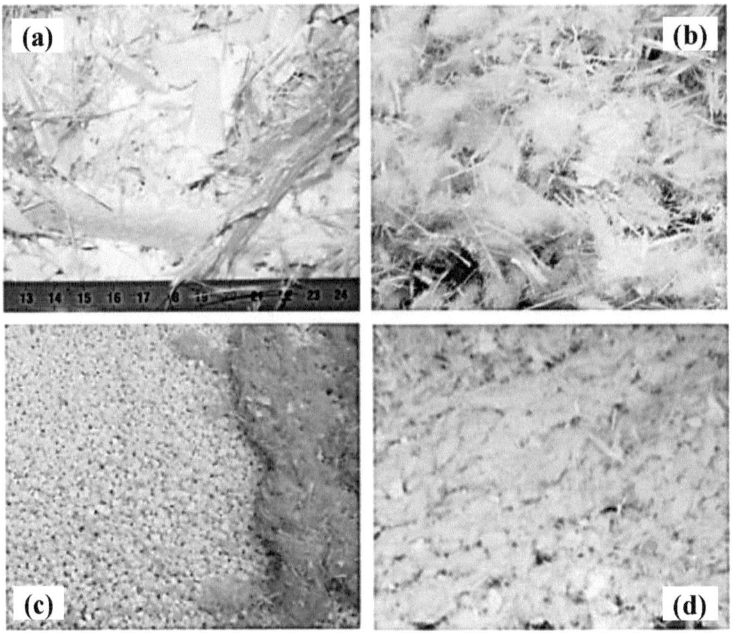

Figure 20.
The results of milling-separation of the glass fiber and acrylic plastic: (a) pre-crushed plastic scrap, (b) separated GFP large fraction, (c) finally milled mixture, (d) separated GFP fine fraction (authors image).

Milling stage	Milling device	Separation method	Separated GFP, wt.%
I	DSA-158	Sieving	16.3
	DSA-2	Sieving	12.2
II	DSL-115	Air classifying	16.5

Table 7.
Results of GFP separation during selective milling by different disintegrators [13].

3.6 Retreatment of tyres and pure rubber

Tyres can be utilized by disintegration in two ways (**Figure 21**):

- preliminary cutting of the tyres to the pieces of 50–150 mm, and size reduction to pieces of 10–30 mm by special devices, or
- direct milling of whole tyres to the rubber powder of size 1–2 mm by special units.

Collars with wires will be removed before cutting-off in both cases.

At the first stage, the pieces of tyres sized 50 to 150 mm can be used in the pyrolysis technology for oil production. The pieces sized 10 to 50 mm can be used as additional fuel in furnaces. Fine powder of fraction 1–2 mm can be used for producing asphalt-concrete of road pavement, which is most beneficial.

At the second stage, further milling of rubber assumes separation of pure rubber from textile and wire fiber. The technology of treatment pure rubber to reclaimed rubber, used instead of caoutchouc, is cost-effective. The ultra-fine rubber powder

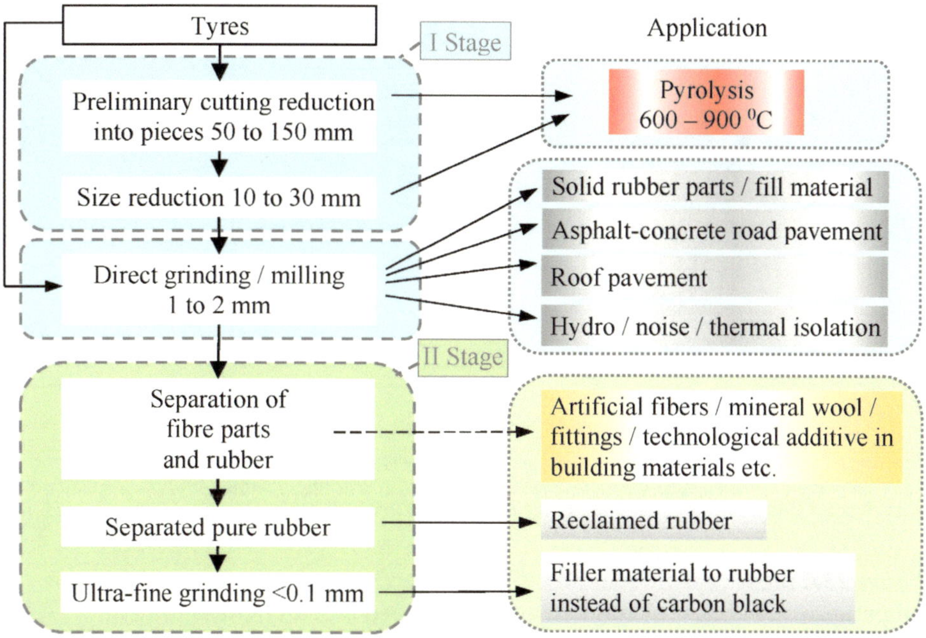

Figure 21.
Principal scheme of utilization of tyres at room temperature (authors image, based on [14]).

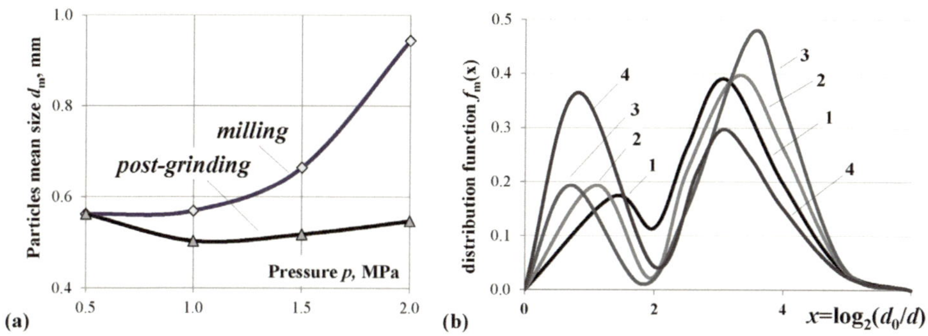

Figure 22.
Dependences of particles size on the feed pressure:(a) of tyre strips in the process of milling and post-grinding; (b) of rubber particles at different feed pressures: 1–0.5 MPa; 2–1.0 MPa; 3–1.5 MPa; 4–2.0 MPa (d_o = 5 mm) [9].

with particles less than 100 μm can be used as carbon black in the production of new rubber. However, this rubber powder is more expensive than carbon black.

Direct milling of whole tyres to the powder of 1–2 mm is more effective. In this case, the disintegrator system consists of two special devices - units for milling and post-grinding.

The granularity of the product depends on the pressure of the tyres against the milling tool. The mean size of the product after milling is shown in **Figure 22a**.

The dependence of particle mean size on the pressure is notable. Post-grinding evens the size of the particles. The granularity of the final product is shown in **Figure 22b**. It can be seen that the distribution function is two-modal.

Comparative grinding of pure rubber at normal and low temperatures was also conducted. The results are shown in **Figures 22** and **23**. The effectiveness of grindability of pure rubber at normal temperature is very low but as it follows from

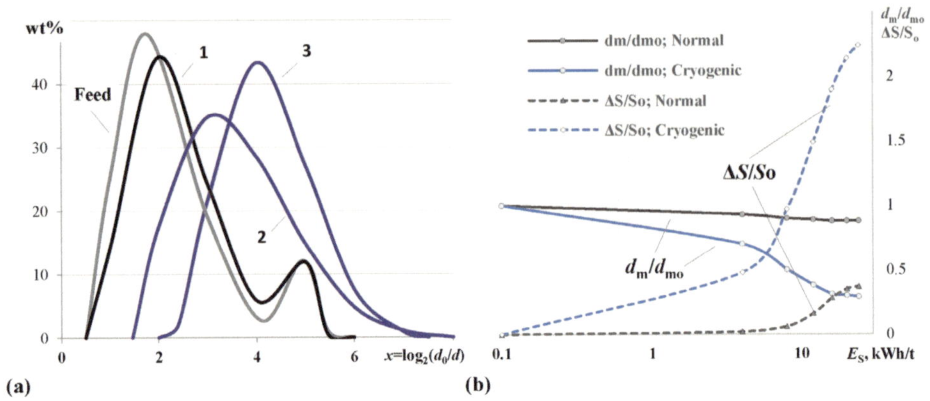

Figure 23.
Dependence of the granularity (a) and specific surface area (b) of ground rubber particles on the specific energy of treatment E_S at normal temperature and cryogenic treatment: 1 – Milling at normal temperature with energy 25 kWh/t; 2 – Cryogenic milling with energy 8.3 kWh/t and 3–27 kWh/t (d_o = 5 mm), (authors image).

Figure 23a, the cryogenic grinding is more effective. Results of normal and low temperature grindings for comparison are grouped together graphically in **Figure 23b**.

As can be seen, the relative mean particle size d_m/d_{m0} and the relative increase in the specific surface area $\Delta S/\Delta S_0$ of rubber particles depend on the specific energy of treatment E_S. **Figure 23** shows that cryogenic grinding is more effective, particularly due to the increase in the specific surface area of rubber particles.

Highly alloyed steel wire found in car tyre collars may have its own value. To find out this value, it was necessary to study first the possibility of separating the wires.

Selective collar size reduction was achieved by using cryogenic grinding in disintegrator DS-158, during which the separation of wire, textile and rubber particles will take place (**Figure 24**).

The amount of rubber in the collars is relatively low, so the main value is probably in the metal. In addition, the amount of liquid nitrogen needed for cryogenic grinding has been determined theoretically and experimentally.

However, the economic efficiency of selective grinding has not been determined. It depends on the market value of alloy steel, amount of tyre collars to be treated and methods used to separate materials. The alternative is disposing collars to landfills, which is currently allowed.

Figure 24.
Cryogenic grinding by the disintegrator DS-158: (a) initial pieces of collar; (b) ground and separated rubber; (c) textile; (d) steel wire (authors image).

Grinding of pure rubber and composites using classical disintegrator mills at normal temperatures is much (by orders of magnitude) less effective than grinding in liquid nitrogen (**Figures 22** and **23**). This is due to high elastic properties of rubber. Under impact, kinetic energy leads to strong elastic deformations of rubber pieces, which preferentially heat the material but do not destroy it.

Therefore, it is rational to use such a grinding method for the following:

- precrushing of rubber agglomerates (**Figure 25**) and

- changing in material properties.

Changing in material properties consists:

a. creation of interstitial composites (e.g., rubber-iron powder);

b. spheroidizing to reduce the surface and improve the free-flow properties of the powders (**Figure 25c**);

c. changing chemical properties under the influence of strong shear shock loads (devulcanization).

The degree of grinding is directly related to the energy consumption of the process. Thus, the feasibility of energy costs for the rubber grinding process is associated with the minimum particle size obtained during grinding. **Figure 25** shows the dependence of the average diameter of rubber particles on the specific grinding energy.

Figure 25.
Dependence of the mean diameter d_m (a) and shape (b, c) of treated at normal temperature pure rubber on the specific energy of disintegrator milling E_S: b – Pure rubber particles before and c – After milling [14].

3.7 Treatment of printed circuit boards

Disintegrator milling as a prospective recycling method to relieve metallic components of current generation of printed circuit boards (PCB) (**Figure 26**).

The PCBs stands out with one of the highest concentrations of the rare and precious metals (RPMs). Therefore, the PCBs can/should be become a sustainable source of the RPMs for future generations of technology.

Traditionally, PCBs are processed using cutting processes (shredder mills) and a combination of low intensity impacts with shear and abrasion (hammer mills).

Both methods have disadvantages. During cutting, the knife must pass through all layers of material, which consist not only of epoxy resins and glass fiber, but also of strong and ductile metals and alloys, as well as ceramics (**Figure 27a, b**). This leads to high wear on the cutting edges. The cutting process also does not effectively

Figure 26.
Initial PCB plates (authors image).

separate into its constituent components. During abrasion and shearing, the largest and most protruding pieces are affected, and these are basically the same metal and ceramic components. As a result, a lot of extra energy is wasted on grinding media wear, unnecessary grinding of metal and ceramic components and unnecessary movement of the entire mass of material.

The high-intensity impact that is used in disintegrator mills generates high stresses in the composite materials, breaking primarily the weakest constituents of them. This should lead to the fact that in the first place the bonds between metallic fraction (MF) and non-metallic (NMF) phase will be destroyed, which will allow further classification to achieve high-quality metal concentrate (**Figure 27c, d**). Thus, due to the selectivity of the impact action, a high level of fragmentation can be achieved – this is the main factor for mechanical enrichment. Such selective disassembly is less energy consuming, relieves the rest of the components of the PCB composite from excessive grinding, and, consequently, transformation into technological emissions.

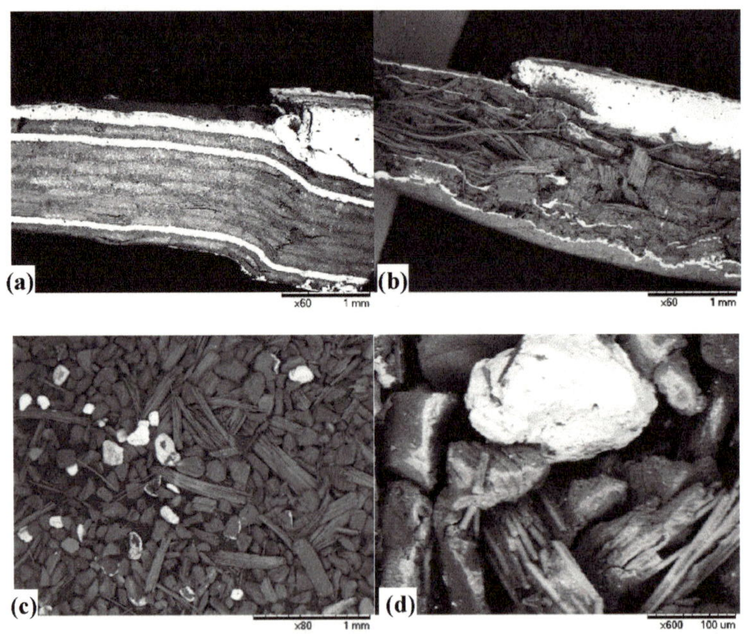

Figure 27.
Structure of the PCBs plate: (a) initial material, (b) after cutting with a shredder and (c, d) ground by the disintegrator DSA-2 (author's image).

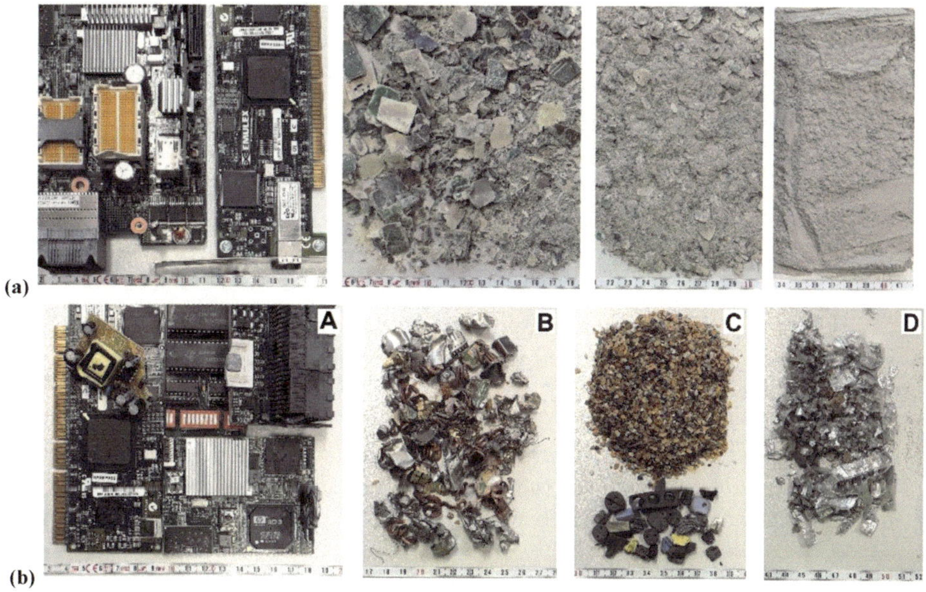

Figure 28.
Example of possible PCB disintegration degrees with existing disintegration systems (a) and PCB milling and separation stages (b): raw PCB (A), metal- (B), plastic- (C) and foil- (D) enriched components after disintegration and separation (author's image).

In this part, the integration of a high-energy impact milling for PCBs recycling comminution state-of-the-art was reviewed from the fragmentation (relieving) process perspective. Modifying PCBs wastes through delamination and fragmentation turning it to MF and NMF mixture and makes it possible to further efficient classification into enriched concentrates (**Figure 28**).

A printed circuit board is a multicomponent metal-plastic multilayer composite material that has a complex structure with brittle and plastic components. The mechanisms for reducing the particle size of plastic and brittle materials are different.

At the pre-crushing stage, large pieces of the composite plates quickly disintegrate into their component parts. The result of milling is a direct fracture. Each separated component is then crushed at its own different speed.

Then each of these components is crushed at a different speed:

- slow reduction in the size of metallic parts as a result of fatigue failure and
- rapid reduction in the size of brittle non-metallic components as a result of direct fracture.

4. Summary

Based on the study of impact milling by disintegrators different polymer materials the following conclusions may be drawn:

1. Impact milling an effective way for retreatment of polymeric materials due to the stresses initiated in materials due be ground. For impact milling DS-series of multi-functional disintegrators and milling systems operating at direct, separation and selective modes were developed.

2. Impact milling is effective way for treatment as brittle as well soft plastic wastes. The grindability of ABS, PMMA, HDPE plastics, based on different fracture mechanisms (direct fracture, low cyclic fatigue), and characteristics of ground product were clarified. Disintegrator retreatment of blends of plastics (e.g. WEEE components) enables to ground and separate plastics of different mechanical properties due to the considerably different dynamics of size reduction.

3. Retreatment of multicomponent materials (e.g. polymeric-metallic PCBs) enables to relieve the valuable metallic components, grind and separate plastic components; by tyres pretreatment (at room and cryogenic temperatures) ways/possibilities were studied and different components (rubber, textile, wire) will be separated.

4. The potential areas of application retreated plastics wastes are shown. The recovered polymer materials can be used in the same process in which they were obtained; valuable metals will be separated and enriched, and used in new PCBs production.

Acknowledgements

This work was supported by the Estonian Research Council grants PRG643 and PRG1145 and by the Environmental Investment Centre Foundation grant KIK19019.

Author details

Priit Kulu and Dmitri Goljandin*
Tallinn University of Technology (TalTech), Tallinn, Estonia

*Address all correspondence to: dmitri.goljandin@taltech.ee

References

[1] Pivnenko K, Jakobsen L, Eriksen M, Damgaard A, Astrup T. (2015). Challenges in Plastics Recycling.

[2] Hopewell J, Dvorak R, Kosior E. Plastics recycling: challenges and opportunities. Philos Trans R Soc Lond B Biol Sci. 2009;364(1526):2115-2126. DOI:10.1098/rstb.2008.0311

[3] Goodship V. Plastic recycling. Sci Prog. 2007;90(Pt 4):245-268. doi: 10.3184/003685007X228748

[4] Rudolph N, Kiesel R, Aumnate C. (2020). Understanding Plastics Recycling: Economic, Ecological, and Technical Aspects of Plastic Waste Handling. DOI: 10.3139/9781569908471.

[5] Ignatyev IA, Thielemans W, Vander Beke B. Recycling of polymers: a review. ChemSusChem. 2014;7(6):1579-1593. DOI:10.1002/cssc.201300898.

[6] Yokoyama T, Inoue Y. Chapter 10 Selection of Fine Grinding Mills. Handbook of Powder Technology, 2007, Vol. 12, 487–508. DOI: 10.1016/S0167-3785(07)12013-3

[7] Tümanok A, Kulu P. Treatment of different materials by disintegrator systems. Proc. Estonian Acad. Sci. Eng., 5/3, 1999, 222–242.

[8] Tamm B, Tymanok A. Impact grinding and disintegrators. Proc. Estonian Acad. Sci. Eng., 1996, 2, p.209–223.

[9] Goljandin D, Kulu P. Disintegrators and Disintegrator Treatment of Materials. Germany: Lap Lambert Academic Publishing; 2015. 162.

[10] Tymanok A. About the energy of material to be treated in multistep rotor milling device. Proc. Techn. Univ. Tallinn, No 393, 1976, p. 139–147.

[11] Kers J, Majak J. (2008). Modelling a new composite from a recycled GFRP. Mechanics of Composite Materials, 44 (6), 623–632. DOI: 10.1007/S11029-009-9050-4

[12] Kers J, Kulu P, Goljandin D, Mikli V. Reuse of reinforced acrylic plastic waste in new composite material development. In: Proceedings of the 5th International Conference of DAAAM Baltic "Industrial engineering - adding innovation capacity of labour force and entrepreneur", 20 22 April 2006, Tallinn, Estonia; 2006. Ed. Küttner, R. Tallinn: Tallinn University of Technology; p. 267–271.

[13] Kers J, Kulu P, Goljandin D, Mikli V. Reprocessing technology of composite plastic scrap and properties of materials from recycled plastics. Proceedings of the Estonian Academy of Sciences Engineering, 2007, 13 (2), p. 105–117.

[14] Lapkovskis V, Mironovs V, Kasperovich A, Myadelets V, Goljandin D. Crumb Rubber as a Secondary Raw Material from Waste Rubber: A Short Review of End-of-Life Mechanical Processing Methods. Recycling; 2020 Dec 1 5(4):32. Available from: http://dx.doi.org/10.3390/recycling5040032

Chapter 8

Effect of Environmental Aging on Tensile Properties of Post-Consumer Recycled (PCR) Polycarbonates

Rashed Islam, Syed Zameer, Jacki Laiz, David Chen, Anthony Yu and Ayyana Chakravartula

Abstract

In this chapter, tensile properties of different grades of post-consumer recycled (PCR) polycarbonate (PC) plastics have been compared with conventional or virgin PC before and after different aging conditions. 50 and 75% recycled PCs showed comparable yield strength (~57 MPa), maximum tensile strength (~70 MPa) and maximum strain (~190–200%) before aging, when compared to virgin PC of same melt flow rate (MFR of ~10 g/10 min). From the fractography analysis (optical and scanning electron microscopy) of the both virgin and 50% recycled PC, it is evident that the fracture morphologies are very similar and they are indicative of ductile failure. It is observed that with the presence of temperature and humidity (60°C 90% RH) aging, tensile strength starts to drop over time but most importantly both 50% and 75% PCR grades showed similar aging behavior compared to virgin PC (10–13% strength degradation after 500 hours of aging). Reliability modelling showed comparable B10, Weibull Alpha and Weibull Beta values between Virgin PC and PCR grades after different aging conditions. Fractography analysis of fresh and aged 75% PCR also showed ductile features.

Keywords: Post Consumer Recycled, Tensile Strength, Melt Flow Rate, Elongation at Break, Weibull Analysis

1. Introduction

Plastics are a versatile class of materials which can be found in products ranging from single-use packaging to components used in automotive and durable goods. Unfortunately, plastics are also identified as an environmental pollutant due to poor recycling rates and poor end-of-life waste management [1, 2]. This has led to many organizations, such as the Ellen MacArthur Foundation, to promote a circular economy for plastic [3].

The circular economy for plastic is based on three tenets: reducing the use of plastics, reusing a product multiple times and recycling at end of life. The first two components are directed more toward plastic packaging but recycling at end of life can be applied to all types of products and industries. Post-consumer recycled (PCR) resin is the recycled product of waste created by consumers and is defined by

IntechOpen

ISO 14021 [4]. It is commercially available from multiple suppliers, but not all grades are created equal. However, there needs to be a demand for recycled plastics for the circular economy to work properly. Regrettably, there is a misperception that recycled plastics are always inferior to virgin plastics and therefore can only be used in "down-cycled" applications. As engineers and scientists interested in both the production of high-quality products and environmental sustainability, we questioned the validity of this misperception and initiated this study to answer the following questions:

- Can recycled plastics be used in consumer hardware?
- Does the use of recycled plastics have any influence on the reliability of the product?
- Are there any manufacturing constraints associated with the use of recycled plastics?

Although there are many factors associated with completely answering the questions above, the factor which we will be evaluating in this paper is strength. Strength of plastics can be quantified by measuring tensile, bending, shear, compression, flexural, and impact properties, among others. Each molded plastic resin's mechanical properties can vary depending on molecular weight, crystallinity, molding parameters, and additives/fillers. The mechanical properties can be defined by testing specimens in compliance with ASTM [5, 6] or ISO [7] testing standards. The advantages of mechanical property testing following ASTM or ISO testing standards are

A. Standard specimens can be obtained from material supplier

B. Useful for comparison of different grades/suppliers

C. Studying the effect of environmental conditions.

Since a plastic's strength degrades over time as it ages in the environment, it is crucial to take this degradation into account when designing parts. Therefore, the aging behavior of plastics needs to be understood thoroughly. As real-time aging can take a significant time (i.e., years), most aging studies are performed using accelerated aging tests utilizing temperature, temperature and humidity, temperature cycling, UV exposure, and/or chemical exposure [8]. The Arrhenius-based accelerated life model is the most popular model to understand the acceleration of aging due to temperature. The rate of reaction, *R*, based on the Arrhenius model can be explained by

$$R(T) = Ae^{-\frac{E_a}{K.T}} \tag{1}$$

where A is a material constant; E_a is the activation energy (eV), K is the Boltzmann's constant (eV/K), and T (K) is the absolute temperature. Based on this model, the acceleration factor (AF) can be calculated as

$$AF = exp\left[\frac{E_a}{K}\left(\frac{1}{T_u} - \frac{1}{T_t}\right)\right] \tag{2}$$

where,
T_u is unit localized temperature during field usage (K).
T_t is unit localized temperature in the test (K).

E_a is activation energy.
K is Boltzmann's constant (8.617385 x10–5 eV/K).
The acceleration factor for temperature and humidity-based acceleration is derived from Peck's Model:

$$AF = \left(\frac{RH_u}{RH_t}\right)^{-n} exp\left[\frac{Ea}{k}\left(\frac{1}{T_u} - \frac{1}{T_t}\right)\right] \quad (3)$$

where,
RH_u is relative humidity in field (%).
RH_t is relative humidity in the test (%).
n is a constant.
Finally, the acceleration factor for aging due to thermal cycling comes from the modified Coffin-Manson model:

$$AF = \left(\frac{\Delta T_{test}}{\Delta T_{field}}\right)^{m} \quad (4)$$

where,
ΔT_{test} = temperature difference in TC test (°C);
ΔT_{field} = temperature difference in field usage (°C);
m = Coffin-Manson exponent.
Any acceleration-based model depends on the distribution of the data that best fits. The life of a consumer product follows a bathtub curve under variable stress and can be explained better by the Weibull distribution, for which the probability density function can be explained as

$$f(t) = \frac{\beta}{\alpha}\left(\frac{t-\gamma}{\alpha}\right)^{\beta-1} e^{-\left(\frac{t-\gamma}{\alpha}\right)^{\beta}} \quad (5)$$

where β is the shape parameter or slope of the curve, α is the characteristic or Weibull life, and α is the location parameter. This is true for 3 parameter Weibull distribution. Typically, γ is considered to be zero and hence the equation can be derived as

$$f(t) = \frac{\beta}{\alpha}\left(\frac{t}{\alpha}\right)^{\beta-1} e^{-\left(\frac{t}{\alpha}\right)^{\beta}} \quad (6)$$

In this chapter, the scope of the research is to focus on the performance variation of different grades of PCR compared to conventional or virgin PC after thermal cycling and high temperature & high humidity aging. The reason behind picking thermal cycling and high temperature & high humidity aging is because these two are the most common environmental exposures for consumer hardware. The effects of chemical, UV and solar aging will be reported at a later date.

2. Experimental procedures

2.1 Samples

The materials studied were commercially available formulations obtained from one resin vendor. All formulations were identical except for PCR content as specified in **Table 1**. The samples were provided by the resin vendor in the form of ASTM D638 Type V Tensile Bars.

After clearly marking the samples and dimensional measurement (width and thickness), they were divided into 7 different groups (**Table 2**) with each group having ~5 ASTM tensile bars of virgin PC, 50% and 75%.

2.2 Tensile measurement

The first group of ASTM tensile bars with no preconditioning were tested in a universal tensile testing machine with a tensile pull rate of 2 mm/min. 5 samples of virgin PC, 50% PCR and 75% PCR, all with white resin and melt flow rate (MFR) of 10 g/10 mins. Next 50% PCR tensile bars (black) with MFR of 10 and 15 were tested. **Figure 1** shows a schematic of a tensile bar and a set-up image of the tensile test. From the load–displacement curves, yield or necking point, the EAB (elongation at break) and UTS (ultimate tensile strength) were recorded.

2.3 Aging conditions

For studying the effect of aging, two different batches were prepared. One batch of ASTM tensile bars were kept inside an environmental chamber with temperature cycling from −20–60°C. Transfer time of 7 minutes and soak time of 23 minutes were set so that the length of 1 cycle is equal to an hour. At each interval (100, 200 and 300 cycles), 5 tensile bars were taken out of the chamber and tensile-tested in a universal tensile testing machine. Profile for thermal cycling is shown in **Figure 2(a)**. Similarly, the second batch of samples were placed in an environmental chamber for high temperature and high humidity exposure. The profile was set so that peak temperature and humidity reached 60 ± 2°C and 90 ± 3%RH, with ramp up and ramp down time of 2 hours according to the high temperature and humidity profile shown in **Figure 2(b)**. As in the thermal cycling test, at each interval (168, 336 and 500 cycles), 5 tensile bars were taken out of the high temperature and high humidity chamber, dimensions measured (width and thickness) and tensile tested. After

Samples	MFR (g/10 min)	Color
Virgin PC	10	White
50% PCR	10	White
75% PCR	10	White
75% PCR	10	Black

Table 1.
List of samples.

Group #	Aging condition	Check point
1	No Aging	N/A
2	−20–60°C	100 Cycles
3	−20–60°C	200 Cycles
4	−20–60°C	300 Cycles
5	60°C 90%RH	168 Hours
6	60°C 90%RH	336 Hours
7	60°C 90%RH	500 Hours

Table 2.
Different aging condition.

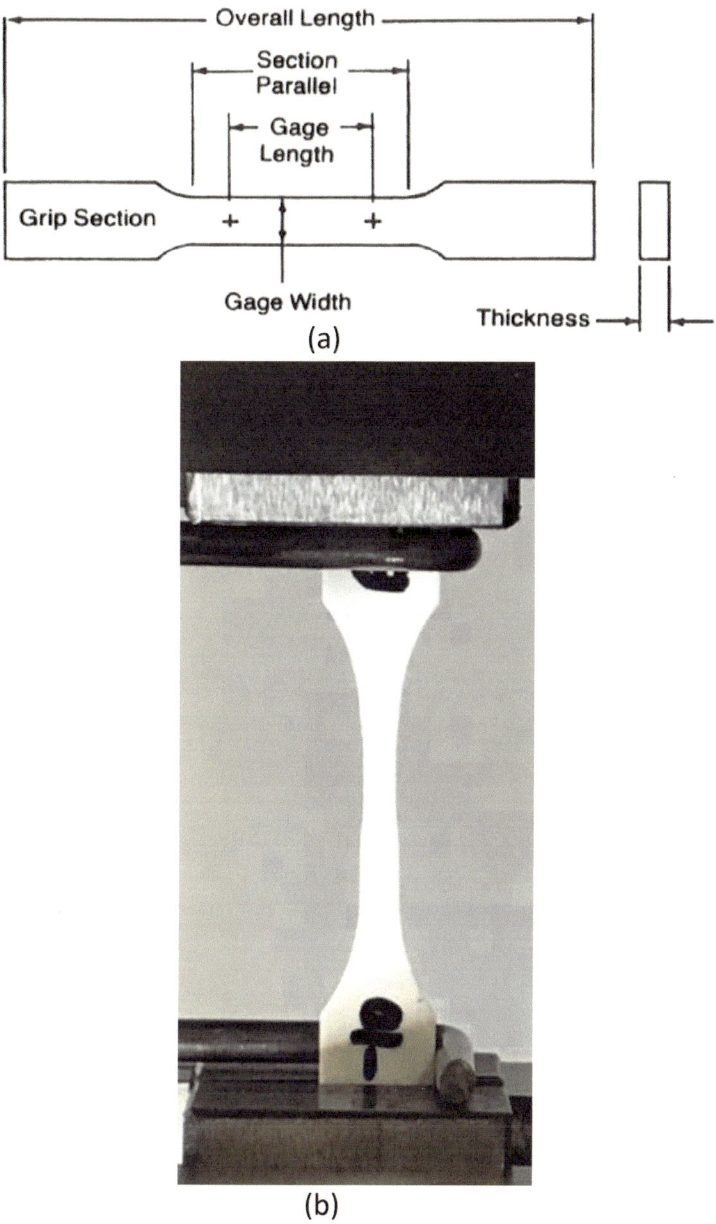

Figure 1.
(a) Schematic of an ASTM tensile bar (type V)[6] and (b) experimental set-up.

tensile testing for all the samples were completed, the fracture surfaces were inspected in optical microscopy and with a scanning electron microscope.

3. Results and discussions

3.1 Dimensional stability after aging

First, the dimensions in widths and thicknesses of each tensile bar were compared before and after aging. **Figure 3(a)** and **(b)** shows the % change in width and thickness after each interval (168, 336 and 500 Hours) of high temperature and high

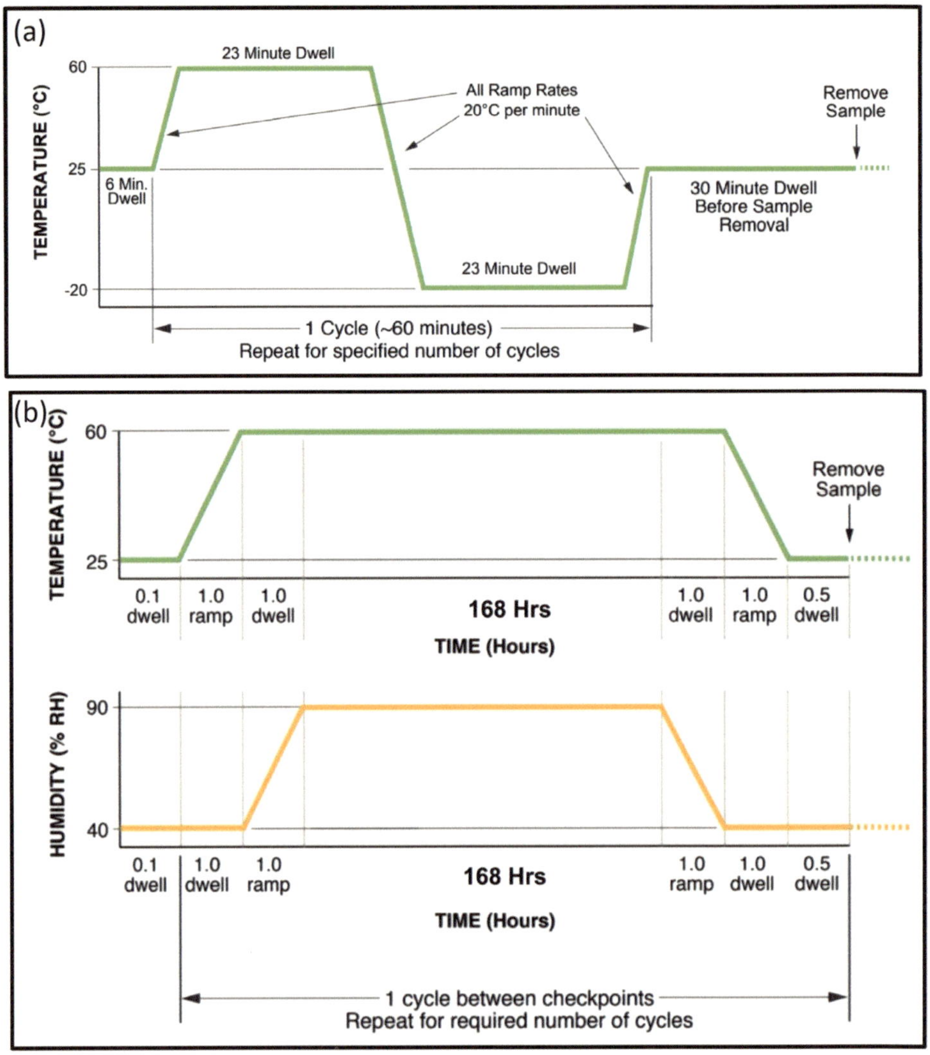

Figure 2.
Profile for (a) thermal cycling test and (b) high temperature high humidity test.

humidity aging. From the figures, it is visible that the dimensions were very stable after aging. Expansion in width and thickness for all the tensile bars were within 0.25%, which were within the measurement error.

3.2 Strength characteristics in unaged condition

Second, we will compare the tensile properties of PCR with virgin or conventional PC at time zero and without any aging. Here breaking strengths of 50 and 75% PCR grades were compared with conventional PC. **Figure 4** shows the load-extension curve of three grades of PC and PCR-PC.

From **Figure 5(a)**, both the PCR grades have comparable tensile strength (at breakage) when compared with a virgin PC of the same MFR. The strength is around 67, 65 and 71 MPa for Virgin PC, 50% PCR and 75% PCR respectively. For comparison, it was made sure the resin color is white for all the grades of PC and PCR. In **Figure 5(b)**, tensile strain at breakage for Virgin PC, 50% PCR and 75%

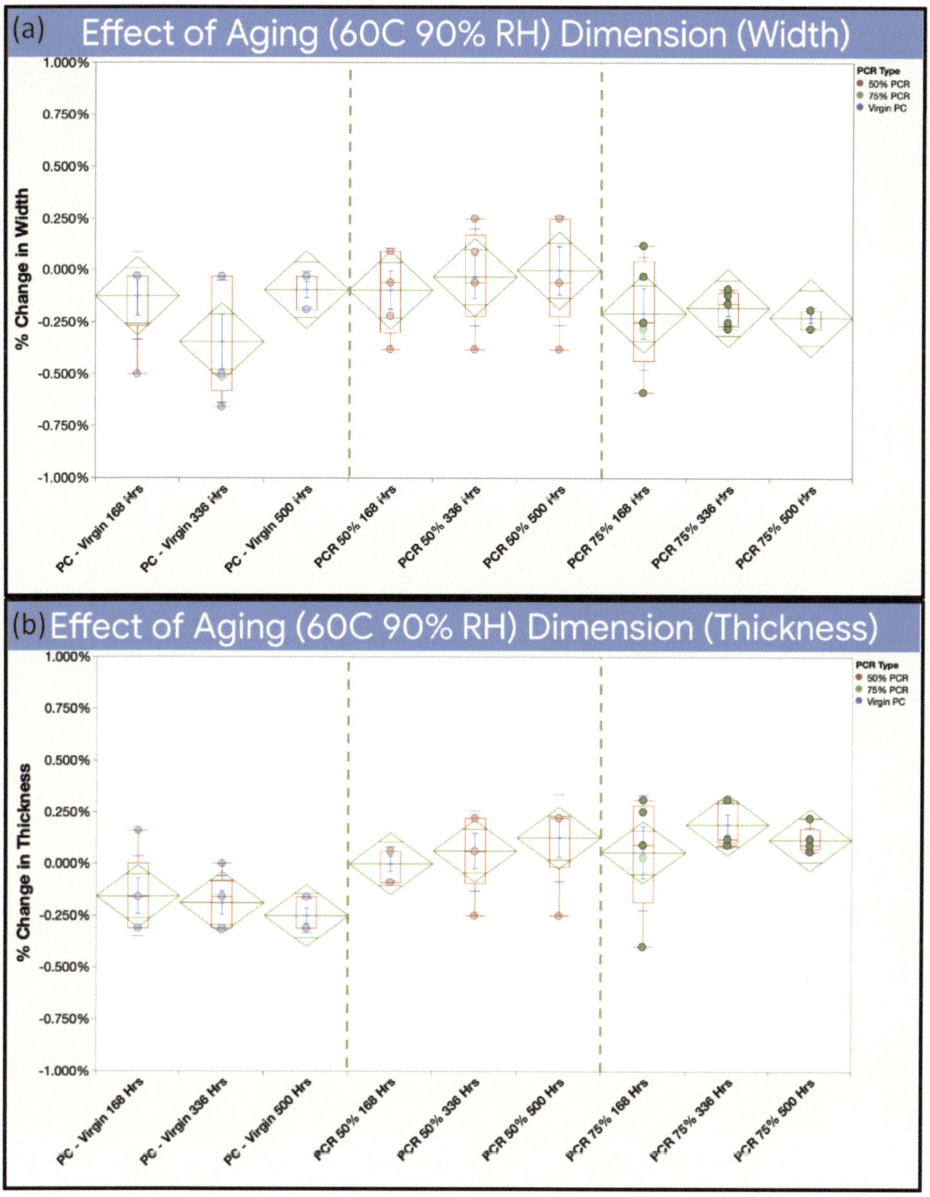

Figure 3.
% change in dimension after aging: (a) width and (b) thickness.

PCR have been compared. The strain follows a very similar pattern as the strength in **Figure 4(a)**. Strain value of 210%, 200% and 213% were recorded respectively for Virgin PC, 50% PCR and 75% PCR. The strain at breakage confirms the ductile behavior for all these three grades of polycarbonates.

3.3 Strength characteristics after aging

Secondly, we studied strength degradation with aging. First, all the three grades were compared at 100, 200 and 300 cycles of the -20° to 60° C thermal cycling. Based on the data recorded in **Figure 6(a)**, temperature cycling does not have a strong effect as the strength did not degrade appreciably for any of the grades.

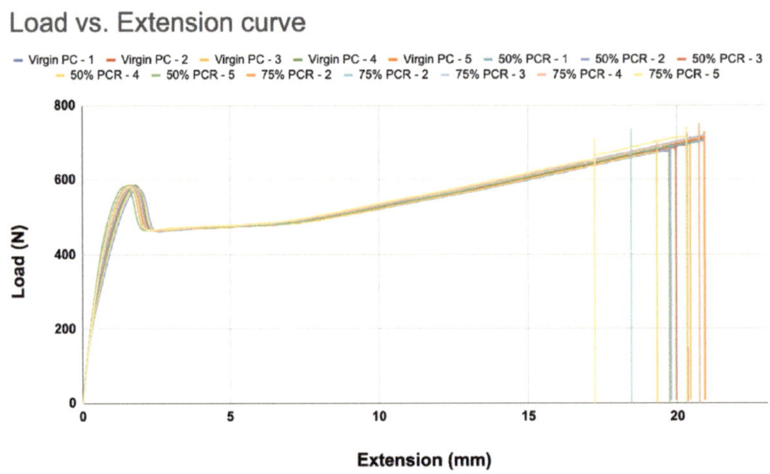

Figure 4.
Load - extension for virgin PC, 50% PCR and 75% PCR before aging.

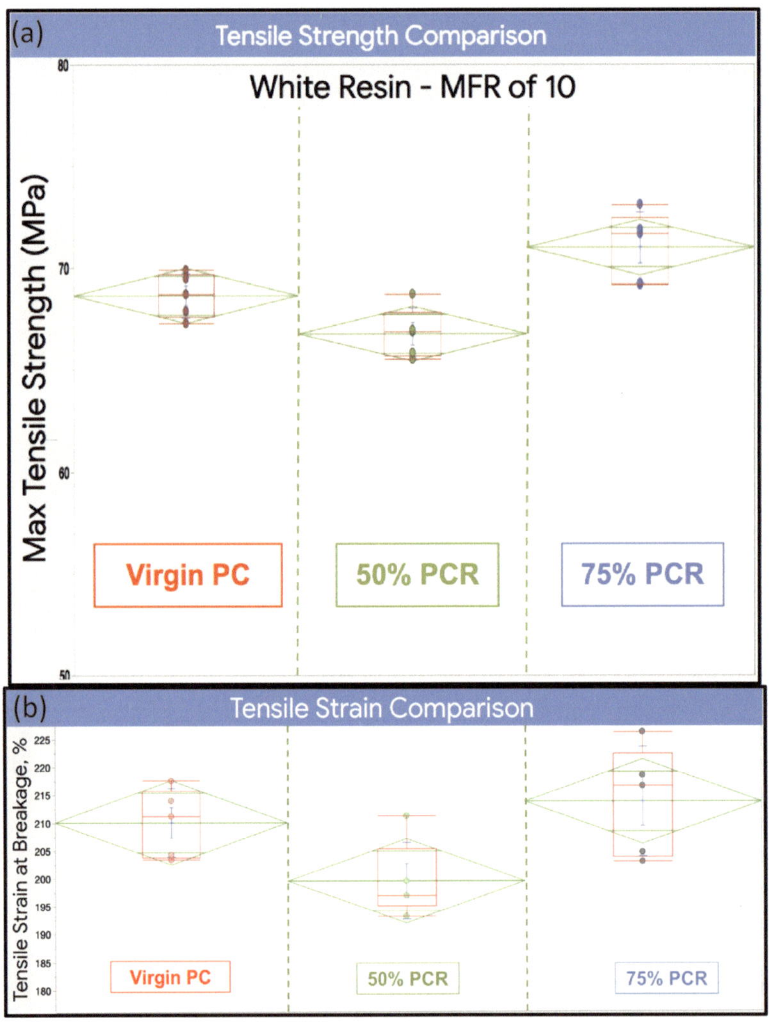

Figure 5.
Comparison between virgin PC and two grades of PCR for (a) tensile strength, (b) tensile strain at breakage.

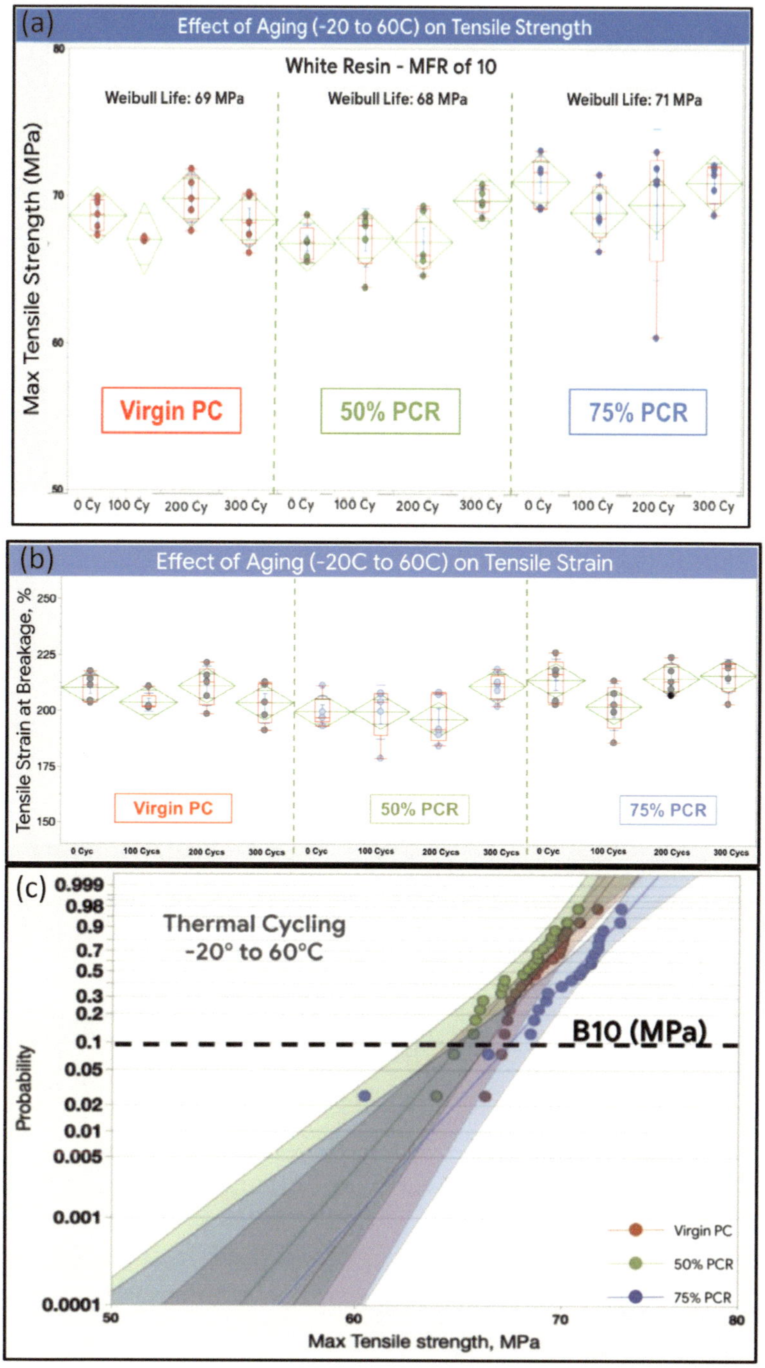

Figure 6.
Effect of thermal cycling on (a) tensile strength, (a) tensile strain and (c) life distribution of virgin PC, 50% PCR and 75% PCR.

Similar results were observed in strain at breakage 6(b). None of the grades show any significant reduction in strain and no pattern was observed for all three grades - which indicates that the tensile bars retained the ductile behavior. The reliability performance was evaluated by B10 (where 10% of the population fails), β (slope of the curve), and *α* (characteristic life). **Figure 6(c)** and **Table 3** shows that the B10

Sample type	Thermal cycling		
	B10	**Weibull α**	**Weibull β**
Virgin PC	66 MPa	69.5	47.7
50% PCR	65 MPa	68.5	42
75% PCR	67 MPa	71.2	40.2

Table 3.
Reliability data after thermal cycling.

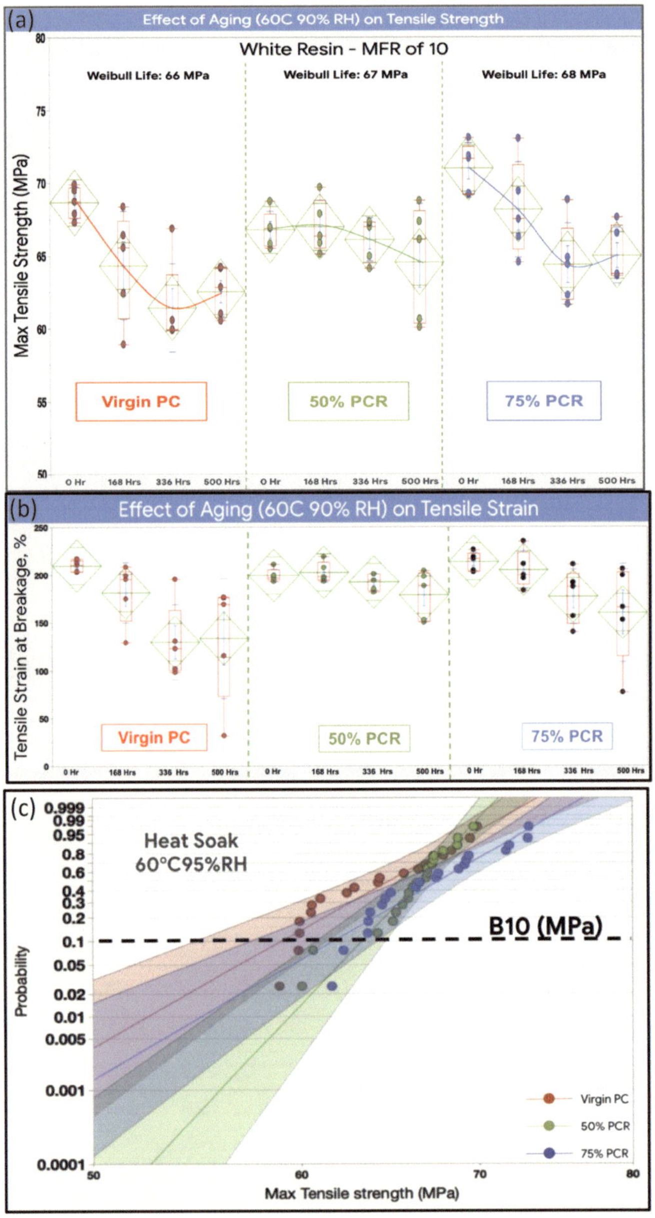

Figure 7.
Effect of heat soak on (a) tensile strength, (a) tensile strain and (c) life distribution of virgin PC, 50% PCR and 75% PCR.

Sample Type	High Temperature and High Humidity		
	B10	Weibull α	Weibull β
Virgin PC	60 MPa	67	20
50% PCR	63 MPa	67.1	28
75% PCR	61 MPa	68.8	20

Table 4.
Reliability data after high temperature and high humidity.

life, Weibull α and Weibull β are very similar for Virgin PC, 50% and 75% PCR respectively [9, 10].

But when the humidity is added to the temperature absolute humidity (gm/m^3) increases from 25.63 gm/m^3 (for 60° C - 20% RH) to 115 gm/m^3 (for 60° C - 90% RH) then the aging effect is more pronounced. The effect in aging is clearly visible in **Figure 7(a)** and **(b)** by the decrease in tensile strength and tensile strain at breakage for all the three grades of PC and PCR. But most importantly, both virgin PC and two grades of PCRs were found to have similar aging behaviors. Literature shows that the molecular weight reduces due to the hydrolysis of carbonate groups [11–15], which leads to the decrease of the tensile strength [16–18]. Prolonged exposure to high heat and humidity eventually induces a ductile to brittle transition [19]. The reduction in tensile strength and ductility are consistent with environmental degradation. Other mechanisms discussed in literature include a decrease in molecular weight due to hydrolysis and formation of small, disk-shaped microcavities or microcracks [11, 14, 15] that cause reduced strength and elongation. In this manuscript, the dominant mechanism has not been verified. In the fractography section, the fracture surface of the unaged and aged samples was analyzed to see whether this hypothesis is in line with the observed features for both conventional PC and PCR grades.

When we take these data into account and model the reliability of destructive degradation (**Figure 7(c)**), we found a comparable B10 (where 10% unit fails) value between virgin PC and two grades of PCR PC after thermal cycling aging. In temperature- and humidity-exposure, the B10 value is understandably lower than the temperature cycle aging but more importantly, virgin PC and PCR show comparable B10 and Weibull alpha values (**Table 4**).

4. Fractography

The purpose of the fractography was to evaluate whether the fracture surfaces added any additional information to the mechanical properties already measured. Four sample pairs were chosen to judge key comparisons:

- Effect of PCR (on unaged, white samples): Conventional PC vs. 50% PCR (white, unaged, MFR = 10 g/10 min)

- Effect of aging (on white samples): Unaged vs. aged (white, MFR = 10 g/10 min)

- Effect of aging (on black, 75% PCR samples): Unaged vs. aged (black, 75%PCR, MFR = 10 g/10 min)

The effect of PCR was gauged by comparing two unaged, white samples with the same MFR. One sample was a conventional PC, while the other was 50% PCR.

From **Figure 8**, both samples have ductile features such as hackle lines, chevron marking (V shaped features) [20] visible across their fracture surfaces. This is consistent with the fact that there were no mechanical property or molecular weight differences between these two samples. The elongation at break and tensile strength between unaged conventional PC and 50% PCR were comparable. Extensions at max load were 20.19 and 19.25 mm and max tensile strength were 67.3 and 65.52 MPa for those specific unaged conventional PC and 50% PCR samples.

The effect of aging on white PCs was determined by comparing two white conventional PC samples with the same MFR. One sample was unaged, while the other was aged for 500 hours at 60°C 90% RH.

From **Figure 9**, the differences in the fractography are more obvious. While the unaged sample has large regions of ductility, the aged sample shows a smoother surface, which indicates a more brittle material. As well, there is an obvious smooth initiation site in the middle of the right half of the surface, which indicates a slower crack growth. This sample still has chevrons and ductile features. The sample on the right (tensile strength: 63.32 MPa) had degraded mechanical properties as

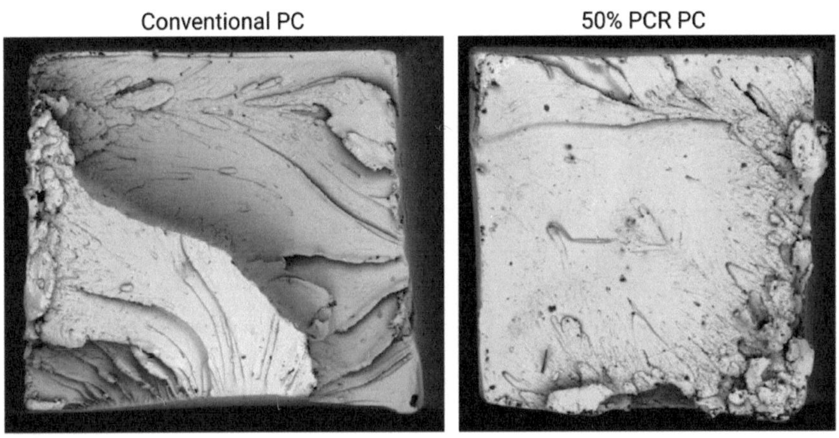

Figure 8.
A comparison of conventional and 50% PCR PC fracture surfaces.

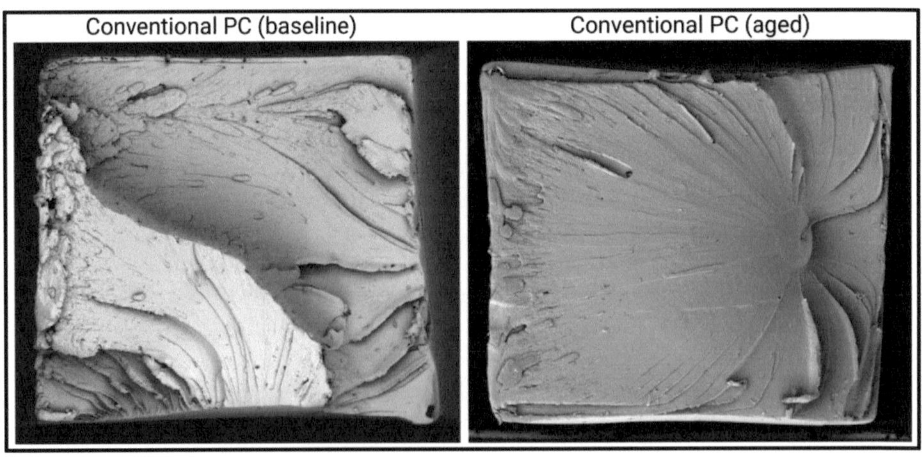

Figure 9.
A comparison of unaged and aged white conventional PC fracture surfaces.

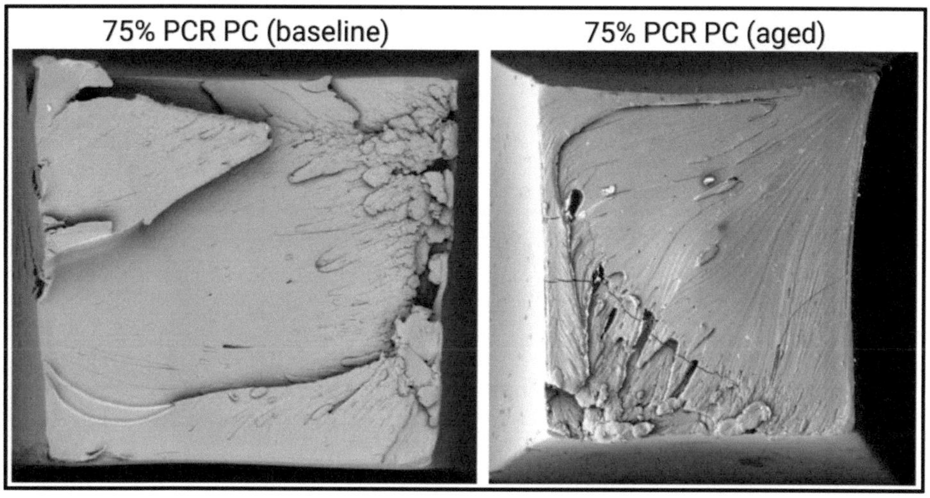

Figure 10.
A comparison of unaged and aged black 75% PCR PC fracture surfaces.

compared to the unaged sample (69.72 MPa), and this is consistent with the differences in the surface morphology.

Finally, the effect of aging on black PCR samples was evaluated. One sample was unaged, and the other was aged for 500 hours at 60°C 90% RH.

Similarly, to the comparison between aged and unaged white samples, the black 75% PCR samples have differences between the aged and unaged samples (**Figure 10**) that agree with the differences in mechanical properties between these two sample groups. The unaged sample has large ductile features, whereas the aged sample has more drawing.

5. Conclusions

In this paper, a comparison between conventional PC and two grades of PCR (50 and 75%) were conducted in terms of the effects of various aging strategies on mechanical strength and fractography. Unaged samples for all three grades with MFR of 10 g/10 min showed comparable strength (67–70 MPa) and the fracture surfaces of conventional and 50% PCR grades indicate ductile failures. As the MFR was increased from 10 to 15 g/10 min, strength is reduced, which is expected in any polymer resin. After thermal cycling at -20° to 60° C, no degradation was observed, whereas after high temperature and high humidity aging (60°C 90%RH) - strength reduction was observed and the degradation was consistent for all the three grades. This can be attributed to the degradation due to environmental aging. In fractography comparison between unaged and aged samples for both conventional PC (white resin) and 75% PCR (black resin), it was also observed that ductile features were less in aged samples. Overall, no significant differences were found between virgin PC and post-consumer recycled PC which indicates that recycled PC can be used in consumer hardware if molding conditions are controlled.

Author details

Rashed Islam[1*], Syed Zameer[1], Jacki Laiz[2], David Chen[1], Anthony Yu[1] and Ayyana Chakravartula[1]

1 Product Integrity Engineering, Google Devices and Services, Google LLC, USA

2 Materials Engineering, Google Devices and Services, Google LLC, Mountain View, CA, USA

*Address all correspondence to: rashed@google.com

References

[1] https://www.statista.com/statistics/959995/flow-plastic-materials-worldwide/

[2] The OECD for the G7 Environment, Energy and Oceans Ministers, "Improving Plastics Management: Trends, policy responses, and the role of international cooperation and trade, September 2018. (https://www.oecd.org/environment/waste/policy-highlights-improving-plastics-management.pdf)

[3] https://www.ellenmacarthurfoundation.org/explore/plastics-and-the-circular-economy

[4] International Standard ISO 14021 - 2016, "Environmental labels and declarations — Self-declared environmental claims (Type II environmental labelling)"

[5] ASTM standard - D638-14, Standard Test Method for Tensile Properties of Plastics.

[6] ASTM D638: THE DEFINITIVE GUIDE TO PLASTIC TENSILE TESTING - How to Perform a Tensile Strength Test on Plastics According to ASTM D638. https://www.instron.us/en-us/testing-solutions/by-standard/astm/multiple-testing-solutions/astm-d638

[7] ISO 527-2: Tensile Testing for Plastics: The Definitive Guide to Performing ISO 527-2 Tensile Testing on Plastics

[8] Ho-Huu C, "Influence of physical aging on the thermal and mechanical behaviors of glassy polycarbonate, Thesis of Doctorate of Applied Science, University of Sherbrooke (Quebec), Canada , 2003.

[9] Kahlen S, Wallner GM, and Lang RW, Solar Energy. 2010; 84 (5): 755-762.

[10] Senden DJA, Engels TAP, So¨ntjens SHM, Govaert LE, Journal of Materials Science. 2012; 47:6043–6046.

[11] Robeson LM, Crisafulli ST, Journal of Applied Polymer Science. 1983; 28: 2925.

[12] Gardner RJ, Martin JR, Journal of Applied Polymer Science. 1979; 24:1269.

[13] Ghorbel I, Thominette F, Spiteri P, Verdu J, Journal of Applied Polymer Sciencc. 1995; 55:163.

[14] Joseph EA, Paul DR, Barlow JW, Journal of Applied Polymer Science. 1982; 27:4807.

[15] Golovoy A, Zinbo M, Polymer Engineering and Science. 1989;29:1733

[16] Flory PJ, Journal of American Chemical Society. 1945;67:2048.

[17] Gent AN, Thomas AG, Journal of Polymer Science: A2 Polymer Physics. 1972;10:571.

[18] Bersted BH, Anderson TG, Journal of Applied Polymer Science. 1990; 39: 499.

[19] Vincent PI, Polymer, 1960;1:425

[20] Jansen JA, Technimet S, Fractographic Characterization of Polycarbonate Failure Modes. In: Proceedings of Plastics: Annual Technical Conference (ANTEC 2004); May 16 - 20, 2004, at Chicago's Navy Pier, Chicago, Illinois.

Chapter 9

Performance Analysis and Modeling of Microplastic Separation through Hydro Cyclones

Fabio Borgia

Abstract

The filtering hydro cyclone is a solid–liquid separation device, generally conical in shape. The hydro cyclone allows the separation of microplastics from water, to facilitate micro-recycling. To test the capabilities of a hydro cyclone at separating microplastics from water, Rietema's standard sizes, mathematical and computational fluid dynamics (CFD) modeling were used. The results show that, even dough the mathematical model in unreliable when considering parameters out-side standard operation conditions, hydro cyclone microplastic separation can be achieved at 98% efficiency. Particles reach the outlet on average in 1.5 s for a flow velocity of 2 m/s, and denser microplastics end up in the underflow.

Keywords: hydro cyclone, water treatment, microplastics, filtration, separation

1. Introduction

Plastics are the most used and versatile material of the modern age, and inevitably global production of plastics has increased exponentially to meet demand. Of all the plastics ever produced it is estimated that 10% has been released into the ocean [1]. Furthermore about 33% of the plastic produced every year is designed to be single use and is discarded within a year [2]. Plastic waste has become a major issue in the last century, mostly damaging marine environments, and activists have taken action to neutralize this problem. Although microplastics are yet not considered as threatening as larger plastics, they are growing in importance. It is extremely important that both plastics and microplastics are collected from rivers, oceans and waste waters to be recycled to block the environmental disaster these are causing.

Microplastics are the product of mechanical, chemical and thermal degradation of plastic objects, varying in size between 1 μm and 5 mm [3]. These particles can be found on surface waters and land, and from fresh-water to deep ocean sediments, making them susceptible to environmental factors such as heat, erosion and extreme pressures that may degrade them at rates not yet well established [4]. Through degradation they can assume many different shapes and sizes that fall into the following 5 categories [3] (**Table 1**).

The density of microplastics can vary significantly from (10–2300 kg/m^3). The most common plastics like ABS (1030–1210 kg/m^3), PET (1300–1500 kg/m^3) and

Abbreviation	Type	Definition	Size
PT	Pellet	A small spherical piece of plastic.	2–5 mm [5]
FR	Fragment	An irregular shape piece of plastic.	0.2–5 mm [6]
FB	Fiber	A strand of filament of plastic.	1–5 mm [7]
FI	Film	A thin sheet or membrane piece of plastic.	1–5 mm [3]
FM	Foam	A piece of sponge, foam or foam like plastic.	1–5 mm [3]

Table 1.
5 categories of degradation characteristics for microplastics found in the environment.

PVC (1150–1700 kg/m^3) are all denser than fresh and sea waters (997–1025 kg/m^3), this allows for most of the microplastics present in water to be separated using gravity. Physical and chemical hazards related to ingestion of microplastics across a diverse range sizes and types have generated ecological concerns. Further, the impacts on human health of the chemical exposure to plastic debris from seafood consumption, and toxins that adsorb onto microplastic debris from the surrounding water, are currently unknown.

This paper proposes to use hydro cyclones to separate them from fresh and seawaters. Firstly, the paper briefly summarizes the functioning of hydro cyclones, secondly applies them to the microplastics separation, through mathematical and CFD models. And finally concludes that they could be very effective in separating up to 98% of the microplastics from the water.

2. Hydro cyclones

Cyclones are used in industry as dust separators; they isolate the dust from air using differential centripetal forces generated by fast circular flow in a cylinder that allow the denser material to flow outward and downward, while the lighter material to flow inward and upward. Hydro cyclones work by using the same principles as the normal cyclone. Hydro cyclones have a cylinder-conical shape which have a tangential feed inlet into the upper cylindrical section and two outlets along the axis at the top and bottom. Hydro cyclones operate vertically: the fluid enters horizontally through an inlet, tangential to the cylinder, that creates a circular flow into the cyclone. This, in turn, produces a vortex effect, where the denser particles circulate around the outer edge of the chamber reducing their kinetic energy through friction along the cylinder and cone walls and, thus, sinking downward and leaving through the spigot. The lighter particles, due to the small area of the spigot, are taken into the inner part of the vortex that flows upward exiting at the top of the cyclone. These two downward and upward cyclonic flows are called, respectively, underflow and overflow [8]. The product of the separation is at the underflow where the denser microplastics exit as a slurry (**Figure 1**).

Theoretical Advantages of using hydro cyclones to separate microplastics are [4]:

- Optimum separation characteristics at varying operation parameters.
- Long life time for appropriate material selection for each application.
- Simple operation.
- Modular design through basic connections and adaptors.

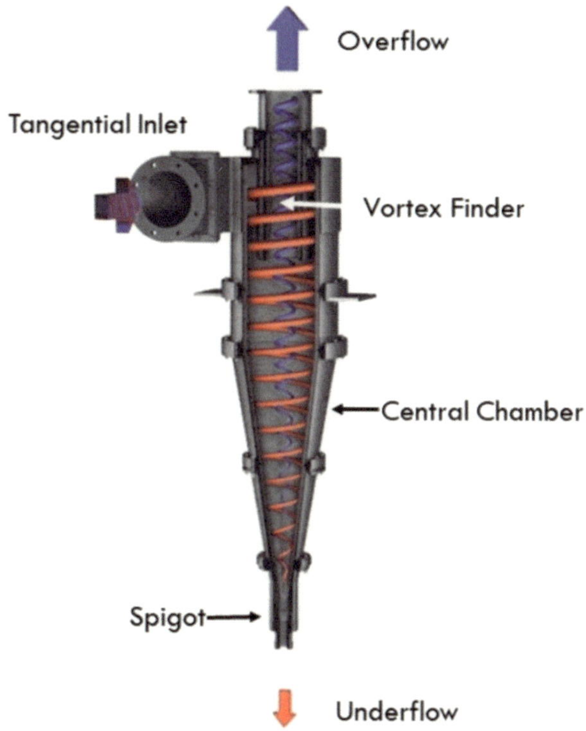

Figure 1.
Diagram of the behavior and characteristics of a hydro cyclone [4].

- Low weight.
- FDA approval.

3. Methodology

Latest investigations classify hydro cyclones based on the particle size of which 50% reports to the overflow and 50% to the underflow, or the so-called $D50_c$ point [9]. All equations used in Sections 3 and 4 have been taken from Richard A. Arterburn et al. [9]. Studies have shown that this classification remains persistent through a range of cyclone diameters and applications. The separation a cyclone can achieve can be approximated using the Eq. (1). The $D50_c$ (base) for a given diameter cyclone is multiplied by a series of correction factors designated by C_1, C_2, and C_3 (Eq. (1)).

$$D50_c(application) = D50_c(base) \times C_1 \times C_2 \times C_3, \tag{1}$$

where $D50_c$ (base) is the micron size that a standard cyclone can achieve under baseline conditions and $D50_c$ (application) is the filtering potential for a particular application.

$$D50_c(base) = 2.84\, xD^{0.66}, \tag{2}$$

where D is the cyclone diameter in centimeters.

The first correlation factor C_1 in Eq. (3) refers to the influence of the concentration of solids contained in the feed. The higher the concentration the coarser the separation. This correlation is a factor of slurry viscosity and particle size and shape. Variables such as liquid viscosity also affect this correlation.

$$C_1 = \left(\frac{(53 - \%_{solids})}{53}\right)^{-1.43}, \tag{3}$$

where $\%_{solids}$ is the percent solid by volume of cyclone feed.

The second correlation C_2 in Eq. (4) is for the influence of pressure drop in the cyclone, measured by taking the difference between feed pressure and overflow pressure. It is recommended that the pressure drop varies between 40 kPa and 70 kPa. This is to limit energy usage as well as equipment wear. As a result, a higher pressure drop would equate in finer separation.

$$C_2 = 3.27 \times \Delta P^{-0.28}, \tag{4}$$

where ΔP is the pressure drop in kPa.

The third correlation C_3 in Eq. (5) corrects the influence of specific gravity on the solids and liquid inside the cyclone. Stoke's law has been used to determine particle diameters which would produce the same terminal settling velocity for a particle of known specific gravity in a liquid.

$$C_3 = \left(\frac{1.65}{(\gamma_s - \gamma_l)}\right)^{0.5}, \tag{5}$$

where γ_s is the specific gravity for solids and γ_l is the specific gravity for liquids.

$D50_c$ (application) may be formulated as the product of the selected minimum size of separation and the associated multiplier for the percentage of solids passing through the overflow (Eq. (6)).

$$D50_c(application) = micronsizexmultiplier, \tag{6}$$

where the multiplier is defined form the **Table 2** below, taken from [9]. The result is that for a identified $D50_c$ (application) micron size, all the particles less than that will go into the overflow and all the particles bigger that that size will discharge to the underflow.

% of solids passing through overflow	Multiplier
98.8	0.54
95	0.73
90	0.91
80	1.25
70	1.67
60	2.08
50	2.78

Table 2.
% of solids passing through overflow and the correspondent multiplier.

Standards	Di/D	Do/D	L/D	Angle
Rietema	0.28	0.34	5.00	20°
Bradley	1/7	1/5	6.8	9°

Table 3.
Rietema and Bradleys standard relations for hydro cyclone dimensions.

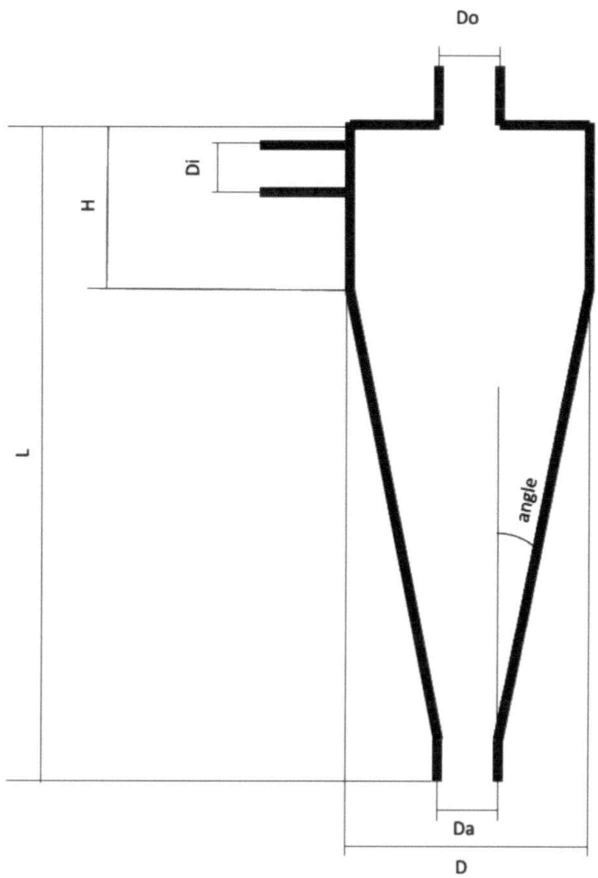

Figure 2.
Hydro cyclone dimension nomenclature and position.

Other cyclone geometric variables such as Di, Do, L and angle, have been found through the standard cyclone dimension relationships of the Rietema and Bradley hydro cyclones [8]. **Table 3** shows the correspondent relations (**Figure 2**).

4. Modeling

In order to create a model of a hydro cyclone that provides an good representation of its behavior and dimensions is critical. The minimum size of microplastics was assumed to be 5 μm; the density of the plastics going inside the hydro cyclone was assumed to be 1500 kg/m^3 (which is the average of densities between the most common plastics and the % that are present in the environment); the % volume of solids (microplastics) in the fluid (water) going inside the hydro cyclone; the pressure drop was considered to be 50 kPa as it is standard for most hydro cyclones;

the % of solids passing through the overflow which relates the multiplier; the Rietema standard cyclone dimension relations where chosen because considered a better fit for the application (**Table 4**).

Initial variables were calculated using the subsequent equations:

$$\rho_{average} = (\rho_{solids} + \rho_{water})/2 = 1248.5\, kg/m^3, \tag{7}$$

$$\gamma_s = \frac{\rho_{solids}}{\rho_{water}} = 1.50, \tag{8}$$

$$\Delta P = P\text{–}P_{drop} = 51kPa, \tag{9}$$

$$D50_c(application) = d_{solids}xMultiplier = 13.9\ \mu m, \tag{10}$$

$$C_1 = \left(\frac{(53\text{–}\%_{solids})}{53}\right)^{-1.43} = 1.03, \tag{11}$$

$$C_2 = 3.27 \times \Delta P^{-0.28} = 1.09, \tag{12}$$

$$C_3 = \left(\frac{1.65}{(\gamma_s\text{–}\gamma_l)}\right)^{0.5} = 1.81, \tag{13}$$

$$D50_c(base) = \frac{D50_c(application)}{C_1 \times C_2 \times C_3} = 26.60, \tag{14}$$

$$D = \left(\frac{D50_c(base)}{2.84}\right)^{\left(\frac{1}{0.66}\right)} = 29.65\, cm, \tag{15}$$

Using the Rietema relations the rest of the hydro cyclone dimensions can be calculated.

$$Di = D \times 0.28 = 8.30\ cm, \tag{16}$$

$$Do = D \times 0.34 = 10.08\ cm, \tag{17}$$

$$L = D \times 5 = 148.28\ cm, \tag{18}$$

$$Da = Do = 10.08\ cm, \tag{19}$$

The dimensions of the apex diameter (Da) are a result of investigations done by [8]. Which result in Da being the equivalent to Do. This is to optimize the flow in the hydro cyclone.

Name	Value	Unit
D_{solids}	5	μm
ρ_{solids}	1500	kg/m^3
$\%_{solids}$	1	%
$\%_{overflow}$	50	%
ρ_{water}	997	kg/m^3
P	101	kPa
P_{drop}	50	kPa
Multiplier	2.78	—
γ_l	1	—

Table 4.
First set of variables for average conditions for microplastic separation.

Name	Value	Equation
C_1	1.348	(11)
C_2	1.087	(12)
C_3	2.846	(13)
$D50_c$ (base)	31.91	(14)
D	39.07	(15)
Di	10.94	(16)
Do	13.28	(17)
L	195.3	(18)
Da	13.28	(19)

Table 5.
Results for particle density 1200 kg/m³.

If the density of plastics going into the hydro cyclone is assumed to be 1200 kg/m^3 (which is the average of densities between the most common plastics between 900 kg/m^3 and 1400 kg/m^3), to achieve the same performance as in the 1500 kg/m^3 case, the dimensions of the hydro cyclone would have to increase as it would take longer to separate the particles. The results are shown in **Table 5**. These show an increase in size for the diameter (D) of 50% and of 30% for the length (L). Although this may seem like a large increase the design can still be manufactured, as the dimensions are relatively similar.

5. Computational fluid dynamics

The CFD simulation was made for the purpose of understanding if the separation of microplastics could be achieved with a hydro cyclone, and in how much time this separation would happen.

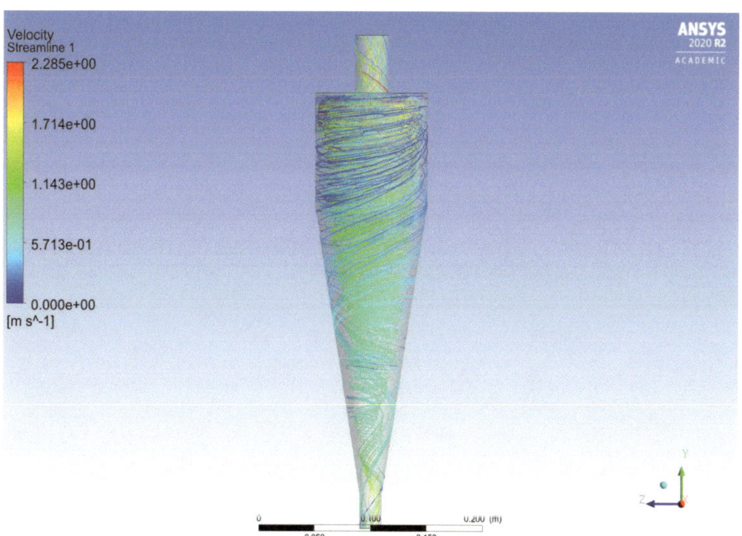

Figure 3.
Water flow simulation in a hydro cyclone through time.

Analysis though CFD was conducted assuming the dimensions indicated in the modeling section. Using Ansys Fluent software.

The CFD simulation was made for the purpose of understanding if the separation of microplastics could be achieved with a hydro cyclone, and in how much time the separation would happen. **Figure 3** illustrates the flow of water inside the hydro cyclone. It can be observe that both underflow and overflow collect water but that most of that exits through the overflow. Also visible is the inner vortex that forms because of the pressure difference.

Figure 4 shows the velocity inside the centre plane of the hydro cyclone. The big red spot on the top left is the velocity at inlet. The two outlets have increased velocities because of the pressure difference. Finally the inner part of the hydro

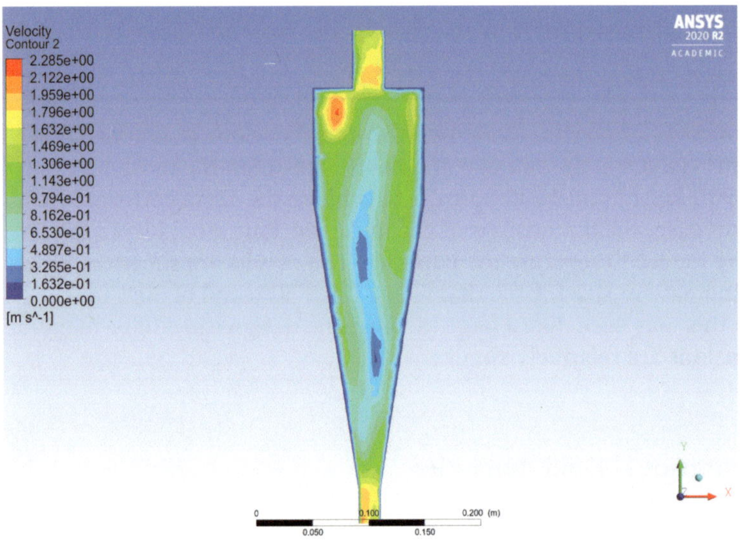

Figure 4.
Velocity gradient for water flow in a hydro cyclone.

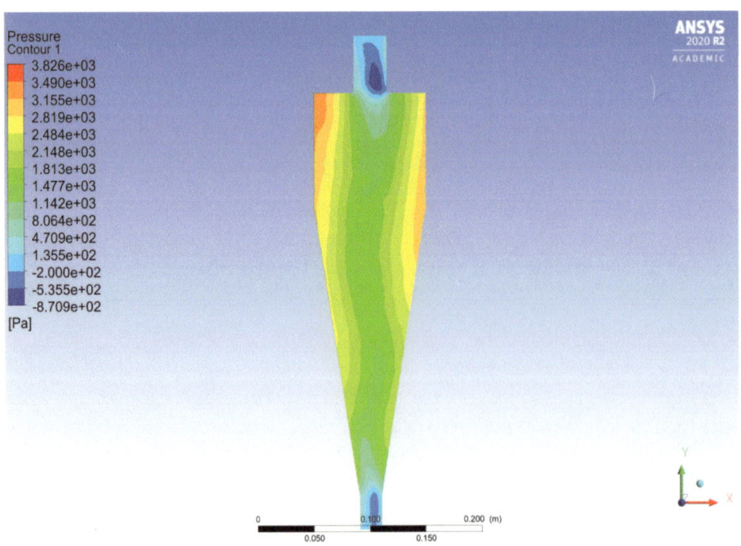

Figure 5.
Pressure gradient for water flow in a hydro cyclone.

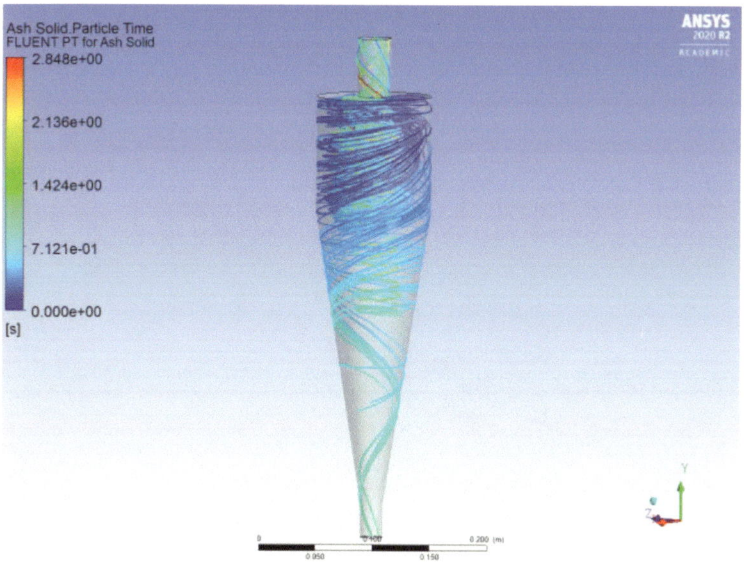

Figure 6.
Particle flow simulation in a hydro cyclone through time.

cyclone is under very low velocities because that is where the inner vortex flows and the velocities at the centre of it are close to 0 m/s.

Figure 5 shows the pressure inside the hydro cyclone and the difference that is created by the two outputs, where the pressure is negative.

Finally microplastic particles of the same physical characteristics where inserted inside a flow of water. **Figure 6** shows the result of this study where less denser plastics exit through the overflow and plastics that have greater density flow through the underflow. On average for a flow of 2 m/s a particle takes 1.5 s to reach any two of the outlets.

From CFD it can be seen that there is a lot of turbulence created near the inlet, where fluid coming inside disrupts the overflow vortex. To counteract this problem the design of the hydro cyclone would have to change to implement a sleave design for the overflow so that the inlet flow does not interact with the inner vortex.

6. Model results

Further studies were made to understand the behavior of some parameters used in the mathematical modeling using Rietema's correlations. The influence of particle dimensions was studied to see the impact on the cyclone diameter (D). **Figure 7** shows this relationship, and it can be deduced that the curve is close to exponential. This means that for particles smaller than 1 μm, the dimensions of the cyclone would be too small to be effective. If the minimum particle size is above 10 μm cyclone dimensions become exponentially bigger, which result in extremely big diameter (D) values. The problem is that Rietema's model used to produce the dimensions is not consistent for larger particle dimensions.

The second study related the % of solids inside the fluid, with the cyclone diameter (D) (**Figure 8**). The relation is linear for values of % solids between 1% and 21% and the change in diameter decreases as the particles increase.

The third study compered the solids density inside the fluid (Graph 3). The result, shown in **Figure 9**, conclude that for densities less then 1100 kg/m^3, the

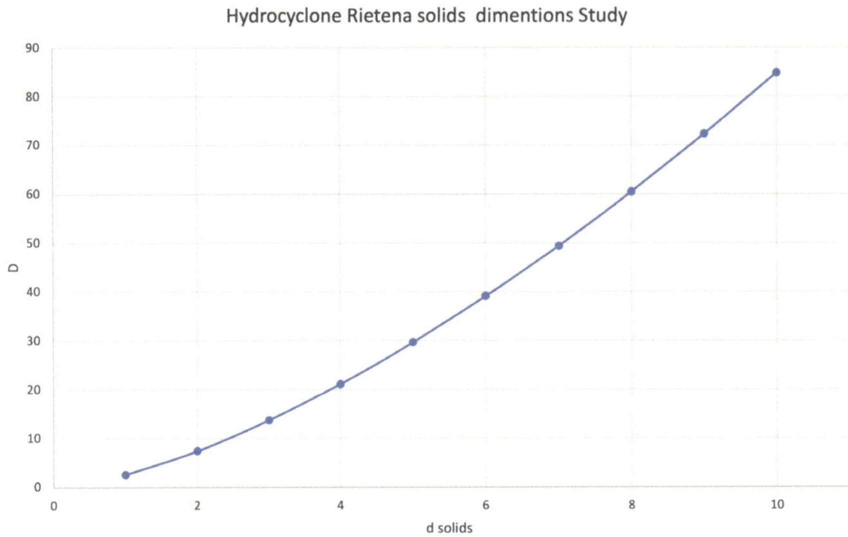

Figure 7.
Hydro cyclone Rietema model, solids dimensions study.

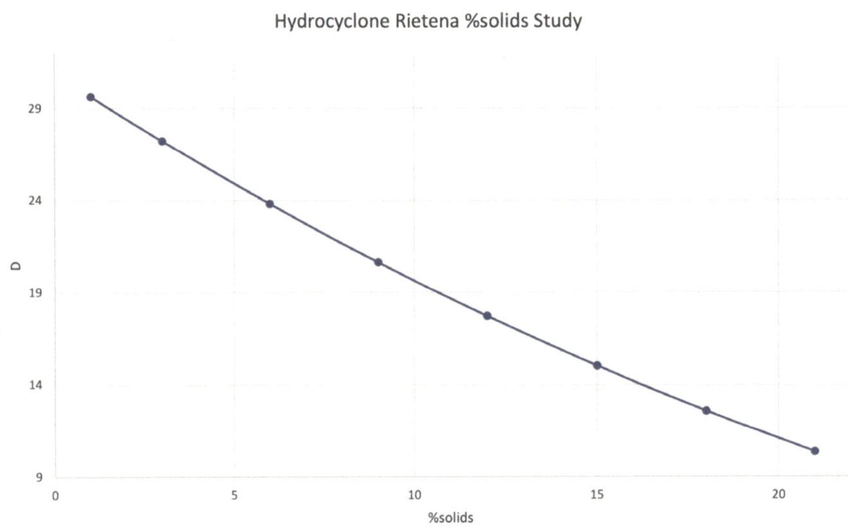

Figure 8.
Hydro cyclone Rietema model, %solids study.

dimension of the diameter (D) become unreliable. For bigger densities the diameter slightly decreases but the variation will become close to none. It can be seen that already from densities of 1200 kg/m^3 the model starts to get defective.

7. Conclusions

The results obtained through analysis and modeling of microplastic filtering hydro cyclones, under standard operational conditions, allow the following conclusions to be drawn.

Rietema's model offers the most consistent results throughout various tests, although becoming unreliable for values that exceed standard conditions. CFD

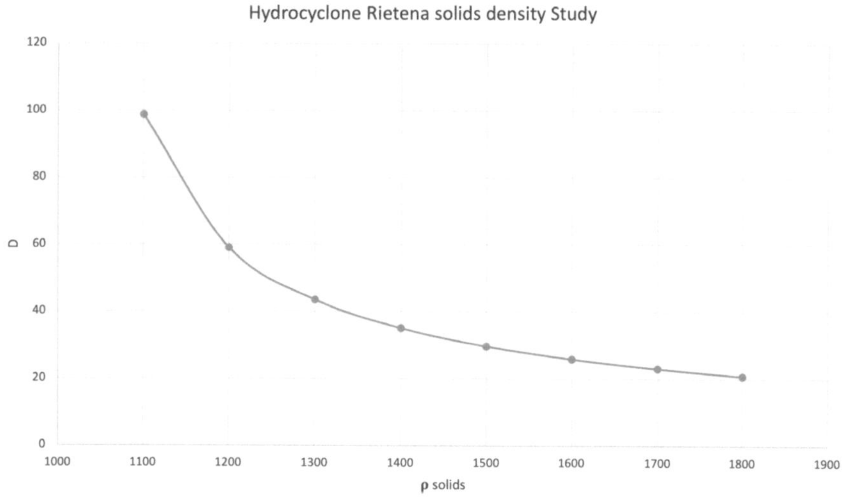

Figure 9.
Hydro cyclone Rietema model, particle density study.

models show that the correct design and dimensioning of the hydro cyclone can separate the flow consistently at a rate of 50% underflow and 50% overflow and the that on average particles take 1.5 s to reach any of the two outlets. The denser microplastics separate before and these reach the underflow in the least amount of time. Particles that reach instead the overflow could be captured by another hydro cyclone to be separated to a greater precision.

Future development could introduce an experimental apparatus to test the theories proposed in this paper. Implementation of this technology could be very useful in cleaning rivers and surface sea water from microplastics and other pollutants without damaging the aquatic flora and fauna. Hydro cyclones could be also mounted on water engine cooling and ballast water tank systems on cargo ships to purify water and prevent corrosion.

Acknowledgements

I wish to express my gratitude and sincere appreciation to the 2nd year students that participated in my group, to bring this idea forward (Benjamin Chesters and Josh McAree).

Nomenclature

angle	Cyclone cone angle	°
C_1	First correlation	-
C_2	Second correlation	-
C_3	Third correlation	-
D	Upper cyclone diameter	cm
Da	Diameter of underflow	cm
Di	Diameter of inlet	cm
Do	Diameter of overflow	cm
$D50_c(application)$	Filtering potential for application	-
$D50_c(base)$	Minimum size of particles	μm

d_{solids}	Dimension of solid particles	kg/m^3
Multiplier	50 % separation multiplier	-
L	Hydro cyclone height	cm
P	Inlet pressure	kPa
P_{drop}	Pressure drop inside the cyclone	kPa
$\rho_{average}$	Average density	kg/m^3
ρ_{solids}	Particles density	kg/m^3
ρ_{water}	Water density	kg/m^3
γ_s	Specific weight of particles	-
γ_l	Specific weight of liquid	-
ΔP	Difference in pressure inside the cyclone	kPa
$\%_{solids}$	Percent of solid particles inside the fluid	%

Author details

Fabio Borgia
University of Brighton, Brighton, UK

*Address all correspondence to: 2borgiafabio@gmail.com

References

[1] Kutz FW, Wood PH, Bottimore DP. Organochlorine pesticides and polychlorinated biphenyls in human adipose tissue. Reviews of Environmental Contamination and Toxicology 1991;20:1–82.

[2] Megharaj M, Ramakrishnan B, Venkateswarlu K, Sethunathan N, Naidu R. Bioremediation approaches for organic pollutants: A critical perspective. Environment International 2011;37:1362–1375.

[3] Christopher Blair, Crawford Brian Quinn; Microplastic Pollutants. Available online: (Accessed on 17/03/2021).

[4] Hydrocyclones. AKW Equipment process and Design. Available online: https://akwauv.de/hydrocyclones-aka-vortex/ (Accessed on 17/03/2021).

[5] Karlsson, T., Arneborg, L., Broström, G., Almroth, B., Gipperth, L. and Hassellöv, M., 2018. The unaccountability case of plastic pellet pollution. Marine Pollution Bulletin, 129 (1), pp. 52-60.

[6] Lehtiniemi, M., Hartikainen, S., Näkki, P., Engström-Öst, J., Koistinen, A. and Setälä, O., 2018. Size matters more than shape: Ingestion of primary and secondary microplastics by small predators. Food Webs, 17, p. e00097.

[7] Li, Y., 2006. Biomechanical Engineering of Textiles and Clothing. Boca Raton: CRC Press.

[8] L. G. M. Vieira, E. A. Barbosa, J. J. R. Damasceno and M. A. S. Barrozo; Performance analysis and design of filtering hydrocyclones. Available online: (Accessed on 17/03/2021).

[9] Richard A. Arterburn; The Sizing and Selection of Hydro cyclones. Available online: https://www.911metallurgist.com/wp-content/uploads/2016/02/sizing-select-cyclones.pdf (Accessed on 17/03/2021).

Section 2

Valorization of Waste Materials

Chapter 10

Industrial Re-Use of Composites

Albert Ten Busschen

Abstract

The amount of obsolete composites is increasing on a global scale, for example yacht hulls from a growing leisure industry and large rotor blades from wind energy production. Until now it has not been possible to recycle or disassemble thermoset composites into their original constituent parts of fibre reinforcement and resins. Subsequently a new method of re-use has been developed. This method involves machining the obsolete composite product into strips or flakes for re-use as reinforcing elements which, when combined with fresh resin and fibre, enable the production of a brand new component. This, in effect, preserves and re-uses the mechanical properties of the original obsolete composite. This method has been proven in manufacturing retaining walls, also guide beams for canals, crane mats and bridge decking, all using the strips or flakes from end of life composite products. For use on an industrial scale, a positive business case is imperative. In order to prove the industrial technology, new products have to contain a sufficiently high percentage of re-used composites in combination with automated processing. This has been achieved with "push-pultrusion" which is in essence a further development of the long established pultrusion process.

Keywords: composite, thermoset, fibre reinforced, end-of-life, rotorblade, re-use

1. Introduction

By definition a composite material is a material that is build up from two or more different components. For example, concrete is a composite material because it is build up from cement, sand and aggregates. However, with the name 'composites' often a category of materials is indicated that can be described as fibre reinforced plastics. In the following, 'composites' will refer to these fibre reinforced plastics. Fibres provide the reinforcement of the plastic, and the result is a strong but light material. These properties, when combined with the freedom to mould them into curved shapes has resulted in composites being used in many applications, notably: wind energy, aerospace structures, transportation, building and infrastructure, sport, yachting and various marine applications.

Industrial production of composite products started during the 1939–1945 World War and involved the use of epoxy or polyester resins reinforced with glass fibres. In the 1980's carbon and aramid fibres became available allowing the manufacture of stiffer, stronger and lighter structures. After the millennium, other reinforcement fibres became commercially available: high-oriented PE, basalt and natural fibres (e.g. flax and hemp). Despite these material innovations, glass fibre remains the volume material due to its excellent price to performance ratio.

Polyester and epoxy resins are thermosets. This means that they are initially liquid and solidify after a chemical cross-linking reaction. This liquid state enables the impregnation of the fibres and in this wet form they can be easily positioned into a mould without using pressure or elevated temperatures. The thermoset resin then solidifies and results in a strong fibre reinforced plastic product that can be subsequently removed from the mould. This combination of materials shows very good resistance to water or other corrosive environments. This results in composites having a long service-life with little or no maintenance. Even in outdoor applications a service-life of composite products is between 60 to 100 years [1].

Since the 1980's thermoplastic polymers have also been used for manufacturing composite products. Initially they were used only in short-glass fibre reinforcement for injection moulded parts (e.g. casings for tools). Typically the thermoplastic polymers PP, PA6, PA66, PBT and PET are used for these products. Thermoplastic composites require both a high temperature and pressure to achieve fibre impregnation. When producing composite parts using the injection moulding process, the high viscosity of the thermoplastic melt means that the impregnation of the fibres can only be achieved by a compounding step in an extrusion process. This results in a reinforcement with a short fibre and with such high pressures required, the size of the products is limited because of the need to use heavy steel moulds and high closing forces.

The commercial use of long-fibre thermoplastic composites has increased in the last twenty years. The production is a two-step process because of the difficulties in impregnation with the viscous thermoplastics, similar to the problem with the short-fibre thermoplastics. The long fibres are first impregnated with a melted thermoplastic polymer into plates or tapes, followed by a second step in which the pre-compounded materials are reheated and shaped into the desired form. This is achieved by hot-press moulding or laser-assisted tape laying. When cooled the product becomes solid. Long-fibre reinforced thermoplastics require PP or high-performance thermoplastics such as PEI, PPS and PEEK. Long-fibre thermoplastic composites are relatively small in volume compared to the total composite market but they are growing [2].

2. End-of-life thermoset composites

Despite their longevity, thermoset composites do eventually come to an End-of-Life (EoL) stage. This can be because of esthetical reasons, damage, or the end of their *guaranteed* structural safety. Generally the composite material itself is still viable. Particularly in the case of rotor blades from modern windmills that are guaranteed for safe use for a period of 20–25 years. When rotor blades are decommissioned (**Figure 1**), the composite material still has very good properties.

The major volume of EoL thermoset composites consists of boat hulls and windmill rotor blades, and these waste streams are expected to increase in the coming years [4]. Boats are a fashion leisure product and as such are periodically replaced and, as mentioned, the dismantling of windmills occurs when the period of guaranteed of structural safety ends or, as rotor blade size increases, for reasons of efficiency.

3. The composite recycling challenge

Thermoset composites are very hard to recycle into the original components (fibres, resin, fillers and core materials). To separate the components, the cross-linked resin must be decomposed because it cannot be melted. Decomposition can be achieved by burning or by dissolving in a chemical substance that can

Figure 1.
Obsolete windmill rotor blades [3].

depolymerize the thermoset polymer. These methods have been extensively investigated since the 1990's, but to date there is no industrial method that is financially viable.

The burning method to regain the fibres is only a partial (caloric) recycling of the material because the resin and organic core materials are not being recycled. In addition, the fibres that are regained from this process are of a very limited value. The glass fibres experience a dramatic loss of fibre strength as reported by Thomason et al. [5]. The coupling agent on the glass fibres (binder) is also destroyed by the burning process. With regard to carbon fibres the situation is slightly more positive because the burning process does not affect their strength. In both cases, however, the after-burning result is not a suitable material for general industrial use.

Two development programmes were set up in the 1990's using the burning process to regain glass fibres from composites in automotive applications, mainly from Sheet Moulding Compound (SMC) and Bulk Moulding Compound (BMC) parts. One programme from the automotive industry in Germany was developed by ERCOM Composite Recycling GmbH that started in 1990 [6]. This development did not lead to a successful industrialisation and stopped in 2004. The other development was the VALCOR-process from the automotive industry PSA in France. This development also did not lead to success and was duly stopped.

Grinding the composite products into a filler was also investigated. Although the resulting filler can be re-used in new products, this method did not lead to a positive business case. This was because virgin fillers that are available on an industrial scale (e.g. limestone, talc, sand) have a very low price level in the range of € 0,10 to € 0,20 per kg. The processing costs to grind the composite products lead to a much higher price level, so competition with traditional fillers is not possible. Over the past years several companies have developed composite grinding methods but they have not been a lasting success.

The burning method was further developed into the 'cement-kiln route' [7]. In this method, EoL composite is fed into a cement oven and the organic components are burned off providing the caloric value to heat the oven. The inorganic components, especially the glass fibre, remain as a filler in the cement. Although not technically a recycling of the composite, some useful components are retained in the form of energy and the fibre remainders that can serve as a filler. The cement kiln route has been accepted in Europe from 2012 to present day as a recognised method for the recycling of composites [8]. This method, however, is expensive: the EoL composite has to be processed into small pieces and to be brought to the cement oven in Bremen, Germany and then there, a gate fee of € 160, –per metric ton has to be paid.

To improve composite recycling, several initiatives in the last 20 years were undertaken to recycle composites into their original components but none have resulted in an industrial process yet. Comprehensive overviews have been given of these initiatives and have been described [9, 10] and presented by the ACMA [11].

4. Principle of structural re-use

In 2015 the Professorship for Polymer Engineering (Windesheim University of Applied Sciences) began to develop the principle of the structural re-use of EoL thermoset composite products [4, 12]. This method is based on keeping the composite structure of the End-of-Life product intact but machining it into smaller parts, e.g. strips or flakes. These smaller parts must have an oblong shape enabling it to act as reinforcing elements once embedded in resin while making new products. In this manner the good properties of the EoL composite, mechanical strength, stiffness and water resistance, remain unimpaired and can be used in the new product. An illustration of the use of EoL composite strips and flakes as reinforcing elements in new composite products is given in **Figure 2**.

To embed the strips or the flakes, additional virgin resin is required. This method implies that the new products will be relatively heavy and can only be simple in shape e.g. profiles, beams or plates. This combination of properties: high strength, high water resistance, relatively heavy and straight-shaped are extremely suited to infrastructural applications. Thermoset composites materials have outstanding resistance to outside conditions [1] and therefore will result in products with a long service life. The method of structural re-use of EoL thermoset composite products has been successfully proven in infra-structural demonstrators like retaining walls, guiding structures that funnel boats into canal locks, crane-mats and bridge decks as will be shown further on in this chapter.

5. Structural re-use in practice

The method starts with the machining of the product into smaller parts, prior to the making of strips or flakes. In **Figure 3** an example is shown where an obsolete polyester boat is broken up into large panels.

In the second step, these panels are sawn into long strips or shredded into flakes, see **Figure 4**.

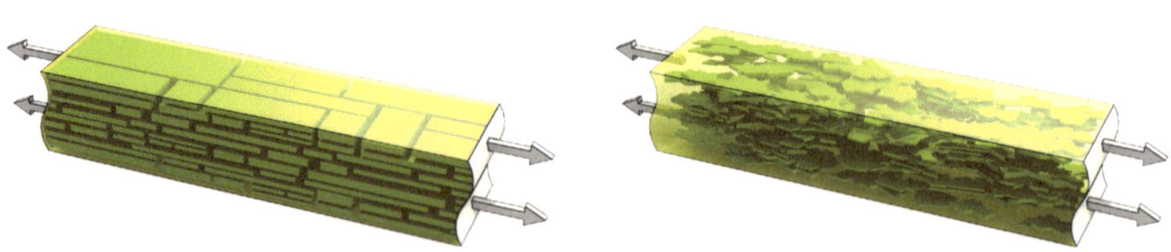

Figure 2.
New composite profiles reinforced with EoL composite strips or flakes [4].

Figure 3.
Breaking up of a polyester boat into panels [4].

Figure 4.
Panels sawn into strips or shredded into flakes [4].

The processing of heavy and large windmill rotor blades into strips or flakes involves more effort than a thinner polyester boat hull. Abolished rotor blades are currently in the region of 30 meter long. Future lengths of abolished windmill rotor blades will be much longer still because currently windmills are installed with rotor blades of over 100 meter length. Abolished rotor blades must therefore be pre-cut preferably on location into transportable sizes, e.g. by mobile waterjet cutting ('cold cutting'), by concrete breaking tools or by diamond blade cutting. These pieces can then be transported to a location where they can be reduced in size still further before eventual shredding.

When rotor blades are shredded, a different material is obtained compared to shreds from boats, because a large part of this composite consists of high-oriented reinforcement. After the shredding process, the material is more needle-like, see **Figure 5** left, than when products with a more random reinforcement are shredded, such as boat hulls, the latter results in a more flake-like product, see **Figure 5** right.

During the process of machining the composite product into strips or flakes, other side products are obtained. First of all, dust is formed during cutting and shredding. But also other components occur originating from the original product such as adhesives, coatings, core materials (foam and wood) and metal parts. Also contaminations that were present on the product can be found back such as dirt, oil and specifically for boat hulls: anti-fouling and growth of shells. Generally these other components are not very harmful for the properties of the end product, provided their percentage is limited. Moreover, the harmful substances will not leach out because it is completely embedded in virgin ('fresh') resin. The company CRC in The Netherlands is specialised in industrial processing of EoL thermoset composites into grades of flakes that can be re-used for making new products.

Processing the strips or flakes into a new product can be done by different techniques. All techniques have in common that the re-used material (strips or flakes) is

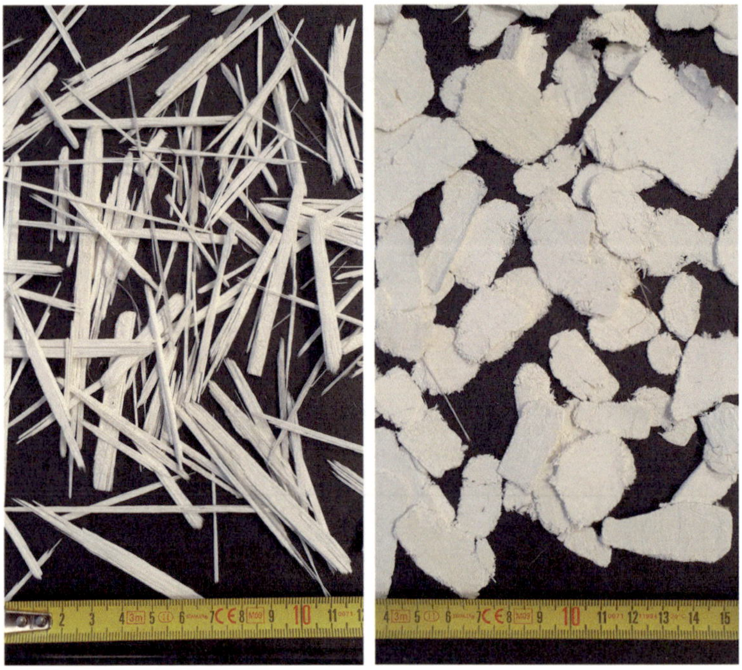

Figure 5.
Shreds from rotor blades (left) and from boat hulls (right).

embedded in virgin resin that binds it together. In most cases also a virgin composite layer is formed around the outside of the product.

6. Mechanical performance of re-used composite

The principle of structural re-use of EoL thermoset composite products is based on the possibility to partially benefit from the strength that is still present in the material. The so-called L/D-ratio of the parts that are re-used (strips or flakes) as reinforcing elements is an important parameter. In this ratio L represents the longest dimension of the re-used composite part and D represents the smallest dimension. Although the role of the L/D-ratio on elastic properties of fibre reinforced materials is described already long ago with elastic models [13], the influence of the L/D-ratio of reinforcing elements on strength of the complete reinforced material or product is more difficult to model.

The influence of the L/D-ratio of reinforcing elements on the strength of the new product that is made with it has been further investigated with a well-defined shape: a rectangular strip. To investigate the effect of the length of strips on reinforcement, a series of panels was produced and tested. To allow for a good comparison, instead of EoL-material, virgin glass mat reinforced polyester laminate with a constant quality was used as a base material to produce the strips.

The base material for making the strips for the investigation was a glass mat reinforced polyester laminate with a thickness of 5 mm. For the glass mat reinforcement 6 layers of continuous glass fibre mat (CFM) of 450 g/m^2 were used (Unifilo U813, OCV). As resin a low-viscous DCPD polyester resin (Synolite 1967 – G 6, AOC) was used. This laminate was made by vacuum infusion resulting in a glass content of 37 wt%. The laminate was made with peel-ply on both sides that was removed after curing for improved adhesion lateron.

Figure 6.
Layers of GRP strips as core in a test panel in preparation for infusion [4].

From the laminate, strips were cut with a diamond-tipped blade to a width of 20 mm, that were used as reinforcing elements in a test panel with the dimensions L x W x T = 1000 x 300 x 24 mm. Incorporating four layers of these strips positioned flatwise resulted in a reinforcing core of 20 mm. A surrounding shell of about 2 mm thickness was created by wrapping the core in an infusion glass mat (Polymat HI-FLOW M03P Core, Scott & Fyfe, consisting of 2x 450 g/m^2 glass mat reinforcement layers). The final panels were produced by means of infusion with the same DCPD polyester resin as for making the 5 mm laminates. The reinforcing strips were used in the core with lengths of 40 mm, 80 mm, 200 mm and 1000 mm, respectively. The photo in **Figure 6** shows the flat-wise incorporation of the fourth layer of 40 mm strips just before it was covered with the infusion glass mat.

To investigate the effect of adhesion between strips and embedment resin, a test panel was produced with 40 mm strips with flat surfaces (by omitting peel-ply layers, resulting in smooth surfaces with relatively bad adhesion). Moreover, a test panel was produced using 1000 mm strips, that was cut from a polyester boat hull without any surface treatment. Finally, a test panel was produced with 200 mm strips placed vertically on their sides instead of flatwise. In all test panels strips were oriented in length direction of the test panel and the strips were placed staggered with respect to the neighbouring strips (both horizontally and vertically). **Table 1** gives the overview of tested configurations.

Length of strips used as reinforcement of core (mm)	**40**	**80**	**200**	**1000**
Virgin GRP strips with peel ply, 4 layers placed flatwise	X	X	X	X
Idem, but without peel-ply (bad adhesion)	X			
Virgin GRP strips with peel ply, 1 layer placed vertically			X	
Strips of 5 x 40 x 1000 mm from boat hull, placed flatwise				X

Table 1.
Combinations of materials and strip lengths tested.

Figure 7.
Bending test of a specimen made with 80 mm strips [4].

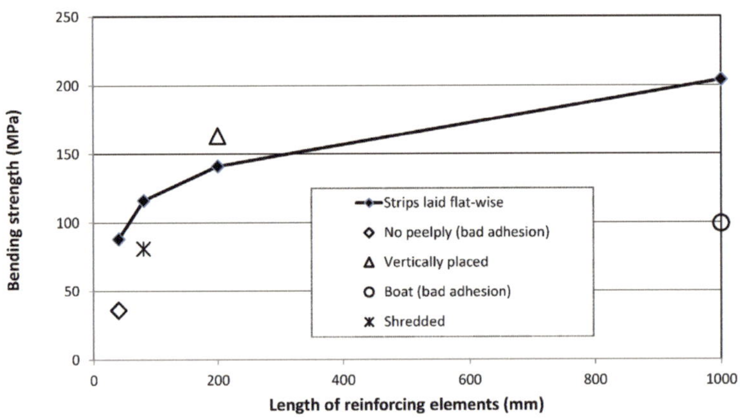

Figure 8.
Mechanical strength as function of strip length for different test materials [4] (markers indicate the mean value of 5 tests).

From each test panel samples were cut in length direction (i.e. in de direction in which the strips are oriented) with the dimensions L x W x T = 360 x 50 x 24 mm. Tests were repeated 5 times. The samples were tested in three-point bending in accordance with ISO 178. The photo in **Figure 7** shows the testing of a sample with four layers of strips of 80 mm in length, placed flat-wise.

The graph in **Figure 8** shows a clear correlation between the bending strength and the length of the strips in the core for flat-wise laid strips with good adhesion. With a 1000 mm strip length the bending strength reaches 204 MPa. The negative effect of bad adhesion on bending strength is seen in the case of the 40 mm strips where the peel-ply has been omitted: the bending strength is as low as 36 MPa, less than half of the strength with 40 mm strips with good adhesion (88 MPa). Also a relatively low strength of 99 MPa is found when using strips cut from a polyester boat-hull, which can be attributed also to bad adhesion. During the test these strips that were obtained from abolished boats, delaminated at the gelcoat side, showing a smooth delamination surface, which is an indication of a relatively bad adhesion. By placing the 200 mm strips vertically, a higher bending strength is observed (163 MPa) as compared to the flat-wise placement (141 MPa).

With the set-up of strips laid flat-wise and a good adhesion with the embedding resin, the effect of the L/D-ratio is considered. It is assumed that the panel strength with strip lengths of 1000 is the maximum attainable panel strength in this set-up (204 MPa). At a strip length of 40 mm (L/D-ratio of 8) only 43% of the maximum attainable strength of the panel is found. With increasing strip length the bending strength of the panel increases in a degressive manner. At a strip length of 200 mm (L/D-ratio of 40) a panel bending strength is found that is 69% of the maximum attainable panel strength. From these considerations it can be concluded that a significant part of the possible maximum panel strength is obtained for a L/D-ratio which is of the order of 50 or higher.

In the graph, also at the location of 80 mm length, the strength of a panel made from flakes is shown. That strength was found to be 81 MPa. The flakes were made by shredding EoL thermoset composite in such a way that a mean flake length of roughly 80 mm was obtained. It must be remarked that the lower strength as compared to the strength of 80 mm strips (116 MPa) can partly be attributed to the lower content of reinforcement (higher resin content).

Shredding may be a more economical way of machining EoL thermoset composites into reinforcing elements than sawing or water cutting, although this has to be investigated further. Shredding is a very promising method because a large quantity of EoL composite products can be machined at relatively low cost. However, the flakes that result from the shredding process must have a quality level that makes them suitable for the re-use in new products, e.g. sufficiently high L/D-ratio, dry, dust free and good adhesion properties with the embedment resin.

7. Demonstrators

The methodology of structural re-use of EoL thermoset composites results in products that are especially suitable for applications in building and infrastructure. Profiles, plates or panels can be made that are mechanically strong and resistant to moisture. The fact that the new products made from EoL composites are not light-weight is generally not a problem for this field of application. In the following section, demonstrator projects are described from re-used EoL composite products.

7.1 Retaining walls

EoL composites profiles were manufactured for retaining walls near the Beatrix lock-gate in Almere, The Netherlands [3, 14]. 80 separate profiles, each with a length of 3.5 metres were produced using two steel moulds. The cross-sectional

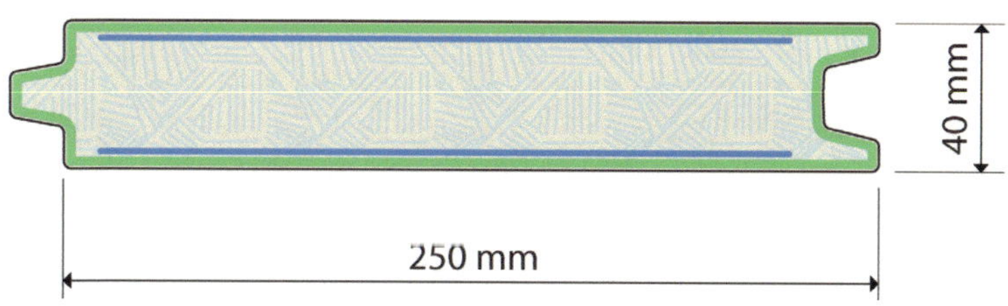

Figure 9.
Cross section of profile for retaining wall built up with a core of EoL composite [3].

Figure 10.
Production of profiles using EoL composite by vacuum infusion under foil [3].

dimensions of the profiles were 40 x 250 mm with a tongue-and-groove detailing. A drawing of the cross-section of the profile is given in **Figure 9**.

The profile has an effective width of 250 mm and a thickness of 40 mm. The outside skin is build up from a virgin glass reinforced polyester laminate containing a 900 g/m^2 random reinforcement (glass mat, green in the picture) and a 900 g/m^2 UD-reinforcement of 200 mm width on both sides (blue in the picture). The core is formed by re-used EoL thermoset flakes that are bound by a polyester resin.

For this demonstrator, the profiles were made by means of vacuum infusion under foil in steel moulds. Into the mould, the glass injection mat was charged (Polymat HI-FLOW M03P Core, Scott & Fyfe, 900 g/m^2) along with the first layer of 200 mm wide UD-glass reinforcement (UNIE840, Selcom 840 g/m^2) on which the EoL flakes and strips were applied. On top of this, a second layer of UD-reinforcement was placed and the glass injection mat was closed around it. Then the vacuum foil was applied. A brown pigmented polyester DCPD resin (Synolite 1967-G-6 of AOC) was injected, giving the profile a wood-like appearance. **Figure 10** shows the charging of the mould with strips and flakes of EoL composite (left) and the product that is infused with brown pigmented resin (right).

Figure 11.
Installation of retaining wall by vibrating profiles with EoL composite in the soil [3].

Figure 12.
Guiding structure with two lowest rows of beams made from EoL composite [3].

The profiles were tested mechanically and the same bending strength was found as for the same profile geometry made of azobé wood, that are commonly used for these applications. The profiles were installed in 2017 by vibrating into the ground without any damage being incurred on the profiles, see **Figure 11**. Inspections and tests on profiles withdrawn from the ground in 2019, two years after installation, showed no signs of degradation nor loss of strength.

7.2 Guiding beams

Guiding structures are placed near bridges or lock gates in canals to guide ships and prevent damaging the infra structure. Traditionally, these structures consist of a steel frame with horizontal guiding beams. These beams are normally made of tropical hard wood (e.g. azobé) and typically have dimensions of 200 x 200 x 4000 mm. In 2019 guide beams were made from re-used EoL thermoset composite to replace the lowest two rows of four guiding structures at Groningen Seaports in Delfzijl, The Netherlands [3]. Because the lowest two rows of guiding beams in a guiding structure are located around and under the water level, this is the location where normally tropical hardwood beams suffer most from fungi attack and re-used EoL thermoset composite will be more durable. **Figure 12** shows a photograph of one of the four guiding structures. The four rows above the water level are made from tropical hardwood.

The strength needed to resist a possible ship collision for a single beam is defined for a single beam as 440 kN when mounted at a support distance of 1800 mm. To achieve this, additional layers of UD-glass reinforcement had to be incorporated into the beam. A prototype beam was tested in three-point bending with a support distance of 1800 mm and a maximum force of 515 kN was recorded. After this successful test, 112 meters of guiding beams were produced using a steel mould with an RTM injection process. The beams have been installed in Delfzijl in October 2019.

7.3 Crane mats

On building sites crane mats are used to obtain a stable work area on which heavy cranes and other machinery can operate. Generally crane mats are composed

Figure 13.
Crane mats in use at a building site [3].

of beams of tropical hard wood. The choice for this material is based on the requirements on strength, wear resistance and durability. The crane mats depicted in **Figure 13** have outer dimensions of 1 x 5 m and are composed of five azobé beams with dimensions 200 x 200 x 5000 mm that are assembled using five steel bars.

In cooperation with Welex (manufacturer of crane mats, based in The Netherlands), a crane mat was made from EoL thermoset composite beams. RTM-infusion was used for the production of the beams. In the mould first a glass injection mat (900 g/m^2), 2 layers of quadraxial glass reinforcement (1200 g/m^2) and 10 layers of UD glass reinforcement (840 g/m^2) were applied after which the core was built up from flakes of EoL composite in combination with fire-dried sand (1–2 mm grain). The five holes of 30 mm diameter for the assembly of the crane mat were created in the beams using tubular inserts. The crane mat was tested under various severe operating conditions. After the filling of the core with resin, the reinforcement layers from the bottom were folded over the top before closing the mould and starting the injection process. The EoL composite crane mat performed very well, showing good resistance to wear by vehicles and was easy to clean, see **Figure 14**.

Figure 14.
Crane mat made of EoL composite tested in practice [3].

Figure 15.
Dinzer bridge with deck profiles of re-used composite: Installation and final result [15].

7.4 Bridge deck profiles

In a project to renew the Dinzer bridge over a canal in Friesland (The Netherlands) deck profiles were applied made from re-used EoL thermoset composite. It was one of the requirements that the profiles should resemble the original hard wood profiles by their cross-sectional dimensions (thickness 95 mm, width 245 mm) and were able to withstand heavy traffic loading.

The profiles were designed with an outside layer of virgin glass fibre reinforced polyester and a core from re-used EoL thermoset composite flakes embedded in polyester resin. Before starting production, prototypes of the profiles were mechanically tested in order to verify the mechanical loading capacity. These tests were successful and showed that the beams fulfilled the requirements for Dutch infra structural design [15]. The required profiles were produced using a resin casting method in a steel mould.

The profiles were successfully mounted on the new bridge structure of the Dinzer bridge and give off the appearance of a traditional deck with wood deck profiles, see **Figure 15**. The big advantage of these profiles made of re-used composite is that the expected service life is much longer and maintenance is lower.

8. Design rules

In general, for building applications in Europe the Eurocode is used. There is, however, no specific Eurocode dedicated to composites. In The Netherlands a specific recommendation for design with composites is available [16] but only for composites made from virgin materials. Nevertheless, the general methodology from the Eurocode can be followed irrespective of the origin of the raw materials from which it is built. This is described in the European standard EN 1990 [17]. In this standard, in Annex D (Design assisted by testing) it is described in what manner material properties can be determined using a test programme.

Based on extensive testing of EoL thermoset composites [18] to date the following conclusions can be made for the design with these materials made with re-used EoL thermoset composites:

- Design for long-term loading (creep) and fatigue should be avoided for parts solely consisting of re-used EoL composites in a product. For these load types continuous virgin fibre reinforcement must be incorporated into the product to ensure resistance to creep and fatigue.

- For design on stiffness and strength (with exclusion of creep and fatigue) design formulas can be used [16], with the conversion factor set to 0.9. It is required that re-used EoL composite is compatible with the virgin resin in which it is embedded (e.g. good adhesion) and that the virgin resin is resistant to the user conditions of the new product and is properly cured.

9. Business case

A business case consists of several components that determine whether or not it is attractive for industry. Most critical factor is economic profitability. Can products be sold for a price that is attractive for the market and leaves a profit for the manufacturing company? There is also an increasing demand for products that are sustainable, this aspect must also be taken into account as non-sustainable products could be excluded in the future. In the Netherlands from 2023 the policy of the government is to purchase only sustainable products.

Product sustainability can be achieved by using raw materials with a low carbon footprint, for example: recycled or re-used materials. The use of materials with a low carbon footprint alone is not enough to ensure sustainability. The energy needed for production and the CO_2 production that is related to the production of the new product must also be included.

The economical profitability of structural re-use of EoL thermoset composites for retaining walls was considered. During the production of the infra-structural demonstrators it had become apparent that manual production was too costly for economic profitability. Trials were carried out with our industry partners using the automated production technique of pultrusion, with a very positive outcome. Following that success, the pultrusion company Krafton in The Netherlands installed a compounder to mix re-used EoL composite flakes with resin and injects this into the core of pultrusion profiles. This is an efficient continuous production process that involves very little labour.

A financial tool was developed to analyse the profitability of a factory that produces these profiles with a pultrusion-based continuous process at a production speed of 15 m/hour [18]. Costs for the production are based on raw material costs, energy consumption, labour costs, depreciation of machinery, rent of production

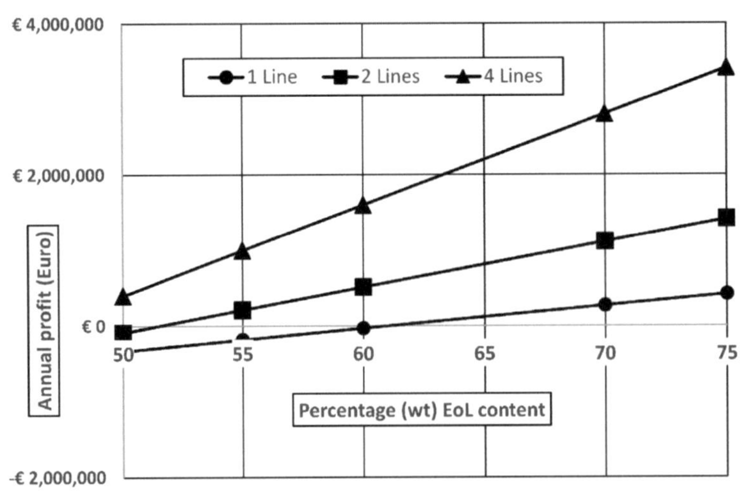

Figure 16.
Annual profit for a model production plant for profiles with EoL composite core [3].

space and overhead. Based on price discussions with local councils and water municipalities in The Netherlands, a factor of 1.3 was applied to the sales price of identical profiles made of tropical hard wood (azobé).

A factor of 1.3 means that a product made of re-used EoL thermoset composite can be sold for a 30% higher price than the traditional tropical hard wood profile. This higher sales price was found acceptable by the water municipalities based on the longer life span in wet conditions and the circular characteristics of the product.

Variables in the tool are the weight percentage of EoL material in the profile and the number of production lines. **Figure 16** shows the annual profit of the model factory as a function of weight percentage of EoL material in pultruded profiles, for production facilities with 1, 2 and 4 production lines respectively.

From the analysis it can be seen that already with one production line the facility becomes profitable when the content of EoL material in the profiles reaches at least 62% by weight. With more production lines the profitability becomes higher as the number of persons working in the pilot plant weighs heavily within the calculation. From the trials by the industry partners using a pultrusion set-up it was found that an EoL content of 70% by weight is possible in the profile under consideration.

For the same profile also the CO_2 footprint was analysed. For this, the ECO-Calculator of EuCIA was used [19]. This tool evaluates the CO_2 footprint of composite products 'from cradle to gate', which means it considers the effect of the raw materials used and the production process. Using this tool the CO_2 emission per kg of product was calculated for two percentages of EoL composite material content (50 wt% and 70 wt%, respectively). Moreover, the CO_2 emission per meter profile was analysed when the profile with the same mechanical performance was made using only virgin raw materials, either as a profile made with an RTM-process with a PET-foam core or as a hollow profile with shear webs inside made with a pultrusion process. Cross-sections of the four profiles are presented in **Figure 17**.

The results of the analysis is given in the graph in **Figure 18**. Obviously the amount of CO_2 for the production of a meter profile is strongly related to the percentage of re-used EoL thermoset composite used. This is mainly connected to the amount of virgin resin that is used to embed the EoL composite flakes. Comparing virgin based profiles with EoL composite material containing profiles, the carbon footprint of the latter becomes advantageous when the amount of re-used EoL thermoset composite is at least 70 wt% or higher.

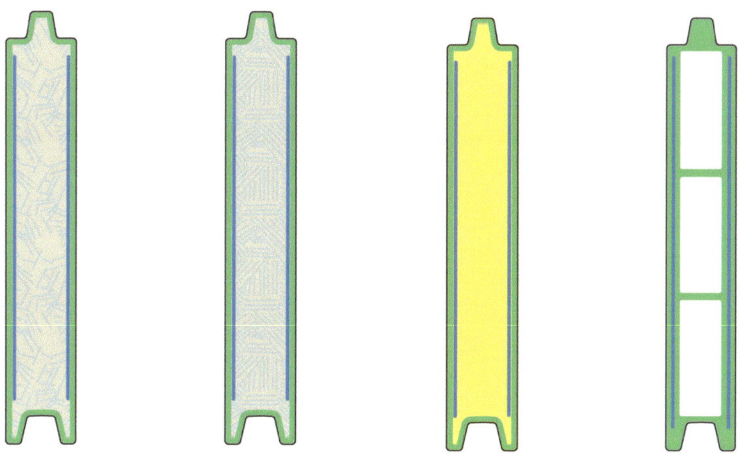

Figure 17.
Profile types with identical mechanical performance analysed for their CO_2 – footprint. From left to right: 50% EoL, 70% EoL, virgin RTM, virgin pultrusion.

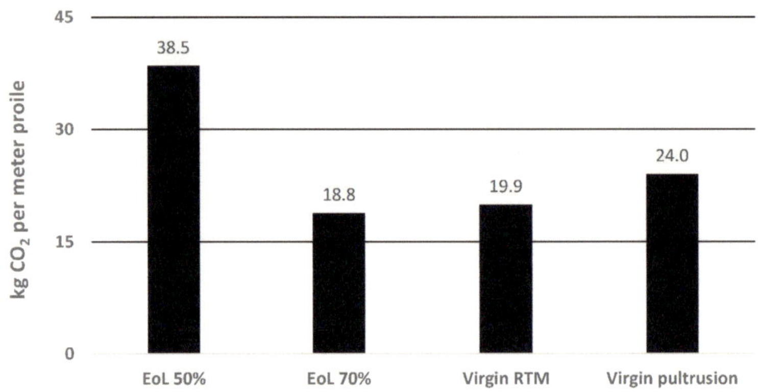

Figure 18.
Graph of CO_2 footprint per meter profile for different profile build up [3].

10. Conclusions

The methodology of structural re-use of EoL thermoset composites as developed by Windesheim and partners offers the possibility of industrial re-use of previously non-recyclable thermoset composites. From the various infrastructural demonstrators it could be concluded that strong, robust and water resistant products can be made. A set of design rules was developed according to the Eurocode to evaluate structural behaviour of products made with re-used EoL thermoset composites. It was shown that industrial processing of these materials can be achieved using a pultrusion type process. The lower labour costs of such a process and the high percentages of EoL-content in the new products can lead to an automated production and a profitable sales of products. It was found that the percentage of EoL material in the new product is the key parameter for both profitability and low CO_2 emission overall.

Author details

Albert Ten Busschen
Windesheim University of Applied Sciences, Zwolle, The Netherlands

*Address all correspondence to: a.ten.busschen@windesheim.nl

References

[1] Halliwell, S. Fibre Reinforced Polymers in Construction: Durability BRE Report IP10/03, September 2003 ISBN 1 86081 635 5

[2] Composites Marked Report 2017 – Market Developments, Trends, Outlook and Challenges Publication of AVK, German Federation of Reinforced Plastics, 2017

[3] Busschen, A. Ten industrial re-use of composites reinforced Plastics, volume 64, No. 3, May/June 2020

[4] Busschen, A. Ten structural re-use of end-of-life thermoset composites reinforced Plastics, volume 61, No. 3, May/June 2017

[5] Thomason, J., Jenkins, P., Yang, L. Glass Fibre Strength – A Review with Relation to Composite Recycling Fibers 2016, 4, 18

[6] Marsh, G. Facing up the recycling challenge. Reinforced Plastics, Volume 45, No. 6, June 2001

[7] Composites Recycling – Where are we now? Publication of Composites UK, Ed. S. Job et al., 2016

[8] Composites Recycling Made Easier. EuCIA-Publication, 2012.

[9] Job, S. Recycling Glass Fibre Reinforced Composites – History and Progress Reinforced Plastics, Volume 57, No. 5, September/October 2013

[10] Koelega, J. Recycling Wind Turbine Blades Graduation Thesis, MSc Industrial Ecology, TU-Delft, March 2019

[11] Update on Recycling ACMA Global Composite Recycling Workshop Paris, March 7, 2016.

[12] Busschen, A. ten Revolutionary re-use of polyester boats the Report, Issue 81, September 2017

[13] Cox, H.L. The elasticity and strength of paper and other fibrouw materials. British Journal of Applied Physics 3: 72-79, 1952.

[14] Busschen, A. ten Remanufacturing of EoL Thermoset Composites into Retaining Walls 3rd International Composites Conference (ICC), Stuttgart, 19 September 2017

[15] Bijker, S., Tromp, L., Siereveld, G. Duurzame aanpak Dinzerbrug: Dekplanken van vezelversterkt kunststof Civiele Techniek (Dutch magazine on civil engineering), No. 4, 2019

[16] CUR Recommendation 96: 2017 Vezelversterkte kunststoffen in bouwkundige en civieltechnische draagconstructies second reviewed edition, CROW 2018, ISBN 978-90-5367-645-5

[17] European standard EN 1990: Eurocode – Basis of structural design Annex D (informative): Design assisted by testing December 2005

[18] Busschen, A. Ten, Topp, M.D.C., Bouwmeester, J., Schreuder, P., Bosman, P., Roetman, E. Industrieel hergebruik van end-of-life thermoharde composieten. Report of professorship for polymer engineering, Windesheim. LKT-RE-106851-1903, September 2019.

[19] EuCIA Eco Impact Calculator https://ecocalculator.eucia.eu

Chapter 11

Phytomass-Derived Multifunctional Activated Carbon as a "Wonder-Material": A Paradigm Shift of Filth-to-Wealth

Palanichamy Kalyani, Thakku Rangachari Banuprabha, Chinnamayan Sudharsana and Nazim Anvarsha

Abstract

Activated carbon (AC) is a wonder-material that finds multifarious applications such as catalytic supports, removal of pollutants, electrodes in energy gadgets, gas storage etc. Surface area, chemical constituents and pore structures are a few traits required in the ACs which largely depend on the source of the precursors and processing methodologies adopted. In this context, the idea of recycling phytomass for producing ACs has attracted researchers seeing that the inexpensive and renewable nature of the phytomass can reduce the overall cost of producing ACs with diversified features and that it does not add CO_2 to the atmosphere leading to global warming (plants release only the same amount of CO_2 as they consumed while growing). Further, phytomass after their life possess no value but their conversion into ACs would be an economically profitable option leading to inexpensive ACs. As a consequent of these advantages this chapter has been planned and designed to provide certain interesting multifunctional aspects of low-cost phytomass derived ACs. The chapter is expected to provide research insights oriented towards identification of unexplored phytomass or wastes which could lead to carbon with novel properties tunable to the applications. Filth-to-wealth or in other words, recycling of wastes provides a strategy categorized under circular-bioeconomy, which is the want of the hour.

Keywords: Phytomass, Activated carbon, Adsorbent, Electrocatalysts, Supercapacitor, Antibacterial, Bio-sensors, Circular-bioeconomy, Recycle

1. Introduction

Activated carbon (AC), no wonder, is regarded as a wonder-material since it is a very vital active material in scores of applications such as catalytic supports, removal of pollutants, electrodes for battery and capacitors, gas storage etc., and these applications require carbon powders with specific functionalities like surface area, chemical constituents, pore structure etc. Since last decade, materials scientists are attracted towards biomass and more specifically, phytomass which have the source from the mass (biomass) of living and dead plants for producing value-added materials. It is interesting to note that a total of 82% of the biomass is of plant origin and crop residues include more than half of the world's agricultural phytomass [1]. Moreover, a study indicate that the annual global production of wood-derived biomass is around 4.6 Gt out of which 60% is utilized for energy production, 20% for industrial 'round wood' manufacture and the remaining 20% will be primary production loss that remains in-field to decay. Further an estimated 80% of forest tree mass is being lost as waste [2]. Generally it is known that starch, cellulose, hemicellulose, fructose, glucose, amino acids, lignins, lipids, organics, inorganics etc. are the chief constituents of phytomass [3] and thus they are apparently rich in carbon content for producing AC powders and are deemed to be low-cost alternative for commercial carbons hitherto utilized for the above-said applications. Obviously, the omnipresent and plentiful agricultural discards effectively offer a secondary, inexpensive and renewable source of carbon. Having seen the basic information on phytomass, it is also pertinent to understand the present scenario of the landfills which have already become flooded with non-treatable garbage and that they do not admit solid-wastes anymore and consequently, we propose ways of segregating, recycling/up cycling/reuse of wastes for reducing disposal problems and improving a Nation's economy or in other words, "waste valorization", the processes of treating wastes for beneficial use that may reduce pollution and the concomitant environmental impacts, has now been developed into a trending research among materials advocates. Also, considering the production cost of materials/products, environmental and energy concerns, the process of reuse of phytomass wastes to generate value-added products is of dire need. Simultaneously, huge economy involved in the clearing-off the (phytomass) wastes globally has activated many research groups to recycle various categories of wastes to achieve value-added products by which the current energy emergency may possibly be mitigated. Thus waste phytomass should be seen as unexplored resources as well as zero-cost source of essential environmental services and not as wastes at all. Hence the production of AC materials, especially from these cheap and natural bio-precursors is a highly attractive research theme in the science of functional materials. Consequently this chapter has been planned to provide recycling based content having a great potential in reducing environmental impact, climatic issue and initiate circular bio-economy model. So this chapter has been designed as a review and summarize the key research reported on six interesting applications where the various zero-cost (waste) phytomass had served as the precursors for producing low-cost ACs. For consistency, the chapter has been sectioned individually under the following headlines based on the applications of phytomass-derived environmental friendly AC as;

1. Adsorbents

2. Electrocatalyst support for hydrogen gas generation from water splitting

3. Supercapacitor electrodes

4. Lithium-ion battery anodes

5. Electrochemical sensors

6. Antibacterial agents

Initially, certain fundamental aspects of AC are also quickly presented for the benefit of the readers. Hence the fundamental research results discussed in the chapter would not only exemplify the multi-dimensional features and applications of phytomass-derived AC but also hints the variety and variability in carbon sources available and break through the conventional idea of obtaining high-performing active carbon too.

2. Fundamental aspects of activated carbon

Activated carbon (AC) refers to a wide range of carbonized black colored materials of high surface area and high degree of porosity [4]. AC has many applications, in addition to those mentioned earlier, like in the environment and industry for the removal, retrieval, separation and modification of various compounds in liquid and gas phases etc. [5].

2.1 Production of AC through physical activation

Physical activation is a commercially adopted two-step process that involves carbonization (pyrolysis) of carbon rich precursors in a neutral atmosphere followed by thermal treatment of the resultant mass in an atmosphere of oxidizing gases such as steam, CO_2, CO_2-N_2 or CO_2-air mixtures in the range of 800–1100°C. This method has the ability to produce AC powders of porous structure and hence physical activation is considered as an inexpensive and green because it is chemical-free. Nevertheless, long activation time, low adsorption capacity of prepared AC and high energy requirements are the main disadvantages [6].

2.2 Production of AC through chemical activation

Chemical activation, often known as wet oxidation, is usually suggested for organic precursor materials containing cellulose, such as wood, sawdust and phytomass. In this method, the precursors are activated at high temperatures in the presence of certain activating chemicals. In the first stage of activation, the raw material will be saturated or impregnated with oxidizing and highly dehydrated chemicals (activating agents) [7]. After impregnation, the suspension is dried and the remaining mixture will be heated for a given duration. Depending on the activating material and the properties of the final product, activation may require temperatures ranging from 400 to 900°C, when the cellulose is expected to break down and char. Ultimately, AC is obtained from the repeated washing of the resultant char.

Chemical activation agents, as mentioned earlier, are dehydrating agents that influence pyrolytic decomposition and inhibits the formation of bitumen, increase the carbon content and with subsequent changes in the thermal degradation of precursors result in the development of the porous structure of the carbon materials. These activating agents penetrate into the carbon structure creating porous network like structure in the AC, thereby increasing its surface area. It is to be mentioned that carbon particle size distribution, porous nature and surface area are the three important aspects that decides the final applications of the AC [8].

In physical activation, carbonization and activation phenomena occur in two separate furnaces while in the chemical activation these processes occur simultaneously in a single furnace. It is to be mentioned that the correct selection of the activation parameters like the amount of impregnation, weight ratio of the activating agents to dry precursors, temperature, final temperature of carbonization, carbonization time and activation atmosphere (space) are important to the quality and physical characteristics of the final AC produced by chemical activation. In fact, chemical activation bestows more porous structure to the ACs than physical activation. Further, chemical activation is more economical as it requires a lesser processing temperature, time and yields higher carbon efficiency. Activated agents react with carbon matrices of the organic precursors and liberate various gases to form a porous structure. However, the need for repetitive washings to remove the unused activating agent from the final product at the end of activation process is one of the disadvantages of this method. In addition, toxic washings produced causes water pollution and therefore require secondary treatments. Different types of chemicals have different reactions with precursors and thus affect the features and nature of the ACs produced. **Figure 1** gives a rough scheme for producing AC by physical and chemical activation methods [3].

2.3 Description on activating agents

Effective activators so far used are alkaline compounds such as potassium hydroxide (KOH), sodium hydroxide (NaOH), calcium chloride ($CaCl_2$) and potassium carbonate (K_2CO_3), acidic compounds like phosphoric acid (H_3PO_4) and sulphuric acid (H_2SO_4), also intermediate metal salts such as $ZnCl_2$. Based on the reactivity and physical nature of the activator, the mixing of the activator and organic precursors could be done by two modes viz., the physical mixing of the activator and precursor (both in dry conditions) and impregnation (solid precursor with melt or fused activator) [9].

H_3PO_4 is widely used in the activation of various lignocellulosic materials. $ZnCl_2$ acts as a dampening agent during activation. K_2CO_3 is known to be a better activator than KOH due to the production with higher yield, higher surface and pore volume and higher adsorbing capacity from aqueous solutions. Activation with alkaline materials such as NaOH and KOH produces ACs with large amounts of surfacial microspores. KOH is being popularly used due to its ability to produce AC with a high surface area, distribution of fine pore size, low environmental pollution, less corrosiveness and cost affordability.

2.4 Precursors for AC

AC can be produced from materials such as wood, coal and some polymers. Wood and coal are relatively economical, but exhaustible. Polymers are the main source of pure carbon whereas it leads to high production costs and the preparation processes require expensive raw materials, enormous time, energy and tedious preparation procedures. AC production costs can be reduced by either choosing a cheap raw material or by applying a proper production method [10]. Nevertheless,

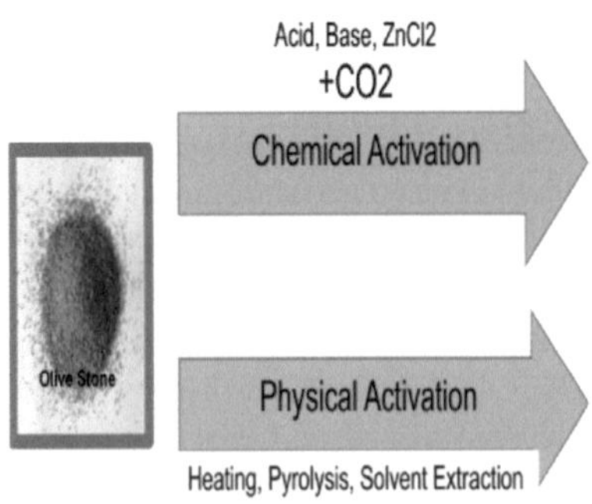

Figure 1.
Scheme showing the preparation of AC by physical and chemical activation methods [6] (Adapted with modification).

it is still a challenge to prepare AC with very specific characteristics, such as a given pore size distribution and using low-cost raw materials processed at low temperature (less energy costs). Therefore, it is necessary to find suitable low-cost raw materials that are economically attractive and at the same time present similar or even better characteristics than the conventional carbons.

The use of waste materials for the preparation of AC is very attractive in view of their contribution to decrease the costs of waste disposal, therefore helping environmental protection [11]. It is already known that any cheap material with a high carbon content and low ash and inorganics can be used as a raw material for the production of AC [12]. Hence the production of AC materials from phytomass has become very much popular in recent years. Literature shows that there have been many interesting research efforts to obtain low-cost AC from a variety of phytomass wastes such as sugarcane bagasse [13], rice straw [14], cotton stalk [15], coconut shells [16], wood [17], nut shells [18], olive seeds [19], apricot stones [20], almond shells [21] and date pits [22] for adsorption studies, for example application of AC.

2.5 Structures of AC

Structure of AC is also considered as an important factor while proposing any new applications. So a short description on the structures of AC is given here. Basically, three important structures have been described.

2.5.1 Porous structure

Generally, ACs show porous characteristics such as specific surface area (SSA), pore volume and pore size distribution and contain up to 15–20% of minerals in the form of ash [12]. The porous structure of AC is presumed to have developed during the carbonization process and further developed during activation when tar, volatile and other carbonaceous materials which might be present in the spaces between the elementary crystallites escape from the precursor. The structure of pores and pore size distribution depends on the nature of the precursors and the activation process. It is believed that during the activation disorganized carbon are removed by exposing the crystallites to the action of activating agent which leads to the development of porous structure. Dubinin [23] classifies pores according to their average width, which represents the distance between the walls of slit shaped pore or the radius of a cylindrical pore, proposed by and officially adopted by the IUPAC. Thus the pores are classified into (i) micropores (diameter (d) < 2 nm), (ii) mesopores (2 nm $< d <$ 50 nm) and (iii) macropores ($d > 50$ nm). **Figure 2** represents a view of these pores.

The micropores form the largest part of the internal surface and are accessible to the adsorptive molecules [25] or electrolyte ions. Generally, micropores contribute at least 90% of the total surface area of an AC, whereas the surface area of mesopores form less than 5% of total surface area and the mesopore volume varies between 0.1 and 0.2 cm^3g^{-1}. The contribution of macropores to the total surface area and pore volume is very small and does not exceed 0.5 m^2g^{-1} and 0.2–0.4 cm^3g^{-1} respectively. SSA and porosity are found out by N_2 adsorption studies.

2.5.2 Crystalline structure

Crystalline structure of AC starts to develop during the carbonization process. The crystalline structure of ACs is different from the graphite structure with respect to the interlayer spacing, which lies between 0.34 & 0.35 Å in ACs and 0.335 Å in

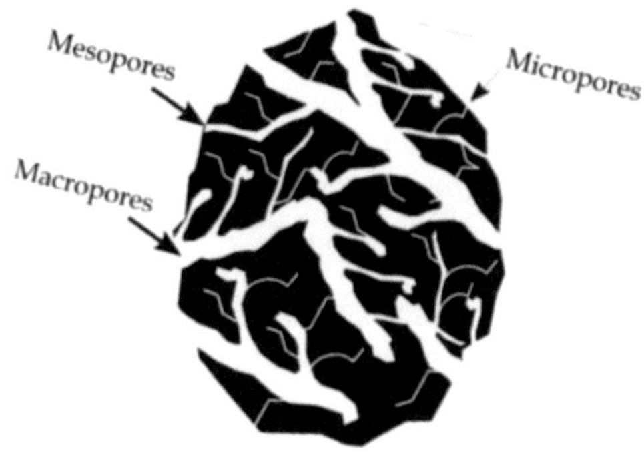

Figure 2.
Schematic presentation of macro, meso and micropores in AC [24] (Adapted with modification).

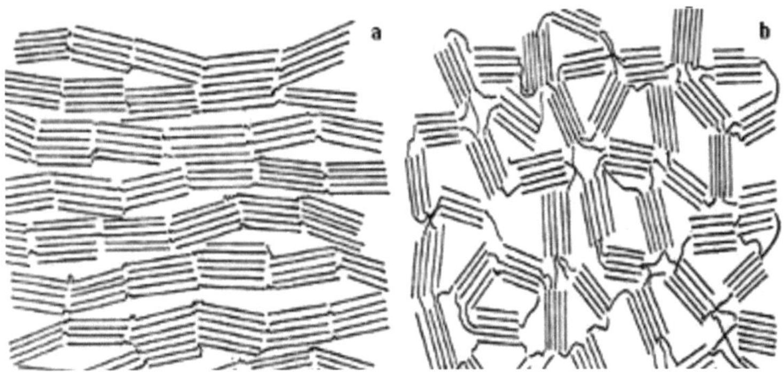

Figure 3.
Scheme of structure of AC (a) graphitized carbon and (b) non-graphitized carbon (adapted with modification from [26]).

graphite. Nevertheless, the basic structural unit of AC is in close approximation with the graphite structure. Based on the graphitizing ability, ACs are classified into graphitizing carbons and non-graphitizing carbons. The above two structures of carbons is schematically presented in **Figure 3**.

Graphitizing carbon may have a large number of graphite layers oriented parallel to each other and is delicate due to the weak cross-linking between the neighbor micro crystallites and has a less-developed porous structure. On the other hand, non-graphitizing carbons are hard due to strong cross linking between crystallites and show a well-developed microporous structure [26]. The formation of non-graphitizing structure with strong cross links is promoted by the presence of associated oxygen or by an insufficiency of hydrogen in the precursors.

2.5.3 Chemical structure

In addition to the porous and crystalline structure discussed above, the AC surface has also chemical structure. It is well established that the adsorption capacity on AC is determined by its porous structure and is strongly influenced by the chemically bonded heteroatoms like oxygen, nitrogen, sulphur and halogens [12, 27, 28]. These heteroatoms are obviously derived from the phytomass precursors and involve in the structure of AC during carbonization process or they may be chemically bonded to the surface during activation [29]. The heteroatoms are likely to be bonded to carbon atoms of the corners and edges of the aromatic sheets or to the carbon atoms at defect positions to form carbon–oxygen, carbon-hydrogen,

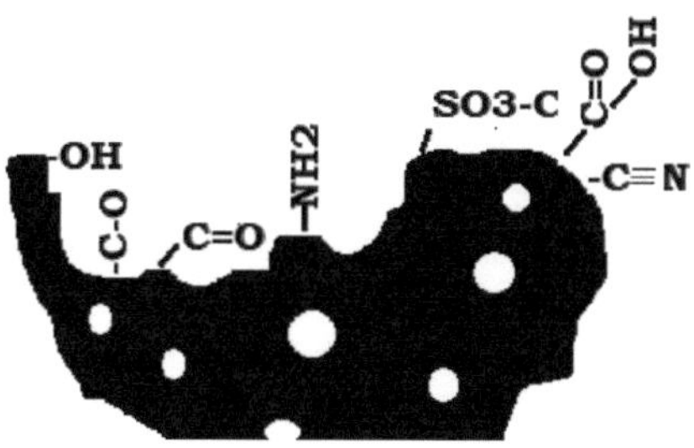

Figure 4.
Model of various organic functional groups on the activated carbon [30] (● = carbon matrix; ○ = pores).

carbon-sulphur, carbon-nitrogen and carbon-halogen surface organic compounds (see **Figure 4**), known as surface groups or surface complexes [28, 31].

Ultimately, the organic hetero functional groups greatly influence the properties and nature of the phytomass-derived AC.

3. Multifunctional aspects of phytomass-derived activated carbon

It can thus be seen that due to the increasing demand of AC, there is a strong need for the sorting out of new precursors for AC which should be cost effective than the commercially available ACs. Although, a variety of raw materials were explored for the preparation of AC in earlier studies, scientists are still exploring new materials depending on their availability and suitability for producing AC with multi-functions. Additionally, application of phytomass carbon electrodes stands as an important class of technology where 3R principles are followed. Thus remarkably, the utilization of phytomass as raw material for the preparation of AC has increased in recent years in view of the foregoing facts on AC. In the following sections, interesting multifunctional aspects of phytomass-derived AC will be deliberated.

3.1 Application of phytomass-derived AC as an adsorbent

The use of charcoal or AC for the adsorption (removal) of pollutants in air or toxic ions and dyes from contaminated water is best known for over 80 years and further adsorption properties, mechanisms, kinetics and theories of adsorption are also well established. But the potential of economically cheaper and renewable phytomass-derived AC as adsorbent was understood since the past decade though the search for better AC is still going-on due to the exponential demand for treatment of industrial effluents.

There are enormous sorptive studies with biomass-derived activated carbons and to cite a few; Tura and Tesema [32] have removed methylene blue using AC derived from *Delonix regia* seed pods. They have shown that adsorption of methylene blue is mainly pH dependent and that maximum adsorption takes place in slight neutral pH. Electrical conductivity and total suspended solid was found to decrease after adsorption which, indicates the decrease of ions from methylene blue dye proving the removal of color i.e. the dye. In yet another special work, Sekaran et al. [33] have reported the preparation of mesoporous-activated carbon from rice husk by precarbonization at 400°C, chemical activation using phosphoric acid at various temperatures and have immobilized Bacillus sp. in the mesoporous-activated carbon for the degradation of sulphonated phenolic compounds in wastewater. *Delonix regia* derived AC has also been utilized to remove Hg [34], Pb and Ni [35].

Good amount of work has been reported on fluoride removal from water. Fito et al. [36] have done a very significant work using H_2SO_4 activated *C. edulis* stem derived AC for the removal of F^- from aqueous solutions. Stem of the *Vitex negundo* plant [37], $CaCl_2$-modifed *Crocus sativus* leaves [38] and bark of *Morinda tinctoria* [39] are further interesting works involving F^- removal. **Figure 5** is a picture where adsorbed molecules in the pores of the AC are shown.

Above reports are just bits from a massive published literature. However, readers can have a detailed outlook from Jorge Bedia et als' report [41] where they have made an excellent exhaustive review on the synthesis and characterization of biomass-derived carbons for adsorption of emerging contaminants from water. All the above read reports convey the need for commercially viable and potential activated carbon-based adsorbents.

3.2 Application of phytomass-derived AC for preparing electrocatalyst for hydrogen gas from water electrolysis

It is well known that, noble metals like Pt and Ru based electrocatalysts are employed for producing hydrogen by electrolyzing water and usually electrocatalysts are fabricated by supporting or loading fine Pt or Ru particles on quality carbon powders (the carbon is called catalyst support), such that more number of active sites will be available for efficient and complete electrolysis. Falling in that line, an innovative and ever first attempt has been reported by the authors of the present chapter on adopting a zero-cost green precursor viz., grass biomass, by converting the grass biomass into a biochar and attempting to produce an electrocatalyst with platinum for generating hydrogen gas through electrolysis of water [42].

Cleaned turf grass blades were chosen as the phytomass of producing AC for supporting Pt particles to finally utilize as electrocatalyst for hydrogen gas generation through electrolysis of water. The procedure involved an activation of grass blades with $ZnCl_2$ followed by heat treatment at 250°C. As an initial trial, 1% Pt was supported over the grass-derived AC powder to result in Pt@G-AC. After various physical characterization studies, Pt@G-AC powder was assessed for catalytic activity in 1 M sulphuric acid solution for H_2 generation through linear sweep and cyclic voltammetric studies. Encouraging results were obtained suggesting that grass can be considered as a renewable alternative for producing carbon supports

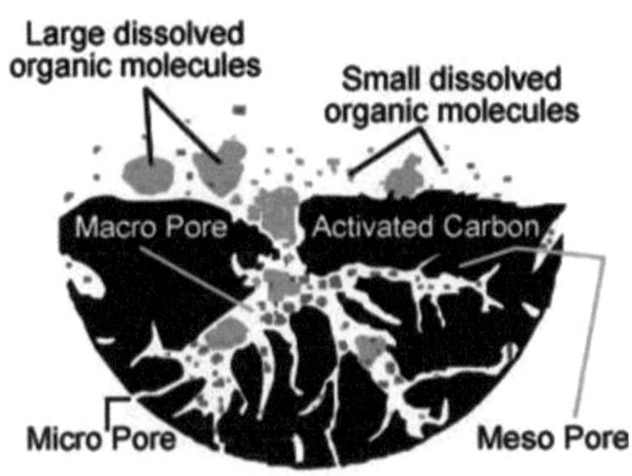

Figure 5.
Image showing the adsorption of molecules in the pores of AC [40] (adapted with modification).

for electrocatalysts but also paves way for the production of low-cost carbon for other applications like adsorbent for color, odor and hazardous pollutants and electrode materials as well, as will be highlighted in the subsequent sections. **Figure 6** gives LSV plot of G-AC with and without 1% Pt electrodes in 1 M H_2SO_4. Since no other reports are available on this particular application it seems that intensive research is this area is highly required and hence it is expected to pick up in the future.

3.3 Phytomass-derived AC for supercapacitor electrodes

Energy storage devices are the key components for a successful and sustainable world and supercapacitors (SCs) are one among them. SCs are able to supply considerable amount of power over a short time with extended cycle life. They offer a higher specific power density than most batteries and a higher energy density than conventional capacitors [43]. SCs are very useful in load leveling applications where a sudden boost of power is needed in a fraction of a second. More importantly, they do not release any heat during their operation and have a very long lifetime thus reducing the cost of maintenance. Also they do not release any hazardous substances that can damage the environment and their performance does not degrade with time. Hence for these reasons, SC is considered as a versatile technology that plays a prime role in partly fulfilling the energy demands of present and the future.

It is well known that certain physical features of the electrode materials determine the performance characteristics of the energy systems [44]. SCs making use of phytomass-derived porous carbon has the advantages like production of low cost carbon electrode components, environmental friendliness and good capacitive performance. Thus the search and research for advanced electrode materials is sought after and obviously very recently, phytomass-derived AC is providing unprecedented opportunities for researchers to design and fabricate innovative electrode materials for high performing SCs.

As far as research on phytomass-derived AC as electrode material for SCs is concerned, a wealth of information related to its preparation methodology, physical properties and electrochemical properties are available in the open literature in the form of reviews and research communications. The authors of this chapter have reported ample number of interesting work and to cite a few; on papaya seeds [45], onion peels [46] and recently on banyan prop root [47] for the possible application as electrode for SCs.

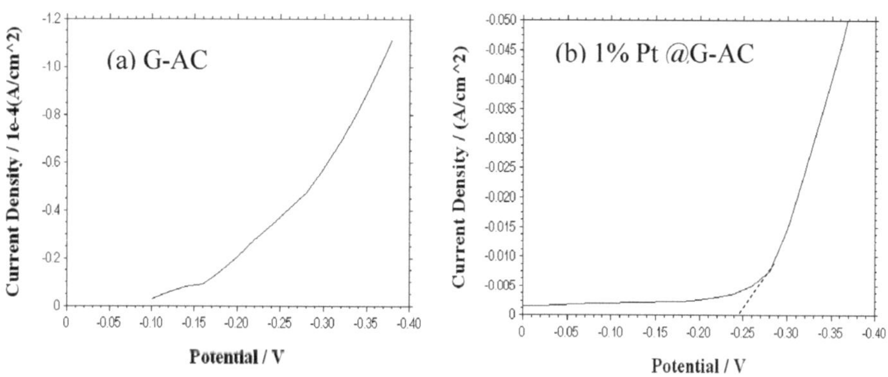

Figure 6.
LSV plot of (a) G-AC and (b) 1% Pt@G-AC electrodes in 1 M H_2SO_4 [42].

One important concept in the charge storage mechanism is the pseudocapacitance which is dealt as follows. In an ideal double layer capacitor, energy is stored in electric double layer and no charge transfer occurs across the interface between the electrode and the electrolyte. However, it is possible that some redox reactions (faradaic) can still happen due to the existence of various heteroatoms like O, N & S present in the form of organic functional groups on the phytomass-derived AC. The capacitance arising from these faradaic reactions is called pseudocapacitance. Therefore the total capacitance is a combination of capacitance contribution from electrostatic charges and faradaic charge transfer redox reactions [48] which is given in equation 1.

$$C_{total} = C_{dl} + C_f; \tag{1}$$

where C_{total} is the total capacitance, C_{dl} is the electrical double layer capacitance, C_f is the pseudocapacitance. Thus by introducing functional groups onto carbon material, pseudocapacitance can be enhanced [49]. The faradaic charge transfer processes at the electrode involving N [50], S [51] and O are given below [52].

$$C* = NH + 2H^+ + 2e^- \leftrightarrow C*H\text{–}NH_2 \qquad O = S = O + 2e^- + H_2O \leftrightarrow\ >SO + 2OH^-$$

$$C*\text{–}NHOH + 2H^+ + 2e^- \leftrightarrow C*\text{–}NH_2 + H_2O; \quad >SO + e^- + H_2O \leftrightarrow\ >SOH + OH^-$$

$$O = Ph = O + 2H^+ + 2e^- \leftrightarrow HO - Ph - OH; \qquad R = O + H^+ + e^- \leftrightarrow R - OH$$

$$>CO + H^+ \leftrightarrow\ >COH^+{}_{(ads)}; \qquad >CO + H^+ + e^- \leftrightarrow\ >COH$$

where C* stands for the carbon structure, Ph and R respectively indicate phenyl and aliphatic groups. Therefore an electrochemical capacitor is called "supercapacitor" or "ultracapacitor". Further understanding on this topic can be had from references [53–55].

Having reviewed various aspects of capacitors and significant reported research, it is still relevant to search for newer and cheaper electrode materials and that too if the electrode materials could be derived from greener sources and waste phytomass then it will be a welcoming suggestion for the current scenario of energy crisis.

3.4 Application of phytomass-derived AC as lithium-ion battery anodes

Research reports on phytomass-derived AC as anodes in lithium-ion batteries seem not very much abundant as available for adsorption and supercapacitor electrodes studies. However, a few noteworthy studies for application as electrodes in lithium-ion battery (LIB) anodes are presented here. Zhang et al. [56] have produced carbons with a high surface area rice straw. They report that the hierarchical porous network with large macroporous channels and micropores within the channel walls enable the porous carbons to provide the pathways for easy accessibility of electrolytes and fast transportation of lithium ions These porous carbons which show a particular large reversible capacity are proved to be promising as anode materials for high rate and capacity LIBs. Bhardwaj et al. [57] synthesized carbon nanomaterials by pyrolysis of tea leaves and used as anode in LIBs. The highest specific capacity reported was 64 mAh g^{-1}. Zhang et al. [58] used pinecone hull and activated at 800°C under CO_2 atmosphere to obtain microporous carbon. This served as the anode for lithium secondary batteries and retained a discharge capacity of 357 mAh g^{-1} and coulombic efficiency of 98.9% was reported to be achieved at higher current density of 10 mA g^{-1}.

Hwang et al. [59] obtained disordered carbon materials by pyrolysis of coffee shells at 800 and 900°C with KOH and $ZnCl_2$ porogens. The first lithium insertion capacity was 524 & 603 mAh g^{-1} for the untreated samples pyrolyzed at 800 & 900°C respectively, while obtained 1150 & 1200 mAh g^{-1} for the KOH treated coffee shells pyrolyzed at the same temperature. Carbon powders of distinct and interesting morphologies were synthesized by pyrolyzing soapnut seeds, jack fruit seeds, date seeds, neem seeds, tea leaves, bamboo stem and coconut fibers, without using any catalyst. These carbon materials were utilized as the anode in LIBs [57]. Amongst the various precursors, carbon fibers obtained from soapnut seeds and bamboo stem, even after 100 cycles, showed the highest capacity of 130 & 93 mAh g^{-1} respectively. In yet another work, Stephan et al. [60] treated banana fibers with pore forming substances such as $ZnCl_2$ and KOH. The BET surface area of the untreated carbon was 36 m^2 g^{-1} and increased to 686 m^2 g^{-1} & 1097 m^2 g^{-1} for the carbons after treatments with KOH and $ZnCl_2$, respectively. On employing these porogen treated carbons in LIBs, the specific capacities for the $ZnCl_2$ treated sample was found to be 3123 m^2 g^{-1} while it was 921 m^2 g^{-1} for the KOH treated sample and for the untreated carbon, the specific capacity was extremely low as 625 mAh g^{-1}.

A comprehensive list of phytomass-based AC utilized in LIBs has been provided in Ref. [55]. Thus the foregoing reports clearly suggest that the phytomass-derived AC has ample scope for investigation for anode materials in advanced energy systems like LIBs.

3.5 Electrochemical sensors

Although a complete description on sensors is out of scope of this chapter, a bird's eye view is worthwhile. A sensor is a device which provides a usable output in response to a physical quantity and converts it into a signal suitable for processing (e.g. optical, electrical, mechanical). Transducer is the active element of a sensor. A biosensor is an analytical device used for the detection of a chemical substance that combines a biological component with a physicochemical detector. The sensitive biological element, for e.g. tissue, microorganisms, organelles, cell receptors, enzymes, antibodies, nucleic acids etc. is a biologically derived material or biomimetic component that interacts with, binds with or recognizes the analyte under study [61]. The biologically sensitive elements can also be created by biological engineering. The transducer or the detector element, which transforms one signal into another one, works in a physicochemical way such as optical, piezoelectric, electrochemical, electrochemiluminescence etc. resulting from the interaction of the analyte with the biological element to easily measure and quantify. A biosensor typically consists of a bio-receptor (enzyme/antibody/cell/nucleic acid/aptamer), transducer component (semi-conducting material/nanomaterial), and electronic system which includes a signal amplifier, processor and a display. There are three main types of electrochemical sensors namely; potentiometric, amperometric and conductometric. **Figures** 7 and **8** respectively scheme the basic representation and components of a biosensor.

Electrochemical biosensors are an important type in sensor technology and have electrodes which translate the chemical signal into an electrical signal such as conductance, resistance or capacitance of the biosensor surface. Electrochemical sensors are able to detect many biomolecules in the human body such as glucose, cholesterol, uric acid, lactate, DNA, hemoglobin and blood ketones [64]. Thus they have great potential to detect diseases related to imbalances of biomolecules. Mostly, they are widely used for biosensing applications, however studies on biosensing–drug delivery applications are only limited. Enzyme- or protein-based electrochemical biosensors that have drug-release capability can be useful for the

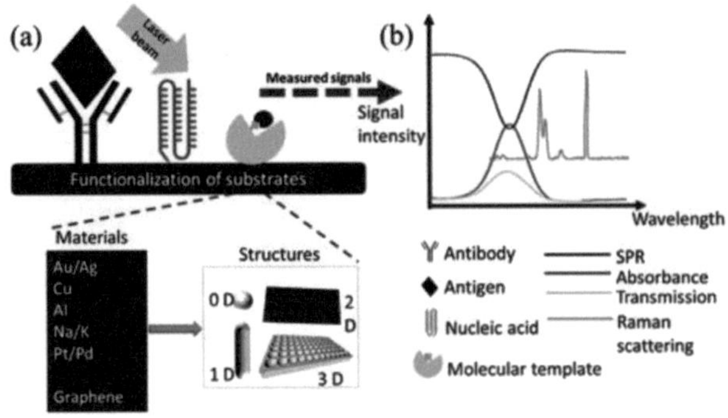

Figure 7.
(a) Basic schematic representation of a biosensor (b) output signal from a sensor [62] (adapted with modification).

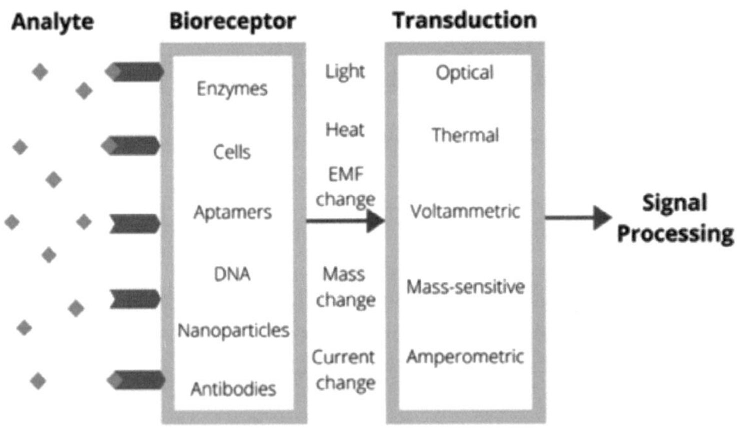

Figure 8.
Components of a biosensor [63] (adapted with modification).

treatment of various diseases. For example, xanthine oxidase enzyme catalyzes the production of hypoxanthine and xanthine and overproduction of these products cause renal failure [65].

High sensitivity, lower detection limits, automation, reduced costs of testing, and development of disposable devices and methodologies capable of working with very small sample volumes are some of the advantages associated with electrochemical biosensors. Also, electrochemical sensors are unaffected by sample turbidity or interference from absorbing and fluorescing compounds like spectroscopy-based techniques; they require comparatively simple instrumentation that requires low power and is portable. Use of electrochemical techniques, over optical and other transduction techniques, exhibits excellent sensitivity and a large linear detection range in a wide range of solvents, electrolytes, temperatures, etc. Electrochemical biosensors can be classified into voltammetric, amperometric, conductometric, impedimetric and potentiometric. Readers can have a complete understanding of various aspects of sensors, types and their applications from references [66–68].

3.5.1 Advantages and disadvantages of biosensors

Biosensor technology has been developed enormously since their introduction in 60s. They offer diverse advantages such as less complicated sensor setup, cheap production of microelectronic circuits and a user-friendly interface with conventional electronic processors [69]. Further electrochemical biosensors are also robust, easy to miniaturize, and offer broad detection limits with the small volumes of analyte (biofluids) requirements even if turbidity or optically absorbing and fluorescing compounds are present. Nevertheless, they have certain disadvantages too that hinder further developments. For example, they do not have the distinct surface architecture that would facilitate high sensitivity in detection and unique recognition of the response to the selected biochemical event [70]. This means that the pH and ionic strength of biofluids can greatly influence the behavior of the biosensors. Hence sensor technology needs to be developed in order to have increased intensity of the signal and improved signal to noise ratio.

3.5.2 Phytomass-derived AC in sensor application

With the impressive progress in the electrochemical sensing technologies and their application in bio-analytical chemistry, it would now be possible to utilize phytomass-based AC to show that high sensitive and low-cost sensor production is commercially feasible in the near future [71]. In this context, a few important research advancements utilizing phytomass-derived AC in sensors has been presented below.

Kim et al. [72] have reported a biomass-derived carbon for the electrochemical determination which involved the initial activation of kelp powder with $ZnCl_2$, followed by a second activation step with KOH. The above AC coated on GCE modified electrode showed high sensitivity, selectivity and a good detection limit for the determination acetaminophen with the detection limit of 0.004 μM. Also the modified electrode showed good result towards acetaminophen in the presence of ascorbic acid and dopamine with the detection limit of 0.007 μM.

Zhang et al. [73] have reported that $ZnCl_2$ activated peel of kiwi fruit based carbon fibre (CF) provided a high sensitivity and selective signaling of ascorbic acid (AA), dopamine (DA), and uric acid (UA) with linear response ranges of 0.05–200 μM, 2–2000 μM, and 1–2500 μM, respectively and its detection limits (S/N = 3) as 0.02 μM, 0.16 μM, and 0.11 μM, respectively and this method was successfully applied to detect AA, DA, and UA in real sample analysis. Wang et al. [74] have shown PBNPs-3D-FKSCs, CuNiNPs-3D-KSCs and CoNPs-3D-KSCs (KSC: kenaf stem carbon, 3D: three dimensional) with honeycomb structure electrodes with good electrocatalytic performances for the reduction of H_2O_2, oxidation of glucose and amino acid.

Oliveira et al. [75] have reported that carbon paste electrode modified with nitric acid activated biochar obtained by the pyrolysis of castor oil cake biomass at 400°C for spontaneous preconcentration of methyl parathion (MP) and for further quantitative determination in drinking water. The electrode showed good sensitivity and limits of detection of MP as 760 μA L $mmol^{-1}$, 39.0 nmol L^{-1}, respectively.

Kalinke et al. [76] for the first time have reported the determination of paraquat (PQ^{2+}) by Differential Pulse Adsorptive Stripping Voltammetry (DPAdSV) using a carbon paste electrode modified (CPME) with biochar obtained from castor oil cake at different temperatures (200–600°C). The best voltammetric response was verified using biochar yielded at 400°C (CPME-BC400). Linear dynamic range (LDR) for PQ^{2+} concentrations between 3.0×10^{-8} and 1.0×10^{-6} $mol L^{-1}$ and a limit of detection of 7.5×10^{-9} $mol L^{-1}$ were verified. The method was successfully applied for PQ^{2+} quantification in spiked samples of natural water and coconut water.

Madhu et al. [77] have reported that PSAC/Co_3O_4 (PSAC: Pongam seed shells derived activated carbon] modified electrodes have potential as nonenzymatic glucose sensor and supercapacitor with ultrahigh sensitivity of 34.2 mA mM^{-1} cm^{-2} with a very low detection limit of 21 nM. Shahzada et al. [78] have reported that sulfur-doped reduced graphene oxide (SrGO) product fabricated using an eco-friendly biomass precursor "lenthionine" through a high temperature doping process have high sensitive electrochemical sensor for detection of 8-hydroxy-2-deoxyguanosine(8-OHdG) molecule. The sulfur-doping amount was regulated and a maximum sulfur content of 2.28 atom% was achieved through controlling the precursor amount. It was homogenous presence of large number of sulfur atoms in SrGO in the form of thiophenic (CSC) bond that produced robust sensitivity (~1 nM), very wide detection window (20–0.002 M).

Ni et al. [79] have modified heteroatom-enriched activated carbon-nickel oxide (HAC-NiO) nanocomposite into NiO-HAC/GCE) and have built a novel glucose sensor which exhibited a wide linear concentration range of 10 μM–3.3 mM and a low detection limit of 1 μM) towards glucose oxidation. Travlou et al. [80] have treated wood-based commercial activated carbon (BAX) and its oxidized counterpart (BAX-O) with melamine and then heated at 450°C in nitrogen. Further oxidation with nitric acid was also done. The carbons were tested for ammonia sensing (45–500 ppm of NH_3). Further the role of the nitrogen functionalities on the electrical performance of the carbons was investigated by testing their selectivity with respect to H_2S sensing. Interestingly, pyridinic groups, acting as p-type impurities were found to be responsible for the observed opposite electrical responses of the melamine impregnated samples upon exposure to NH_3/H_2S. This facilitated H_2S dissociation into H^+ and HS^- ions, speculates the authors. The latter ions, either by providing ionic conductive paths through the carbon matrix or through their oxidation to SO_2 may cause a decrease of the normalized resistance.

Hayat et al. [81] have deposited a TiO_2 modified activated carbon on the surface of screen printed carbon electrodes (SPCEs) and used in the direct oxidation of phenols. Calibration curves showed a high sensitivity and wide linear range for each studied compounds viz., p-nitrophenol, 1-naphthol, catechol and hydroquinone. The authors say that there was no interference of Na^+, K^+, Cl^-, Br^-, Mg^{2+}, Zn^{2+} and NO_4^- ions and show 96% recoveries in real sample analysis.

Koskun et al. [82] have synthesized activated carbon (AC) decorated monodisperse nickel and palladium alloy nanocomposites modified glassy carbon electrode (Ni-Pd@AC/GCE NCs) by in-situ reduction technique and they showed a very low detection limit of 0.014 μM, a wide linear range of 0.01 mM–mM and a very high sensitivity of 90 mA mM^{-1} cm^{-2}. Furthermore, monodisperse Ni-Pd@AC/GCE was utilized to detect glucose in real sample species.

Aparna et al. [83] have reported that the $NiFe_2O_4$-AC-modified glassy carbon electrode (GCE) showed excellent electrocatalytic activity towards DA (dopamine) compared to $NiFe_2O_4$/GCE and AC/GCE. This has been attributed to the synergistic action and the large surface area of the nanocomposite. Differential pulse voltammetry (DPV) was employed for the detection of DA wherein the detection limit of 0.4 μM along with a linear range of 5 μM to 100 μM was realized. Wang et al. [84] utilized DPASV (differential pulse anodic sweep voltmmetry) technique to show that the peak currents have linear relationship with Pb^{2+}, Cd^{2+} and Zn^{2+} concentrations respectively in the range of 0.5–2.25 mg/L, 0.5–4.0 mg/L and 1.0–4.0 mg/L with detection limits of 0.1, 0.3 and 1.0 mg/L (S/N = 3) respectively.

Thus it is hoped that the information provided in this section proves valuable and stimulates further research and developments in the promising field of phytomass-derived AC for flexi-sensors in the near future.

3.6 Application of phytomass derived AC as potential antimicrobial

Natural plant products are always interesting to explore because of their significant antibacterial, antifungal, antiviral and anticancer activity and the presence of elements such as S & N has been proved responsible for their action [85]. Likewise, it has already been mentioned that ACs derived from phytomass too has these heteroatoms present in the form of organic functional groups in its structure to explain the activity observed against certain human pathogens. It is motivating and interesting to know that the application of kitchen soot was earlier in practice as an antimicrobial and was called "old woman's remedy" [86]. Consequently, natural carbon seemed to occupy a paramount place in household medicine. Scanning of literature shows that not many research reports have appeared pertaining to antibacterial activities of phytomass-derived AC and hence a few valid reports are given here to substantiate the importance of phytomass-derived ACs. Yallappa et al. [87] have done a revolutionary research using groundnut shell-based nano carbon, which proved in-vitro antibacterial activity. In-vitro evaluation of antibacterial efficacy using *Passiflora foetida* derived AC against a score of pathogens has been reported by Dheeban et al. [88]. Similarly, Lakshmi et al. [89] have elaborately reviewed interesting reports on AC nanoparticles from biowaste as a new generation antimicrobial agents.

In yet another research, Shamsi et al. [90] have reported a clear zone of inhibition of carbon nanoparticles obtained from sandal wood bark against *B. cereus*, *E. coli*, *C. violeceum* and *P. notatum*. Sheena et al. [86] have reported enormous

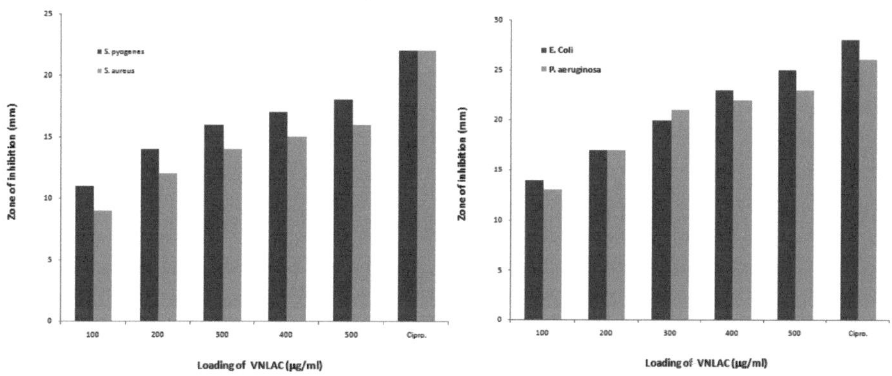

Figure 9.
Antimicrobial activity of Vitex negundo *leaves AC against gram-positive pathogens (left) & gram-negative pathogens (right) [30].*

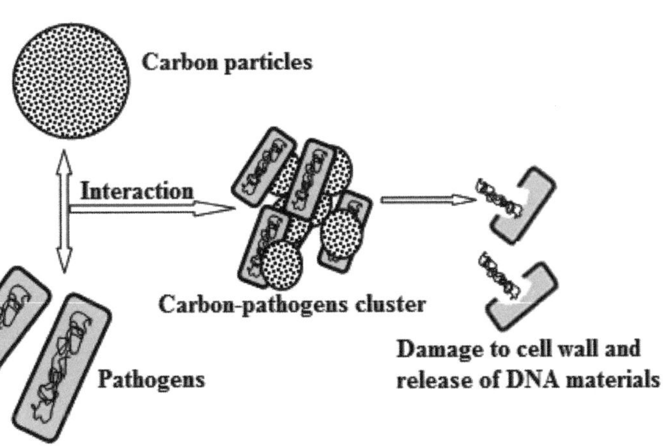

Figure 10.
Scheme of plausible mechanism of antimicrobial activity of AC [30].

antimicrobial activity of carbon nanoparticles isolated from natural sources against pathogenic gram-negative and gram-positive bacteria. Karthik et al. [91] have prepared AC from *Tribulus terrestris* and have proved activity against *E. coli*, *B. subtilis*, *S. aureus*, and *K. pneumoniae*. Anvarsha et al. [30] has recently reported *Vitex negundo*-derived AC and its importance as an antibacterial agent against human pathogens. Zone of inhibition was measured with reference to 5% DMSO and values are plotted in **Figure 9**. A schematic of mechanism of antimicrobial activity of AC particles is shown in **Figure 10**, which is self explanatory.

All these reports conclude that carbon materials produced from the chemical activation of phytomass still has huge potentials against various bacterial strains and that in-depth investigation should be devoted for a commercial success in the near future. Hopefully, the results envisaged through the above studies would certainly attract researchers globally to study various phytomass-derived carbons as novel and affordable therapeutic agents which can effectively inhibit the growth of various strains of microbes.

4. Summary, conclusions and future prospects

The objective of the chapter is to appreciate and provide details of the multifunctional aspects of zero-cost phytomass-derived AC in the areas where the most coveted commercial AC is hitherto celebrating. Fundamental aspects of AC were discussed initially followed by six important applications of the phytomass-derived AC to understand the value-added advantages of phytomass-derived AC. It is well documented that the preparation of carbonaceous materials from phytomass or biomass wastes has also important added advantages such as an effective management of the wastes and lower synthesis costs. Consequently, this chapter is a fair consolidation of research done on six major fields where the novel phytomass-derived AC can find applications equal to commercial ACs now. Nonetheless, researchers are still looking for practical and affordable carbons which can be applied at the commercial scale that leads to improved performance and applications in future.

With the authors' own experience in the title subject and with a global view to realize the suitability and applicability of the synthesized phytomass-derived ACs in the commercial quarters, the problems of significance which may invite research attention has been identified and have been listed below.

- Extensive investigation needs to be carried out to produce ACs with even better surface characteristics with tailored pore-size distribution through different routes such as chemical activation, physical activation, two stage activation or microwave heating.

- Process economy mainly depends on the selection of precursors and methods of preparing AC. Thus the cost analysis should be carried out to evaluate the practical applications.

- The use of phytomass-derived ACs and their modified ones in photocatalysis has recently been reported as a novel application. Hence investigation on this aspect is expected to gain impetus in the near future, which would sure be lending hands to address environmental issues.

- Solid wastes that are disposed by the industries and agricultural sectors are a great environmental concern. Hence research activity is needed to utilize these solid wastes to convert in to useful products, adopting filth-to-wealth concept.

It is hoped that at least some of these research activities will be initiated in the near future with unconditional research collaboration. But with a certain amount of commercial success that has already been achieved and the multitude of R & D efforts that are presently going on worldwide regarding the elimination of safety problems, improvement in performance and reduction of cost etc. phytomass carbon in the above-said fields will see a tremendous leap in the commercial market and will definitely open up new avenues in academic, research and industrial sectors. Advantageously, utilization of phytomass-derived multifunctional AC strengthens the circular-bioeconomic status of a Nation too. Interestingly, the authors of the present chapter has already committed similar type of studies with AC derived from a few other phytomass such as calotropis stem, palm leaves, coconut leaves etc. for the six classes of applications seen in the foregoing sections. Since the physical features and chemical constituents influence the performance tunable for multifarious applications, we have diversified choice of sources of phytomass carbon. Thus the consolidation of research work presented in this chapter opens up avenues for the utilization of various zero-cost phytomass for producing new carbon materials for various novel applications, which obviously proves that the phytomass-derived multifunctional AC is a "wonder-material" and the concept is unquestionably a paradigm shift of filth-to-wealth.

Author details

Palanichamy Kalyani*, Thakku Rangachari Banuprabha, Chinnamayan Sudharsana and Nazim Anvarsha
Department of Chemistry, DDE, Madurai Kamaraj University, Madurai, Tamilnadu, India

*Address all correspondence to: kalyani_1973@yahoo.com

References

[1] Smil V. Crop Residues: Agriculture's Largest Harvest. Crop residues incorporate more than half of the world's agricultural phytomass. BioScience. 1999;49(4) 299-308.

[2] Tripathi N, Hills CD, Singh RS, Atkinson CJ. Biomass waste utilisation in low-carbon products: harnessing a major potential resource. npj/Climate and Atmospheric Science. 2019;2(35) 1-10.

[3] Amthor JS. From sunlight to phytomass: on the potential efficiency of converting solar radiation to phyto-energy. New Phytologist. 2010;188 939–959.

[4] Tadda MA, Ahsan A, Shitu A, ElSergany M, Arunkumar T, Jose B, Razzaque MA, Daud NNN. A review on activated carbon: process, application and prospects. Journal of Advanced Civil Engineering Practice and Research. 2016;2(1) 7-13.

[5] Reza M, Yun S, Afroze S, Radenahmad N, Bakar A, Saidur R, Taweekun J, Azada AK. Preparation of activated carbon from biomass and its applications in water and gas purification, a review. Arab Journal of Basic and Applied Sciences. 2020;27(1) 208–238.

[6] Saleem J, Shahid UB, Hijab M, Mackey H, McKay G. Production and applications of activated carbons as adsorbents from olive stones. Biomass Conversion and Biorefinery. 2019; 9 775–802.

[7] Bergna D, Romar H, Lassi U. Physical activation of wooden chips and the effect of particle size, initial humidity, and acetic acid extraction on the properties of activated carbons. Journal of Carbon Research. 2018;4(4) 66.

[8] Johnson PJ, Williams RS. Activated carbon for automotive applications. Carbon Materials for Advanced Technologies. 1999; 235-268.

[9] Heidarinejad Z, Dehghani MH, Heidari M, Javedan G, Ali I, Sillanpaa M. Methods for preparation and activation of activated carbon: A review. Environmental Chemistry Letters. 2020; 18 393–415.

[10] Lafi WK. Production of activated carbon from acorns and olive seeds. Biomass and Bioenergy. 2001;20(1) 57-62.

[11] Dias JM, Alivim-Ferraz MCM, Almeida MF, Rovera-Utrolla J, Sanchez-Polo M. Waste materials for activated carbon preparation and its use in aqueous–phase treatment: A review. Journal of Environmental Management. 2007;85(4) 833-846.

[12] Bansal RC, Donnet JB, Stoeckli HF. Active Carbon. Marcel Dekker: New York; 1988.

[13] Cronje KJ, Chetty K, Carsky M, Sahu JN, Meikap BC. Optimization of chromium (VI) sorption potential using developed activated carbon from sugarcane bagasse with chemical activation by zinc chloride. Desalination. 2011;275(1-3) 276-284.

[14] Fierro V, Muniz G, Basta AH, El-Saied H, Celzard A. Rice straw as precursor of activated carbons: Activation with ortho-phosphoric acid. Journal of Hazardous Materials. 2010; 181(1-3) 27-34.

[15] Deng H, Li G, Yang H, Tang J, Tang J. Preparation of activated carbons from cotton stalk by microwave assisted KOH and K_2CO_3 activation. Chemical Engineering Journal. 2010;163(3) 373-381.

[16] Azevedo DCS, Araujo JCS, Bastos-Neto M, Torres AEB, Jaguaribe EF,

Cavalcante CL. Microporous activated carbon prepared from coconut shells using chemical activation with zinc chloride. Microporous and Mesoporous Materials. 2007;100(1-3) 361-364.

[17] Gomez-Sarrano V, Cuerda-Correa EM, Fernandez-Gonzalez MC, Alexandre-Franco MF, Macias-Garcia A. Preparation of activated carbons from chestnut wood by phosphoric acid chemical activation. Study of microporosity and fractional dimension. Materials Letters. 2005;59(7) 846-853.

[18] Lua AC, Yang T. Characteristics of activated carbon prepared from pistachio-nut shell by zinc chloride activation under nitrogen and vacuum conditions. Journal of Colloid and Interface Science. 2005;290(2) 505-513.

[19] Stavropoulos GG, Zabaniotou AA. Production and characterization of activated carbons from olive-seed waste residue. Microporous and Mesoporous Materials. 2005;82(1-2) 79-85.

[20] Youssef AM, Radwan NRE, Gawad IA, Singer GAA. Textural properties of activated carbons from apricot stones. Colloids and Surfaces A: Physicochemical and Engineering Aspects. 2005;252(2-3) 143-151.

[21] Bansode RR, Losso JN, Marshall WE, Rao RM, Portier RJ. Adsorption of volatile organic compounds by pecan shell and almond shell based granular activated carbons. Bioresource Technology. 2003;90(2) 175-184.

[22] Girgis BS, EI-Hendawy ANA. Porosity development in activated carbons obtained from date pits under chemical activation with phosphoric acid. Microporous and Mesoporous Materials. 2002;52(2) 105-117.

[23] Dubinin MM. The potential theory of adsorption of gases and vapors for adsorbents with energetically non uniform surfaces. Chemical Reviews. 1960;60(2) 235–241.

[24] Hassan HZ, Mohamad AA, Alyousef Y, Al-Ansary HA. A review on the equations of state for the working pairs used in adsorption cooling systems. Renewable and Sustainable Energy Reviews. 2015;45 600–609.

[25] Moreno-Castilla C. Adsorption of organic molecules from aqueous solutions on carbon material. Carbon. 2004;42(1) 83–94.

[26] Franklin R.E. Crystalline growth in graphitizing and non-graphitizing carbons. Proceedings of the Royal Society.1951;209(1097) 196–218.

[27] Rodriguez-Reinoso F, Molina-Sabio M. Activated carbons from lignocellulosic materials by chemical and/or physical activation: An overview. Carbon. 1992;30(7) 1111–1118.

[28] Valix M, Cheung WH, Zhang K. Role of heteroatoms in activated carbon for removal of hexavalent chromium from wastewaters. Journal of Hazardous Materials. 2006;1359 395–405.

[29] Rodriguez-Reinoso F. The role of carbon materials in heterogeneous catalysis. Carbon. 1998;36(3) 159–175.

[30] Anvarsha N, Kalyani P. Phytomass-inspired carbon and its importance as an antibacterial agent against human pathogens. Future Journal of Pharmaceutical Sciences. 2021;7 48-60.

[31] Castro-Muniz A, Suarez-Garcia F, Martinez-Alonso A, Tascon JMD. Activated carbon fibres with a high content of surface functional groups by phosphoric acid activation of PPTA. Journal of Colloid and Interface Science. 2011;361(1) 307–315.

[32] Tura AM, Tesema SS. Removal of methylene blue dye from wastewater using low cost activated carbon

prepared from *Delonix regia*. International Journal of Chemical and Biochemical Sciences. 2018;13 13-19.

[33] Sekaran G, Karthikeyan S, Gupta VK, Boopathy R, Maharaja P. Immobilization of Bacillus sp in mesoporous activated carbon for degradation of sulphonated phenolic compound in wastewater. Materials Science and Engineering: C. Materials for Biological Applications. 2013;33 735–745.

[34] Shabudeen PSS, Daniel S, Indhumathi P. Utilizing the pods of *Delonix regia* activated carbon for the removal of mercury (II) by adsorption technique. International Journal of Research in Chemistry and Environment. 2013;3(1) 60-65.

[35] Festus AA, Elvis OA, Morayo AB. Equilibrium sorption of lead and nickel from solutions by flame of the forest (*Delonix regia*) pods: Kinetics and isothermic study. Journal of Environmental Protection. 2013;4 261-269.

[36] Fito J, Said H, Feleke S, Worku A. Flouride removal from aqueous solution onto activated carbon of *Catha edulis* through the adsorption treatment technology. Environmental Systems Research. 2019;8(1) 25-35.

[37] Suneetha M, Sundar BS, Ravindhranath K. Removal of fuoride from polluted waters using active carbon derived from barks of *Vitex negundo* plant. Journal of Analytical Science and Technology. 2015;6(1) 1–19.

[38] Dehghani MH, Farhang M, Alimohammadi M, Afsharnia M, Mckay G. Adsorptive removal of fluoride from water by activated carbon derived from $CaCl_2$- modified *Crocus sativus* leaves: Equilibrium adsorption isotherms, optimization, and influence of anions. Chemical Engineering Communications. 2018;205(7) 955-965.

[39] Amalraj A, Pius A. Removal of fluoride from drinking water using aluminum hydroxide coated activated carbon prepared from bark of Morinda tinctoria. Applied Water Science. 2017;7 2653–2665.

[40] http://biomassproject.blogspot.com/2018/08/activated-carbon-for-gold-and-silver.html

[41] Bedia J, Peñas-Garzón M, Gómez-Avilés A, Rodriguez J, Belver C. A review on the synthesis and characterization of biomass-derived carbons for adsorption of emerging contaminants from water. Journal of Carbon Research. 2018;*4*(4) 63-116.

[42] Kalyani P, Anitha A, Darchen A. Activated carbon from grass - a green alternative catalyst support for water electrolysis. International Journal of Hydrogen Energy. 2013;38(25) 10364-10372.

[43] Awasthi GP, Bhattarai DP, Maharjan B, Kim K, Park CH, Kim CS. Synthesis and characterizations of activated carbon from *Wisteria sinensis* seeds biomass for energy storage applications. Journal of Industrial and Engineering Chemistry. 2019;72 265–272.

[44] Rufford E, Hulicova-Jurcakova D, Zhu Z, Lu GQ. Nanoporous carbon electrode from waste coffee beans for high performance supercapacitors. Electrochemistry Communications. 2008;10(10) 1594-1597.

[45] Kalyani P, Anitha A, Darchen A. Obtaining activated carbon from papaya seeds for energy storage devices. International Journal of Engineering Sciences and Research Technology. 2015;4(1) 110-120.

[46] Anitha A, Kalyani P. Capacitive performance of onion peel derived carbon. International Journal of Current Research. 2014;6(9) 8433–8438.

[47] Kalyani P, Banuprabha TR, Velkannan V. Activated carbon from banyan prop root biomass and its application in pseudo capacitors: a strategy towards circular economy for energy. Ionics. 2021;27 1357–1368.

[48] Conway BE, Birss V, Wojtowicz J. The role and the utilization of pseudo capacitance for energy storage by supercapacitors. Journal of Power Sources. 1997;66(1-2) 1-14.

[49] Ismanto AE, Wang S, Soetaredjo FE, Ismadji S. Preparation of capacitor's electrode from cassava peel waste. Bioresource Technology. 2010; 101(10) 3534–3540.

[50] Grimshaw J. Electrochemical Reactions and Mechanisms in Organic Chemistry. Elsevier: Amsterdam; 2000.

[51] Zhao X, Zhang Q, Chen CM, Zhang B, Reiche S, Wang A, Zhang T, Schlogla R, Sua DS. Aromatic sulfide, sulfoxide and sulfone mediated mesoporous carbon monolith for use in supercapacitor. Nano Energy. 2012;1(4) 624–630.

[52] Hu CC, Wang CC, Wu FC, Tseng RL. Characterization of pistachio shell-derived carbons activated by a combination of KOH and CO_2 for electric double-layer capacitors. Electrochimica Acta. 2007;52(7) 2498–2505.

[53] Béguin F, Frackowiak E. Supercapacitors: Materials, Systems and Applications. Wiley-VCH Verlag GmbH and Co. Weinheim: Germany; 2013.

[54] Kalyani P, Anitha A. Biomass carbon and its prospects in electrochemical energy systems. International Journal of Hydrogen Energy. 2013;38(10) 4034-4045.

[55] Kalyani. P, Anitha A. Biomass carbon: Prospects as electrode material in energy systems. Innovations in engineered porous materials for energy generation and storage applications. 2018;218-240

[56] Zhang F, Wang KX, Li GD, Chen JS. Hierarchical porous carbon derived from rice straw for lithium ion batteries with high-rate performance. Electrochemistry Communications. 2009;11(1) 130-133.

[57] Bhardwaj S, Jaybhaye SV, Sharon M, Sathiyamoorthy D, Dasgupta K, Jagadale P, Gupta A, Patil B, Ozha G, Pandey S, Soga T, Afre R, Kalita G, Sharon M. Carbon nanomaterial from tea leaves as an anode in lithium secondary batteries. Asian Journal of Experimental Science. 2008;22(2) 89-93.

[58] Zhang Y, Zhang F, Li GD, Chen JS. Micro porous carbon derived from pinecone hull as anode material for lithium secondary batteries. Materials Letters. 2007;61(30) 5209-5212.

[59] Hwang YJ, Jeong SK, Nahm KS, Shin JS, Stephan AM. Pyrolytic carbon derived from coffee shells as anode materials for lithium batteries. Journal of Physics and Chemistry of Solids. 2007;68(2) 182-188.

[60] Stephan MA, Kumar PT, Ramesh R, Thomas S, Jeong, SK, Nahm KS. Pyrolitic carbon from biomass precursors as anode materials for lithium batteries. Materials Science and Engineering: A. 2003;430(1-2) 132-137.

[61] Kozitsina AN, Svalova TS, Malysheva NN, Okhokhonin AV, Vidrevich MB, Brainina KZ. Sensors based on bio and biomimetic receptors in medical diagnostic, environment, and food analysis. Biosensors. 2018;8(2) 35-69.

[62] Liu J, Jalali M, Mahshid S, Wachsmann-Hogiu S. Are plasmonic optical biosensors ready for use in point-of-need applications? Analyst. 2020;145 (2) 364-384.

[63] Martynko E, Kirsanov D. Application of chemometrics in biosensing: A brief review. Biosensors. 2020;10(8) 100-120.

[64] Akolpoglu MB, Bozuyuk U, Erkoc P, Kizilel S. Biosensing-drug delivery systems for in vivo applications advanced biosensors for health care applications. 2019;249-262.

[65] Harrad LE, Bourais I, Mohammadi H, Amine A. Recent advances in electrochemical biosensors based on enzyme inhibition for clinical and pharmaceutical applications. Sensors. 2018;18 164-189.

[66] Grieshaber D, MacKenzie R, Voros J, Reimhult E. Electrochemical biosensors - sensor principles and architectures. Sensors. 2008;8(3) 1400–1458.

[67] Bakker E. Electrochemical sensors. Analytical Chemistry. 2004;76 3285-3298.

[68] Stradiotto NR, Yamanaka H, Zanoni MVB. Electrochemical Sensors: A powerful tool in analytical chemistry. Journal of the Brazilian Chemical Society. 2003;14 2-17.

[69] Mehrotra P. Biosensors and their applications – a review. Journal of Oral Biology and Craniofacial Research. 2016;6(2) 153–159.

[70] Turner APF. Biosensors: sense and sensibility. Chemical Society Reviews. 2013;42 3184-3196.

[71] Sanati A, Jalali M, Raeissi K, Karimzadeh F, Kharaziha M, Mahshid SS, Mahshid S. A review on recent advancements in electrochemical biosensing using carbonaceous nanomaterials. Microchimica Acta. 2019;186 773-795.

[72] Kim D, Kim JM, Jeon J, Lee J, Oh J, Antink WH, Kim D, Piao Y. Novel two-step activation of biomass-derived carbon for highly sensitive electrochemical determination of acetaminophen. Sensors and Actuators B: Chemical. 2018;259 50-58.

[73] Zhang W, Liu L, Li Y, Wang D, Ma H, Ren H, Shi Y, Han Y, Ye BC. Electrochemical sensing platform based on the biomass-derived micro porous carbons for simultaneous determination of ascorbic acid, dopamine, and uric acid. Biosensors and Bioelectronics. 2018;121 96-103.

[74] Wang L, Zhang Q, Chen S, Xu F, Chen S, Jia J, Tan H, Hou H, Song Y. Electrochemical sensing and bio sensing platform based on biomass-derived macro porous carbon materials. Analytical Chemistry. 2014;86 1414−1421.

[75] Oliveira PR, Kalinke C, Gogola JL, Mangrich AS, Junior LHM, Bergamini MF. The uses of activated biochar for development of a sensitive electrochemical sensor for determination of methyl parathion. Journal of Electro Analytical Chemistry. 2017;799 602-608.

[76] Kalinke CS, Mangrich AS, Marcolino-Junior LH, Bergamini MF. Carbon paste electrode modified with biochar for sensitive electrochemical determination of paraquat. Electroanalysis. 2016;28(4) 764-769.

[77] Madhu R, Veeramani V, Chen SM, Manikandan A, Lo AY, Chueh YL. Honeycomb-like porous carbon−cobalt oxide nanocomposite for high-performance enzymeless glucose sensor and supercapacitor applications. ACS Applied Materials and Interfaces. 2015;7 (29) 15812−15820.

[78] Shahzada F, Zaidic SA, Koo CM. Highly sensitive electrochemical sensor based on environmentally friendly biomass-derived sulfur-doped graphene for cancer biomarker detection. Sensors

and Actuators B: Chemical. 2017;241 716-724.

[79] Ni Y, Xu J, Liang Q, Shao S. Enzyme-free glucose sensor based on heteroatom-enriched activated carbon (HAC) decorated with hedgehog-like NiO nanostructures. Sensors and Actuators B. 2017;250 491–498.

[80] Travlou NA, Ushay C, Seredych M, Rodríguez-Castellón E, Bandosz TJ. Nitrogen-doped activated carbon-based ammonia sensors: Effect of specific surface functional groups on carbon electronic properties. ACS Sensors. 2016;1 591–599.

[81] Hayat A, Rhouati A, Mishra RK, Alonso GA, Nasir M, Istamboulie G, Marty JL. An electrochemical sensor based on TiO_2/activated carbon nanocomposite modified screen printed electrode and its performance for phenolic compounds detection in water samples. International Journal of Environmental Analytical Chemistry. 2016;96(3) 237–246.

[82] Koskun Y, Savk A, Sen B, Sen F. Highly sensitive glucose sensor based on mono disperse palladium nickel/ activated carbon nanocomposites. Analytica Chimica Acta. 2018;1010 37-43.

[83] Aparna TK, Sivasubramanian R. $NiFe_2O_4$ nanoparticles-decorated activated carbon nanocomposite based electrochemical sensor for selective detection of dopamine in presence of uric acid and ascorbic acid. Journal of Chemical Sciences. 2018;130 14-25.

[84] Wang WJ, Cai YL, Li BC, Zeng J, Huang ZY, Chen XM. A voltammetric sensor for simultaneous determination of lead, cadmium and zinc on an activated carbon fiber rod. Chinese Chemical Letters. 2018;29(1) 111–114.

[85] Humaira P, Raedah A, Nadia H, Sayeed M, Asad UK. Synthesis, characterization and biological evaluation of some novel nitrogen and sulphur containing organometallic heterocycles. Arabian Journal of Chemistry. 2017;10(8) 1098–1106.

[86] Sheena V, Sunny K, Soumya J. Antimicrobial activity of carbon nanoparticles isolated from natural sources against pathogenic Gram-negative and Gram-positive bacteria. Journal of Nanoscience. 2013;2013 1–5.

[87] Yallappa S, Deepthi R, Yashaswini S, Hamsanandini R, Chandraprasad M, Ashok Kumar S, Hegde G. Natural biowaste of groundnut shell derived nano carbons: synthesis, characterization and its in vitro antibacterial activity. Nano-Structures and Nano-Objects. 2017;12 84–90.

[88] Dheeban Shankar P, Basker S, Karthik K, Karthik S. In vitro evaluation of antibacterial efficacy using *Passiflora foetida* activated carbon. Research in Pharmacy. 2016;6 1–9.

[89] Lakshmi SD, Avti PK, Hegde G. Activated carbon nanoparticles from biowaste as new generation antimicrobial agents: A review. Nano-Structures and Nano-Objects. 2018;16 306–321.

[90] Shamsi TN, Parveen R, Afreen S, Azam M, Fatma T, Haque QMR, Fatima S. In-vitro antibacterial and antioxidant activities of sandalwood (*Santalum album*). Austin Journal of Biotechnology and Bioengineering. 2014;1(2) 1–3.

[91] Karthik K, Saraswathy CP. Antimicrobial activity of chemically modified activated carbon of *Tribulus terrestris*. Research in Biotechnology. 2015;6(3) 42–46.

Chapter 12

A Zero-Waste Process for the Treatment of Spent Potliner (SPL) Waste

Samir I. Abu-Eishah, Manal D.M. Raheem, Fatma A.S. Aljasmi, Fatima M.O. Alameri, Amna G.R. Alblooshi and Intesar F.R. Alnahdi

Abstract

This work presents a deep analyses of an environmentally friendly process to recover all valuable minerals contained in the spent potliner (SPL) such as graphite carbon and aluminum fluoride (AlF_3) and production of sodium sulfate (Na_2SO_4) and gypsum ($CaSO_4$) when H_2SO_4 is used as the leaching agent. The level of emission of hazardous gases such as HCN (weak acid) and HF are minimized by direct scrubbing of the HCN in aqueous $AgNO_3$ solution to produce a stable silver cyanide (AgCN) product. The HF can be recovered as a liquid by condensation and used within the process and/or in production of metal fluorides such as the highly-soluble potassium fluoride (KF); a main source of fluoride in industry. Almost pure CO_2 gas is also recovered from the process gas streams.

Keywords: aluminum production, spent potliner (SPL), leaching by H_2SO_4, aluminum fluoride recovery, graphite carbon recovery, zero-waste process

1. Introduction

SPL is a hazardous solid waste material produced in the aluminum smelting industry [1]. It is generated when the graphite carbon and the refractory lining of the aluminum electrolytic cell reach the end of their useful life. After about 5 to 8 years of smelter operation, the cathode liner materials deteriorate and affect the aluminum electrolytic cell performance thus need to be replaced. Various factors contribute to cell lining degradation, for example, mechanical stress, electrolyte penetration and side reactions [2].

About 20 to 25 kg of SPL is generated per each ton produced of primary aluminum [3]. Worldwide aluminum production was about 63.6 million tons in 2018, generating about 1.4 million tons of SPL [4], which is a real environmental burden to the aluminum industry, and these figures are subject to increase [5]. In 2018, the United Arab Emirates (UAE) produced 2.64 million tons of aluminum and 29,040 tons of SPL (~11 kg SPL/ton aluminum). This SPL is distributed to the UAE cement industry for use as a feedstock and a fuel alternative [4].

SPL is classified as a hazardous waste by the US Environmental Protection Agency (EPA) since it contains significant amounts of toxic fluoride and cyanide

IntechOpen

compounds (in addition to a trace amounts of polycyclic aromatic hydrocarbons, PAH), which can have adverse impacts on the environment if not adequately disposed. Cyanides are highly toxic and must be destroyed or removed from the SPL before its disposal or reuse. SPL has a high pH value due to the presence of alkali metals and oxides that make it corrosive.

Some of the SPL constituents react with water and produce flammable, toxic and explosive gases such as H_2, NH_3 and CH_4. Thus, SPL disposal is becoming one of the largest environmental concerns and the SPL stored around the world needs to be safely disposed.

Both the aluminum and fluoride species are very valuable materials and need to be recovered, preferably in the form of aluminum fluoride (AlF_3) that can be recycled to the aluminum smelting plant to produce elemental aluminum. The graphite carbon also needs to be recovered and reused at least in manufacturing of cathodes for the aluminum electrolytic cells.

In this work, we are developing an environmentally-friendly process, while properly, safely and effectively disposing the other constituents of the SPL. In this process we aim to recover the aluminum and fluoride species, the graphite carbon, in addition to other side products, that at the end leads to zero-waste. In the discussion below, equations numbering (i) within the text, for i = 1, ... n, stands for the final form of the reactions taking place during the leaching process with H_2SO_4 as well as the equations used in the process analyses.

Also, the numbering appearing in the tables stands for chemical reactions within the cathode (**Table 1**), potential gases that might evolve from the SPL reactive species when hydrolyzed (**Table 4**), other potential reactions (**Table A.5**), and SPL trace constituents' reactions with H_2SO_4 (**Table A.6**).

1.1 SPL compounds generated during the aluminum smelting process

The aluminum smelting process involves electrolysis of alumina (Al_2O_3), dissolved in cryolite (Na_3AlF_6), in a cell having graphite electrodes and linings used to transmit current from the cathodic collector bar and to contain the molten Aluminum product and the alumina-containing electrolyte.

New lining materials of aluminum electrolytic cells are made from clean and virgin graphite materials. The cathode graphite material is typically 15–25% porous, but it gets penetrated by bath materials after the start of electrolysis [6]. Penetration is initiated by the metallic sodium Na(c), followed by the electrolyte [7]. The chemical reactions within the cathode result in the formation of various carbides, nitrides, cyanides, and others within the pot linings (refractory, cathodes, and sidewalls) [8].

The spent cathode contains a lot of fluoride and cyanide. During the extended operation of the electrolytic cell, fluoride is brought in by AlF_3 and Na_3AlF_6 and is absorbed into the cell linings. Cyanides are produced by the chemical reaction between metallic sodium (from cryolite), atmospheric nitrogen penetrating into the cathode carbon through openings in the potshell and through the cathode carbon itself. Indicative examples of the chemical reactions that take place within the cathode are shown in **Table 1** along with their calculated change in the heat of reaction (ΔH_R) and and change in the Gibbs free energy of reaction (ΔG_R) (using HSC Chemistry 6.1 software) at 30°C.

1.2 SPL composition

The SPL composition varies from one plant (or from one cell) to another [9]. Various factors contribute to this variation, some of which include the cell design, cathode materials, side reactions, operation time, shutdown time and electrolyte

#	Chemical Reaction	ΔH_R, kJ/mol	ΔG_R, kJ/mol
$Na_{(C)}$, CO, Na_2CO_3, and NaCN formation reactions:			
1	$6NaF + Al \rightarrow 3Na(c) + Na_3AlF_6$	160.1	111.8[b]
2	$O_2(g) + 2C \rightarrow 2CO(g)$	−221.2	−275.4
3	$3CO(g) + 2Na(c) \rightarrow Na_2CO_3 + 2C$	−814.9	−628.7
4	$2Na(c) + 2C + N_2(g) \rightarrow 2NaCN$	−196.6	−155.1
$NaAlSiO_4$ (Nepheline) formation reaction at low SiO_2/Al_2O_3 ratios:			
5	$6NaF + 3SiO_2 + 2Al_2O_3 \rightarrow 3NaAlSiO_4 + Na_3AlF_6$	−46.9	−43.1
$NaAlSi_3O_8$ (Albite) formation reaction at high SiO_2/Al_2O_3 ratios:			
6	$6NaF + 9SiO_2 + 2Al_2O_3 \rightarrow 3NaAlSi_3O_8 + Na_3AlF_6$	−80.8	−94.4
Reactions that contribute to changes in Na_3AlF_6 (cryolite) ratio:			
7	$Na_3AlF_6 + 2CO(g) + 6Na_{(C)} \rightarrow NaAlO_2 + 6NaF + 2C$	−973.2	−827.4
8	$Na_3AlF_6 + 2Na_2CO_3 + 2C \rightarrow NaAlO_2 + 6NaF + 4CO(g)$	549.4	354.9[c]
9	$2Na_3AlF_6 + N_2(g) + 6Na(c) \rightarrow 2AlN + 12NaF$	−742	−646.6
10	$Na_3AlF_6 + NaCN + 2Na(c) \rightarrow AlN + 6NaF + C(s)$	−275.9	−248.9
Other NaCN consuming reactions:			
11[a]	$2Al_2O_3 + NaCN + 2Na(c) \rightarrow 3NaAlO_2 + AlN + C$	−294.1	−247.3
Additional $NaAlO_2$ formation reactions:			
12	$AlN + 2CO(g) + Na(c) \rightarrow NaAlO_2 + 2C + 5\ N_2(g)$	−1204.4	−1008.1
13[a]	$Al_2O_3 + CO(g) + 2Na(c) \rightarrow 2\ NaAlO_2 + C$	−495.7	−412.8
Al_4C_3 formation reactions:			
14	$4Na_3AlF_6 + 12Na(c) + 3C \rightarrow Al_4C_3 + 24NaF$	−856.1	−650.5
15[a]	$8Al_2O_3 + 3C + 12Na(c) \rightarrow 12NaAlO_2 + Al_4C_3$	−499.9	−343.2
16	$4Al + 3C \rightarrow Al_4C_3$	−215.9	−203.3
17	$2Al + N_2(g) \rightarrow 2AlN$	−636.5	−573.3

[a]*The alumina data are for α-Al_2O_3 since the data for the actual similar β-Al_2O_3 ($Na_2O{\cdot}11Al_2O_3$) compound is not available. The Na (l) data was used in the equations that require Na(c) data, which means that the actual ΔG_R is slightly more negative when Na(c) is on the right side of the equation and slightly more positive when Na(c) is on the left side of the equation.*
[b]*Only -ve at T > 700°C.*
[c]*Only -ve at T ≥ 650°C.*

Table 1.
Chemical reactions within the cathode [6, 8] and their calculated ΔH_R and ΔG_R at 30°C.

composition [10]. Most of the chemical components of the SPL are direct constituents of the electrolytic bath that infuse the carbon cathode and subsequently the refractory lining. While some of the phases are additives to the electrolytic bath, others are the result of side reactions [11].

Typical composition ranges of the SPL constituents are shown in **Table 2**, from which the SPL contains about 6.2 wt% Al, 17.5 wt% F, 39 wt% C (as graphite), and 21 wt% Na [12].

Table 3 shows the main elemental composition of the SPL along with the major phases or compounds of these elements. For example, the major forms of cyanides are identified as sodium cyanide (NaCN), sodium ferrocyanide $Na_4Fe(CN)_6$ and sodium ferricyanide $Na_3Fe(CN)_6$. Fluorides are mostly found in the form of sodium fluoride (NaF). Other reported forms of fluoride include sodium aluminum fluoride (Na_3AlF_6) and calcium fluoride (CaF_2) [15].

Compound	wt% (low)	wt% (high)
NaF	8.0	16.0
Na_2CO_3	6.5	6.5
NaCN	0.1	2.0
$NaAlO_2$	5.0	10.0
C	20.0	40.0
Na_3AlF_6	7.0	14.0
CaF_2	3.0	7.0
$NaAlSiO_4$	3.0	7.0

Table 2.
Predominant SPL compounds and their composition ranges [12].

Element	Composition range, wt%	Major phases/compounds
C	9.6–50	Graphite carbon
Na	7.0–20	NaF, Na_3AlF_6
Al	4.7–22.1	Al metal, α-Al_2O_3, others
F	9.7–18.9	NaF, Na_3AlF_6, CaF_2
Ca	1.1–2.9	CaF_2
Li	0.3–1.1	LiF, Li_3AlF_6
Mg	0.3–0.9	MgF_2
Si	0.0–2.3	Refractory SiO_2, $NaAlSiO_4$
Fe	0.3–2.1	Fe_2O_3
S	0.1–0.3	Gypsum ($CaSO_4$)
Ti	0.15–0.24	TiB_2
CN	0.02–0.44	NaCN, $Na_4Fe(CN)_6$, $Na_3Fe(CN)_6$

Table 3.
SPL main elements [13] and their major phases / compounds [14].

1.3 SPL properties

When the linings are removed from the pot they contain substantial amounts of sodium fluoride and sodium aluminum fluoride. In addition, the SPL contains Al metal, Na metal, Aluminum nitride (AlN), Aluminum carbide (Al_4C_3), and sodium cyanide (NaCN) that absorbs and reacts with atmospheric water (humidity) and emits hazardous gases to the atmosphere. **Table 4** shows potential gases evolved when the SPL is hydrolyzed, i.e. subjected to humidity, along with their calculated ΔH_R and ΔG_R at 30°C. However, some authors claim that reactions 19, 23 and 25 (in **Table 4**) produce Al_2O_3. However, it is well known that Al_2O_3 results from Gibbsite {$Al(OH)_3$} only after it is calcined (at temperatures above 400°C) [17].

Other reactions include those of ionic ferro- and ferri-cyanide with water [18]. For example,

$$Na_4Fe(CN)_6(ia) + 6H_2O \rightarrow 6HCN\uparrow + Fe(OH)_2\downarrow + 4Na^+ + 4OH^-$$

Note: (ia) is used in the HSC database for aqueous electrolyte (neutral), which is formed from undissociated aqueous species (ions).

#	Chemical Reactions[a]	ΔH_R, kJ/mol	ΔG_R, kJ/mol
18	$Al + 3H_2O \rightarrow Al(OH)_3(s) + 1.5H_2(g)$	−419.5	−427.7
19	$2Al + 3H_2O \rightarrow Al_2O_3 + 3H_2(g)$	−819.3	−872.3
20	$Al + NaOH + H_2O \rightarrow NaAlO_2 + 1.5H_2(g)$	−422.1	−453.3
21	$2Na(c) + 2H_2O \rightarrow 2NaOH + H_2(g)$	−295.6	−279.7
22	$AlN + 3H_2O \rightarrow Al(OH)_3 + NH_3(g)$	−147.3	−157.0
23	$2AlN + 3H_2O \rightarrow Al_2O_3 + 2NH_3(g)$	−275.0	−330.8
24	$Al_4C_3 + 12H_2O \rightarrow 4Al(OH)_3 + 3CH_4(g)$	−1686.4	−1658.0
25	$Al_4C_3 + 6H_2O \rightarrow 2Al_2O_3 + 3CH_4(g)$	−1647.1	−1691.8
26	$NaCN + 2H_2O \leftrightharpoons HCOONa(ia) + NH_3(g)$	−49.9	−75.3

[a]*In energy calculations: Na(l) is used instead of Na(c). (ia) is used in the HSC database for aqueous electrolyte (neutral), which is formed from undissociated aqueous species (ions).*

Table 4.
Potential gases that might evolve from the SPL reactive species when hydrolyzed [7, 16] and their calculated ΔH_R and ΔG_R at 30°C.

1.4 Main products and side products of the SPL treatment

Fluoride is the main product of the various SPL treatment processes. Fluorides are used as fluoropolymers (e.g. Teflon), which is utilized as a part of an extensive variety of uses such as cosmetic and reconstructive surgeries, paints, cookware, scratching semiconductor gadgets, cleaning, etching glass and aluminum and in evacuating rust. Aluminum hydroxyfluoride (AlF_2OH) is of particular importance among the produced fluorides. It has a high market value and can be converted to aluminum fluoride (AlF_3), which is one of the important key materials for aluminum metal production and constitutes a major cost in it [19].

Carbon is the main side product recovered during the SPL treatment; over 87% of which is in the form of graphite. Graphite behaves as a non-metal and a metal because it can resist high temperatures and it is a good electrical conductor. Graphite is also good as a refractory material because of its high-temperature stability and chemical inertness thus it is used in the production of refractory bricks. Furthermore, it can be used in production of functional refractories for continuous casting of steel and as lining blocks in iron blast furnaces due to its high thermal conductivity. In high-temperature applications (e.g. arc furnaces), it is used in production of phosphorus and calcium carbide. It can also be used as anode in aqueous electrolytic production of halogens (e.g. chlorine and fluorine), cathode in the aluminum industry, or as a fuel [4]. The other compounds (e.g. CaF_2) can be used as part of the feed in cement production.

2. Recovery of fluoride values from the chemical leaching of SPL

The majority of the chemical leaching processes of the SPL targeted fluoride recovery in the form of metal fluorides such as sodium fluoride (Villiaumite, NaF), calcium fluoride (CaF_2), sodium aluminum fluorides [e.g. cryolite (Na_3AlF_6) and $5NaF.3AlF_3$ complex], aluminum fluoride (AlF_3), aluminum hydroxyfluoride (AlF_2OH) or aluminum hydroxyfluoride hydrate ($AlF_x(OH)_{(3-x)}.xH_2O$, x = 1 or 2) [19]. The most valuable fluoride among these are AlF_3 and AlF_2OH. The AlF_3 is constantly needed in aluminum smelters to maintain the cryolite balance [20].

The AlF_2OH can be easily converted to AlF_3, for example by its reaction with HF [12]. However, NaF has a low market value since it is not consumed as much as AlF_3 in a typical smelter. The CaF_2 is also of low market value and limited quality.

Most of the AlF_3 recovery methods involve very complex and expensive processes mainly because they were not successful in precipitating AlF_3 due to its relatively high solubility in water [21]. Another problem is the AlF_3 meta-stability (200–250 g/L) which can delay its crystallization by several hours [22]. A combination of HF, fluorosilicic acid (H_2SiF_6) and ammonium bi-fluoride (NH_4HF_2) was used to precipitate AlF_3 by [23], however, these acids are highly toxic and/or expensive. In addition, calcination at 500°C to get the final AlF_3 product is required; thus, increasing the energy demand.

Leaching of the SPL CaF_2 and Na_3AlF_6 by $Al(NO_3)_3.9H_2O$ or $AlCl_3.6H_2O$ was tried and found to be very slow (24 h, at 25°C [24, 25]. The SPL fluorides (NaF, CaF_2 and Na_3AlF_6) were leached as fluoride precipitates and the NaF and Na_2CO_3 were removed from the SPL by water washing [26]. 76–86 mol% of the SPL refractory (Na_3AlF_6 and CaF_2) were extracted by using 0.34 M Al^{3+} solution at 25°C in 24 h.

After an initial water wash to leach NaF, followed by a single-leaching step using 0.5 M HNO_3 and 0.36 M $Al(NO_3)_3$ at 60°C [27], a total of 96.3% of the remaining fluoride was recovered along with 100% of the Mg and 90% of the Ca originally present in the SPL as MgF_2 and CaF_2, respectively.

Bishoy [28] subjected the SPL to NaOH leaching first followed by HNO_3 leaching at various combinations of temperatures and liquid/solid ratios. The contribution of the alkali and acid concentrations on the leaching process was found to be 51.80% and 2.61%, respectively. The best combination (2.5 M NaOH, 5 M HNO_3, 4.5-liter solution/kg SPL (or simply, L/S ratio), and 75°C) resulted in only 50.62% leaching of the SPL compounds.

Shi et al. [29] used a two-step alkaline-acidic leaching process to separate the cryolite from SPL and to purify the graphite carbon. Their results showed a recovery of 65.0% of soluble Na_3AlF_6 and Al_2O_3 compounds starting with NaOH leaching. However, they recovered 96.2% of the CaF_2 and $NaAl_{11}O_{17}$ compounds in the following HCl leaching step. By combining the acidic and alkaline leaching solutions, 95.6% of the cryolite precipitates (at pH = 9, T = 70°C, and time = 2 h) with a 96.4% purity.

Parhi & Rath [30] adopted a similar two-step leaching process to recover carbon and cryolite fractions from the SPL. They used HCl for leaching of CaF_2 and $NaAl_{11}O_{17}$ followed by NaOH for leaching of Na_3AlF_6 and Al_2O_3. A maximum leaching efficiency of 86.01% was achieved at (10 M HCl, 1.5 M NaOH, 4.5 L/S ratio and 100°C). The carbon recovery increased from 42.19% to 76.85% after treatment.

Zhao (2012) [31] presented a leaching process using water and H_2SO_4 to recover HF form the SPL. The cake obtained contains graphite powder, aluminum hydroxide {$Al(OH)_3$} and alumina (Al_2O_3) while the filtrate contains fluorides and sulfates.

Cao et al. [32] recovered fluoride and carbon from the SPL by a water washing followed by leaching with aluminum sulfate {$Al_2(SO_4)_3.18H_2O$} solution at 25°C for 24 h. The carbon recovery achieved was 88%. $Al_2[(OH)_{0.46}F_{0.54}].6H_2O$ and $5NaF.3AlF_3$ precipitated at (90°C, pH 5.5, 3 h) with a maximum fluoride recovery of 99.7%. The main products after calcination were AlF_3 and $5NaF.3AlF_3$.

Li et al. [33] employed a two-step leaching process: (1) NaF is leached by water from the imbedded electrolyte, then (2) Na_3AlF_6, CaF_2 and $NaAl_{11}O_{17}$ are leached using acidic anodizing wastewater (H_2SO_4 solution). Then the electrolyte components are precipitated from the mixed filtrates of steps (1) and (2). Most of the NaF in the SPL was dissolved in step (1); the residual electrolyte was mainly cryolite (with ~0.95% NaF). The purity of the carbon recovered was about 95.5% under

(80°C; L/S = 8 L/kg; 300 rpm; 3 h). The cryolite recovery from the mixed filtrate at (75°C; 4 h; pH 9; F/Al ratio of 6:1) was 98.4% while the Na_2SO_4 crystals purity was 92.0%.

The solubility of aluminum hydroxyfluoride at 30–70°C and its precipitation from synthetic solutions was studied by [34]. Their results suggest that when NaOH is used for the pH adjustment, a high F:Al ratio as well as higher pH were problematic because of the competitive co-precipitation of sodium fluoroaluminates hydrates ($NaAlO_2.xH_2O$) [34, 35]. Further, high purity $AlF_2OH{\cdot}H_2O$ crystals were produced at F:Al ratio of 1.6 and pH of 4.9.

Ntuk et al. [34] used two methods of AlF_2OH crystallization: (1) partial neutralization-crystallization for the bulk AlF_2OH and (2) solution evaporation-crystallization for the beneficiation of the very small AlF_2OH particles (< 30 μm), i.e. those below the acceptable size.

A leachate solution containing (AlF_2^+, Na_2SO_4) was mixed with a controlled amount of NaOH (pH 4.5–5.5) and fed to a crystallizer to selectively produce $AlF_2OH.H_2O$, which was then filtered and separated from the Na_2SO_4 solution. Around 76–86% of the fluoride was recovered from the SPL. It should also be noted that AlF_2OH can be easily converted to AlF_3 by its reaction with HF [19].

The main properties of potential leaching acids and the after leaching produced acids are listed in **Table 5**.

2.1 Solubility of SPL constituents in water

Water leaching is a process that can extract a substance by its dissolution in water. Some of the SPL constituents such as NaF, Na_2CO_3, NaCN, and $NaAlO_2$ are soluble in water but with varying degrees and their solubilities mostly increase with the increase of temperature. Other SPL constituents such as $NaAlSiO_4$, Na_3AlF_6, CaF_2, and C are insoluble in water even at high temperatures (say, 100°C). **Table 6** shows the SPL individual constituents' solubilities in water at 25 and 100°C.

The hydrolysis of some of the SPL individual constituents (namely, NaCN, NaF, $NaAlO_2$ and Na_2CO_3) is discussed below.

NaCN when mixed with water or come in contact with aquatic species, the results will be detrimental to the health of that species. When NaCN is hydrolyzed,

	Name	MW, kg/kmol	Boiling point, °C	Density @ 25°C, kg/m^3
Leaching Acid				
HCl(g)	Hydrochloric	36.5	−85.1	1.49
35–37 wt% HCl		—	—	1200
100wt% HNO_3	Nitric	63.0	83.0	1510
68 wt% HNO_3		—	—	1410
100wt% $HClO_4$	Perchloric	100.5	203.0	1768
70 wt% $HClO_4$		—		1664
96–98 wt% H_2SO_4	Sulfuric	98.1	337.0	1840
Produced Acid				
HCN(l)	Cyanic	27.0	25.6	687.6
HF(l)	Fluoric	20.1	19.5	990[a]

[a]*Saturated liquid at 19.5°C.*

Table 5.
Some properties of mineral acids (sought for SPL leaching) and after-leaching produced acids.

Compound	Name	Solubility at 25°C, g/L	Solubility at 100°C, g/L
1. SPL main compounds:			
NaF	Sodium fluoride	41.5	50.5
Na_2CO_3	Sodium carbonate	170	436
NaCN	Sodium cyanide	637	480
$NaAlO_2$	Sodium aluminate	H. soluble	H. soluble
$NaAlSiO_4$	Sodium aluminosilicate	Insoluble	Insoluble
Na_3AlF_6	Cryolite	Insoluble	Insoluble
CaF_2	Calcium fluoride	0.016	Insoluble
C	Graphite	Insoluble	Insoluble
2. Other SPL potential compounds:			
$NaAlSi_2O_6$	Sodium aluminodisilicate	Insoluble	Insoluble
$Na_4Fe(CN)_6$	Sodium ferrocyanide(ia)	H. soluble	H. soluble
$Na_3Fe(CN)_6$	Sodium ferricyanide(ia)	H. soluble	H. soluble
TiB_2	Titanium diboride	Insoluble	Insoluble
Al_2O_3	Aluminum oxide	Insoluble	Insoluble
LiF	Lithium fluoride	Insoluble	Insoluble
Li_3AlF_6	Lithium aluminum hexafluoride	1.12[a]	Very low
MgF_2	Magnesium fluoride	Insoluble	Insoluble
TiB_2	Titanium diboride	Insoluble	Insoluble
Fe_2O_3	Ferric oxide	Insoluble	Insoluble

[a]At 20 °C. H. soluble = highly soluble.

Table 6.
Solubility of the SPL individual constituents in water at 25 and 100°C.

it will produce sodium formate and ammonia gas (for T > 50°C) [36] according to Eq. (1):

$$NaCN + 2H_2O \rightleftharpoons HCOONa(ia) + NH_3(g) \tag{1}$$

where (ia) refers to aqueous electrolyte (neutral) formed from undissociated aqueous species. However, the above reaction (Eq. 1) is very slow [37] although it is spontaneous (ΔG_R = -75.3 kJ/mol at 30°C, see **Table 4**).

When NaCN is dissolved in excess water, hydrated sodium ion $[Na(H_2O)_4]^+$ and a CN^- ion are produced. However, $[Na(H_2O)_4]^+$ is a strong acid conjugate that will not react with water):

$$NaCN + 4H_2O \rightarrow \left[Na(H_2O)_4\right]^+ + CN^- \ (\text{cold}, pH > 7) \tag{2}$$

According to [36], it was stated that when NaCN is mixed with water at room temperature, it can undergo the reaction given by Eq. (3):

$$NaCN + H_2O \rightarrow NaOH + HCN(g) \tag{3}$$

However, this reaction (Eq. 3) is non-spontaneous (ΔG_R = +59.6 kJ/mol, see **Table A.5**) and is not possible at room temperature, but its reverse reaction is possible (spontaneous, ΔG_R = -59.6 kJ/mol) and well known:

$$HCN + NaOH \rightarrow NaCN + H_2O \tag{4}$$

NaF dissolves in water to produce hydrated sodium $[Na(H_2O)_4]^+$ ion and F^- ion:

$$NaF + 4H_2O \rightarrow \left[Na(H_2O)_4\right]^+ + F^- \tag{5}$$

that further reacts with water to form HF(l) and OH^- ion (the strongest base):

$$F^-(l) + H_2O \rightarrow HF(l) + OH^- \tag{6}$$

$NaAlO_2$ is highly soluble in water and decomposes completely in highly alkaline solutions and turns to sodium tetra-hydroxy aluminate $Na[Al(OH)_4]$ or its ionic forms (ΔG_R = -23.8 kJ/mol, see **Table A.5**):

$$NaAlO_2 + 2H_2O \rightarrow Na\left[Al(OH)_4\right] \tag{7}$$

$NaAlO_2$ is claimed by some authors to react with water at high temperature and with time and produce NaOH and $Al(OH)_3$ according to.

$$NaAlO_2 + 2H_2O \rightarrow NaOH + Al(OH)_3 \downarrow (\text{amorphous}). \tag{8}$$

However, this claim is not true since the reaction is non-spontaneous (ΔG_R = +25.6 kJ/mol, see **Table A.5**) and its spontaneity decreases with temperature (more $+\Delta G_R$) regardless of the retention time.

Na_2CO_3 is also highly soluble in water. The kinds of ions produced are as follows:

$$Na_2CO_3 + H_2O \rightarrow 2Na^+ + (CO_3)^{2-} + H_3O^+ + (OH)^- \tag{9}$$

Again, the claim that Na_2CO_3 reacts with H_2O to produce NaOH and CO_2(g) is also not true because it is non-spontaneous reaction (ΔG_R = +131 kJ/mol, see **Table A.5**).

On the other hand, **Table 7** shows the solubilities of the compounds produced after SPL acid leaching and/or during processing. These information are very helpful in devising the separation techniques of these products as discussed below in process description.

2.2 Process selection and the decision matrix

Bishoyi [28] made an extensive comparison to find out the best suitable leaching acid among H_2SO_4, HCl, HNO_3, and perchloric acid ($HClO_4$) while fixing the L/S ratio and observed that H_2SO_4 gave maximum leaching efficiency at 25°C. But as the temperature is increased from 25–100 °C, all of these acids gave rise to almost the same leaching percentage. However, all of the acids undergo complete ionization in water.

The order of decreasing strength of the four acids under investigation is as follows: $HClO_4$ (strongest), HCl, H_2SO_4, and HNO_3 (weakest). At 25°C, the dissociation constant (pK_a) of $HClO_4$, HCl, H_2SO_4, and HNO_3 are -8, -6.3, -3 ($pK_{a,1}$), and -1.4, respectively [38]. The larger the pK_a of an acid, the smaller its extent to dissociate at a given pH (i.e. the weaker the acid). Strong acids have pK_a values $\leq$ -2. Note: $pK_a = pH - \log_{10}[A^-]/[HA]$, [HA] and $[A^-]$ are the molar equilibrium concentrations (mol/L) of the acid and its anionic part, respectively.

On the other hand, the corrosivity of an acid depends on its level of dissociation, its concentration and phase. A vapor phase acid is more corrosive than a liquid

Compound	Name	Solubility at 25°C, g/L	Solubility at 100°C, g/L
(1) Intermediate products			
$Al_2(SO_4)_3$	Aluminum sulfate	H. soluble	H. soluble
$Al(OH)_3$	Aluminum hydroxide	0.001	Insoluble
SiO_2	Silica	Insoluble	Insoluble
$CaSO_4$	Gypsum	Insoluble	Insoluble
(2) Final products			
C	Graphite	Insoluble	Insoluble
Na_2SiO_3	Sodium silicate	H. soluble	H. soluble
$Ca(ClO_4)_2$	Calcium perchlorate	H. soluble	H. soluble
$AlF_x(OH)_{(3-x)}$, (x = 1 or 2)	Aluminum hydroxyfluoride	Soluble	Less soluble
AlF_3	Aluminum fluoride	7.3	17.2
Na_2SO_4	Sodium sulfate	H. soluble	H. soluble
(3) other products			
HF	Hydrogen fluoride	H. soluble	H. soluble
KF	Potassium fluoride	H. soluble	H. soluble
HCN	Hydrogen cyanide	H. soluble	H. soluble
AgCN	Silver cyanide	Insoluble	Insoluble
CO_2	Carbon dioxide	Insoluble	Insoluble

Table 7.
Solubility of the after-leaching SPL products at 25 and 100°C.

Parameter	H_2SO_4	HNO_3	HCl	$HClO_4$
Reactions spontaneity. See **Tables A.1** to **A.4** in Appendix A	All -ve	All -ve	5 -ve 2 +ve	5 -ve 2 +ve
Acid molarity (M) [31, 32]	5	5	10	7.5
pK_a or degree of corrosivity at 25 °C [38]	−3.0	−1.4	−6.3	−8.0
Acid cost, $/kg (2019 prices)	0.2–0.35	0.2–0.25	0.15–0.35	4.0–4.5
L/S ratio [31, 32]	2.5	4.5	4.5	4.5
Optimum temp., °C [31, 32]	50	75	100	100

Table 8.
Values of the decision parameters sought for various leachant acids.

phase acid. In addition, the corrosivity of an acid increases as temperature is increased. **Table 8** shows the values of the parameters used in process selection among the four leachant acids mentioned above.

Table 9 shows the factors affecting process selection (decision matrix), factors weight and fraction among the sought leachant acids. In **Table 9**, F_i = Factor weight/Σ factor weights. Overall score = Σ F_i x Score $_i$. Based on that, the overall score in decreasing order is as follows: H_2SO_4 (highest), HNO_3, HCl, and $HClO_4$ (lowest).

In this work, we have calculated the change in the heat of reaction (ΔH_R) and the change in the Gibbs free energy of reaction (ΔG_R) for the reactions of the

Factor	Factor weight	F_i	Acid Individual Score, $Score_i$			
			H_2SO_4	HNO_3	HCl	$HClO_4$
Spontaneity	20	0.22	100	100	70	80
Acid molarity	10	0.11	80	80	50	50
Degree of corrosivity	10	0.11	50	60	30	20
Acid cost	10	0.11	75	80	90	40
L/S ratio	20	0.22	70	50	70	50
Optimum temperature	20	0.22	80	50	25	25
Overall Score =	90	1.00	78.3	68.9	55.6	46.7

Table 9.
Decision matrix: Factor, factor weight, fraction (F_i), individual and overall scores sought for the leaching acids.

individual constituents of the SPL waste. **Table A.1** to **A.4** in Appendix A show the calculated ΔH_R and ΔG_R at 30°C for the reactions with H_2SO_4, HNO_3, HCl, and $HClO_4$, respectively. Inspection of these values shows that most of these reactions are exothermic (-ΔH_R) and spontaneous (-ΔG_R). We have also calculated ΔH_R and ΔG_R for all other potential reactions of the SPL constituents with H_2SO_4 (see **Table A.5**) as well as for the reactions with H_2SO_4 of potential trace materials that might present in the SPL (see **Table A.6**).

The operating conditions for these acids are as follows: H_2SO_4 liquid at room temperature, liquid HNO_3, HCl gas, and $HClO_4$ gas. The commercial grades of these acids are usually available at 98 wt% H_2SO_4, 68 wt% HNO_3 (pH = 1.2), 34–36 wt% HCl (pH = 1.1), and 70 wt% $HClO_4$. Because of this, the higher the concentration of the acid available for use, the lower the molarity is required for leaching. However, in all cases, an alkali leachant (e.g. NaOH) needs to be used either before or after the acid leaching step. But in this work, we have decided to add NaOH after the acid leaching step.

All of these leaching acids produce the same acid gases (namely, HCN, HF and CO_2), SiO_2 along with the existing graphite carbon. However, H_2SO_4 produces insoluble gypsum ($CaSO_4$) and soluble sodium sulfate (Na_2SO_4) along with other soluble salts that need to be crystallized and separated (i.e. AlF_2OH and/or AlF_3). However, the other leaching acids produce two soluble salts along with AlF_2OH and/or AlF_3 that makes separation more difficult. **Table 10** shows the generated intermediate and final products when H_2SO_4, HNO_3, HCl, or $HClO_4$, are used as the leaching acids. Based on that, the H_2SO_4 as a leachant seems to have more advantages above the other leaching acids, among which is the production of Na_2SO_4; one of the most profitable sodium salts. Thus, in the next discussion we will concentrate on leaching the SPL constituents by H_2SO_4 solution.

Product	Leaching Acid			
	H_2SO_4	HNO_3	HCl(g)	$HClO_4$(g)
Intermediate	$Al_2(SO_4)_3$	$Al(NO_3)_3$(ia)	$AlCl_3$	$Al(ClO_4)_3$(ia)
Soluble or aqueous	AlF_2OH, AlF_3, Na_2SO_4	AlF_2OH, AlF_3, $NaNO_3$, $Ca(NO_3)_2$	AlF_2OH, AlF_3, NaCl, $CaCl_2$	AlF_2OH, AlF_3, $NaClO_4$, $Ca(ClO_4)_2$
Insoluble	C, SiO_2, $CaSO_4$	C, SiO_2	C, SiO_2	C, SiO_2
Gas	CO_2, HF, HCN	CO_2, HF, HCN	CO_2, HF, HCN	CO_2, HF, HCN

Table 10.
Products resulting from SPL treatment as a function of leachant acid.

Lastly, it should be noted that the aluminum salts $Al_2(SO_4)_3$, $Al(NO_3)_3$, $AlCl_3$, and $Al(ClO_4)_3$ behave as acidic or basic solutions in water. For example, in $Al_2(SO_4)_3$, the SO_4^{2-} anion is neutral while the Al^{3+} is not. In the reaction:

$$Al_2(SO_4)_3 + 6H_2O \leftrightharpoons 2Al(OH)_3 + 6H^+ + 3SO_4^{2-} \tag{10}$$

the produced H_2SO_4, which is a strong acid, dissociates in the aqueous phase to form $2H^+$ and SO_4^{2-} ions, and as a result, the solution is considered acidic. For this reason, any of the above-mentioned aluminum salts, if present in the aqueous solution, can behave as acidic leachants for some of the SPL constituents (such as Na_3AlF_6 and CaF_2). This conclusion is used here as a basis for the selection of the SPL acid leaching process.

3. Leaching of the SPL individual constituents by H_2SO_4 solution

The leaching process starts with the dissolution of the water-soluble compounds of the SPL (namely, NaF, NaCN, Na_2CO_3, and $NaAlO_2$) in the H_2SO_4 solution rather than leaching in water followed by the acid. However, leaching of these four compounds in water is possible but it is very slow and requires large vessels.

Leaching reactions of the above-mentioned water-soluble compounds with H_2SO_4 are presented by Eqs. (11) to (14). See reactions R1 to R4 in **Table A.1**.

$$2NaF(s) + H_2SO_4 \rightarrow Na_2SO_4 + 2HF(g) \tag{11}$$

$$Na_2CO_3\ (s) + H_2SO_4 \rightarrow Na_2SO_4 + H_2O + CO_2(g) \tag{12}$$

$$2NaCN\ (s) + H_2SO_4 \rightarrow Na_2SO_4 + 2HCN(g) \tag{13}$$

$$2NaAlO_2\ (s) + 4H_2SO_4 \rightarrow Al_2(SO_4)_3 + Na_2SO_4 + 4H_2O \tag{14}$$

On the other hand, the graphite present in SPL is the only compound that does not react with acids (e.g. H_2SO_4), alkalis (e.g. NaOH) or acidic Al^{3+} solution. However, the reactions of the three other insoluble compounds present in the SPL (namely, $NaAlSiO_4$, Na_3AlF_6, and CaF_2) are explained below.

1. The $NaAlSiO_4$ dissolves in aqueous H_2SO_4 solution and produces the intermediate product $NaAl_3(SO_4)_2(OH)_6$ according to Eq. (15)

$$3NaAlSiO_4 + 3H_2SO_4 \rightarrow 3SiO_2(s) + Na_2SO_4 + NaAl_3(SO_4)_2(OH)_6 \tag{15}$$

$NaAl_3(SO_4)_2(OH)_6$ dissolves in excess H_2SO_4 [39] according to Eq. (16).

$$2NaAl_3(SO_4)_2(OH)_6 + 6H_2SO_4 \rightarrow 3Al_2(SO_4)_3 + Na_2SO_4 + 12H_2O \tag{16}$$

By multiplying Eq. (10) by 2, adding it to Eq. (15), and dividing the result by 3 gives the net result presented by Eq. (17) (similar to that reported by [40]:

$$2NaAlSiO_4 + 4H_2SO_4 \rightarrow Al_2(SO_4)_3 + Na_2SO_4 + 2SiO_2 + 4H_2O \tag{17}$$

2. The cryolite (Na_3AlF_6) does not react with H_2SO_4 spontaneously; it has a high $+\Delta G_R$. However, it reacts spontaneously with concentrated NaOH solution to produce NaF and the intermediate product $NaAl(OH)_4$ according to Eq. (18):

$$Na_3AlF_6(s) + 4NaOH \rightarrow NaAl(OH)_4 + 6NaF(aq) \tag{18}$$

However, both resulting products (NaF and $NaAl(OH)_4$) need to be leached with or neutralized by H_2SO_4 according to Eq. (11) (for NaF) and according to Eq. (19) for $NaAl(OH)_4$:

$$2NaAl(OH)_4 + 4H_2SO_4 \rightarrow Al_2(SO_4)_3 + Na_2SO_4 + 8H_2O \tag{19}$$

An alternative to this two-step leaching process expressed by Eqs. (16) and (18), the Na_3AlF_6 can be leached with an acidic Al^{3+} solution comprised of $Al(OH)_3$ and H_2SO_4, which was found to be more effective than leaching with an acid only or an alkali only [41, 19]. This acidic Al^{3+} solution can be prepared according to Eq. (20):

$$2Al(OH)_3(s) + 3H_2SO_4 \leftrightharpoons Al_2(SO_4)_3(aq) + 6H_2O \tag{20}$$

and the reaction of Na_3AlF_6 with the above solution gives

$$2Na_3AlF_6(s) + 2Al(OH)_3(s) + 3H_2SO_4 \rightarrow 3Na_2SO_4 + 4AlF_3 + 6H_2O \tag{21}$$

However, the $Al_2(SO_4)_3$ (or acidic Al^{3+}) solution is already produced by Eqs. (14) and (17) presented above. Here, the $Al_2(SO_4)_3$ has an *amphoteric* character, i.e. it can both act as an acidic and a basic solution in the aqueous phase. Thus, the Na_3AlF_6 reacts (spontaneously) with the present acidic $Al_2(SO_4)_3$ solution to give Na_2SO_4 and AlF_3 according to Eq. (22):

$$2Na_3AlF_6(s) + Al_2(SO_4)_3(aq) \rightarrow 3Na_2SO_4(aq) + 4AlF_3(aq) \tag{22}$$

3. The reaction of CaF_2 with H_2SO_4 is less spontaneous (very small -ΔG_R that decreases with temperature) and gives $CaSO_4$ and HF according to Eq. (23), (which is not required at this stage of leaching):

$$CaF_2(s) + H_2SO_4(aq) \rightarrow CaSO_4(s) + 2HF(aq) \tag{23}$$

However, CaF_2 can react (spontaneously) with the solution presented by Eq. (19) according to Eq. (24).

$$3CaF_2(s) + 2Al(OH)_3(s) + 3H_2SO_4 \rightarrow 3CaSO_4(aq) + 2AlF_3(aq) + 6H_2O \tag{24}$$

But again, CaF_2 can also react (spontaneously) with the acidic $Al_2(SO_4)_3$ produced by Eqs. (14) and (17) to give $CaSO_4$ precipitate and aqueous AlF_3:

$$3CaF_2(s) + Al_2(SO_4)_3\ (aq) \rightarrow 3CaSO_4(s) + 2AlF_3(aq) \tag{25}$$

4. Process description

In this work, we propose a process for leaching of the main constitutes of the SPL waste by H_2SO_4 solution. The combination of **Figures 1, 2** and **3** constitute the process flow diagram (PFD) of the proposed leaching process. Note: The numbers in red color beside the stream numbers on these figures, are the stream input temperature (30°C) or the calculated temperature using heat of mixing and reaction thermochemical data along with the energy balance equations. Most of the acid leaching reactions are exothermic (-ΔH_R) except those appearing in bold numbers in the ΔH_R column of **Table A.1** in particular.

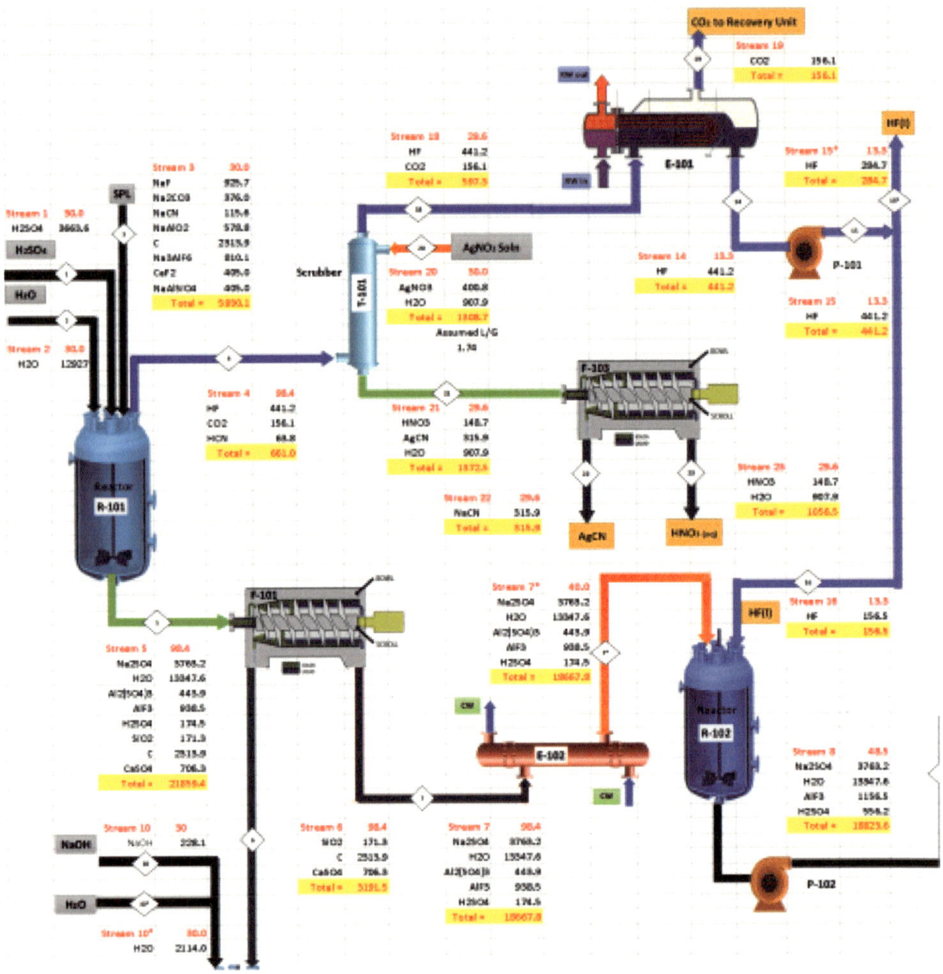

Figure 1.
Process flow diagram and material balance for the SPL treatment.

The collected SPL waste first passes through crushing and grinding steps. The resulting SPL fines are fed to an agitated semi-batch reactor filled with a pre-prepared H_2SO_4 solution. To ensure that all the SPL particles are sufficiently exposed to the solution, a 2.5 M H_2SO_4 (with 5 wt% excess) is used along with a recommended L/S ratio of 2.52 liters of H_2SO_4 acid solution per kg of SPL [19]. The reactor contents should be kept under agitation for 2–4 h. A 40,000 tons of SPL is assumed to be processed annually (or 5930 kg/h based on a stream factor of 0.77). However, a total of 220 working days per year (batch-wise operation, 22 working days per month, and allowing 2 months for shutdown and maintenance, i.e. stream factor = 0.6) is suggested elsewhere [19].

Considering the composition ranges of the SPL main constituents reported in [12] and presented in **Table 2**, the composition, the mass and molar flow rates based on the SPL upper composition limit are given in **Table 11**.

The products generated during processing are classified into three categories or streams: (1) gaseous stream (HCN, HF and CO_2), (2) insoluble products stream (graphite, gypsum and SiO_2), and (3) soluble products stream (aluminum fluorides and sodium salts, mainly, Na_2SO_4). Details on processing of each of these streams are given below and demonstrated in **Figures 1, 2** and **3** generated by the authors.

DOI: http://dx.doi.org/10.5772/intechopen.99055

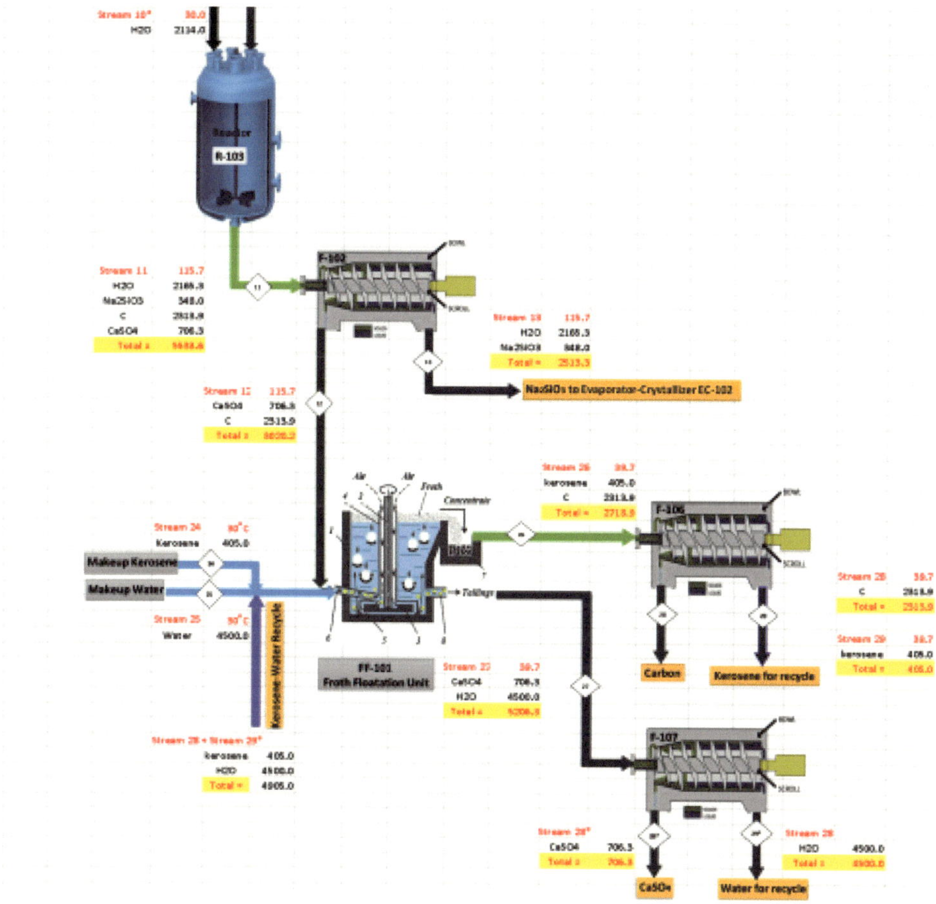

Figure 2.
Process flow diagram and material balance for the SPL treatment … continued.

1. During the leaching step, a gas stream (mainly, HCN, HF and CO_2) leaves reactor R-101, cooled (not shown on the PFD) and then sent to a gas emission-control scrubber (T-101) where the HCN gas is scrubbed by its reaction with a silver nitrate ($AgNO_3$) solution sprayed at the top. See **Figure 1**. This reaction is spontancous and exothermic. As a result, silver cyanide (AgCN) is produced according to Eq. (26). See reaction R8 in **Table A.1**.

$$HCN(g) + Ag(NO_3(aq) \rightarrow AgCN(s) + HNO_3(aq) \quad (26)$$

The AgCN is insoluble in water, but it is slightly soluble in aqueous HNO_3. The AgCN, is separated from the aqueous solution via filter F-103. The AgCN salt is stable at ambient conditions and is very valuable in gold extraction. However, it is highly toxic by ingestion and its contact with skin and eyes can cause severe irritation. It has a LD_{50} oral (rat) of 123 mg/kg.

Note: It should be mentioned that no reaction will take place between aqueous $AgNO_3$ used in Eq. (26) and HF(l), HF(g) or CO_2, since these reactions are non-spontaneous at temperatures ≤90°C.

The HF can be recovered as a liquid from the HF-CO_2 gas mixture by cooling/condensation in E-101 to below its condensation temperature (at its partial pressure in the gas stream). The remaining gas from E-101 is sent to a CO_2

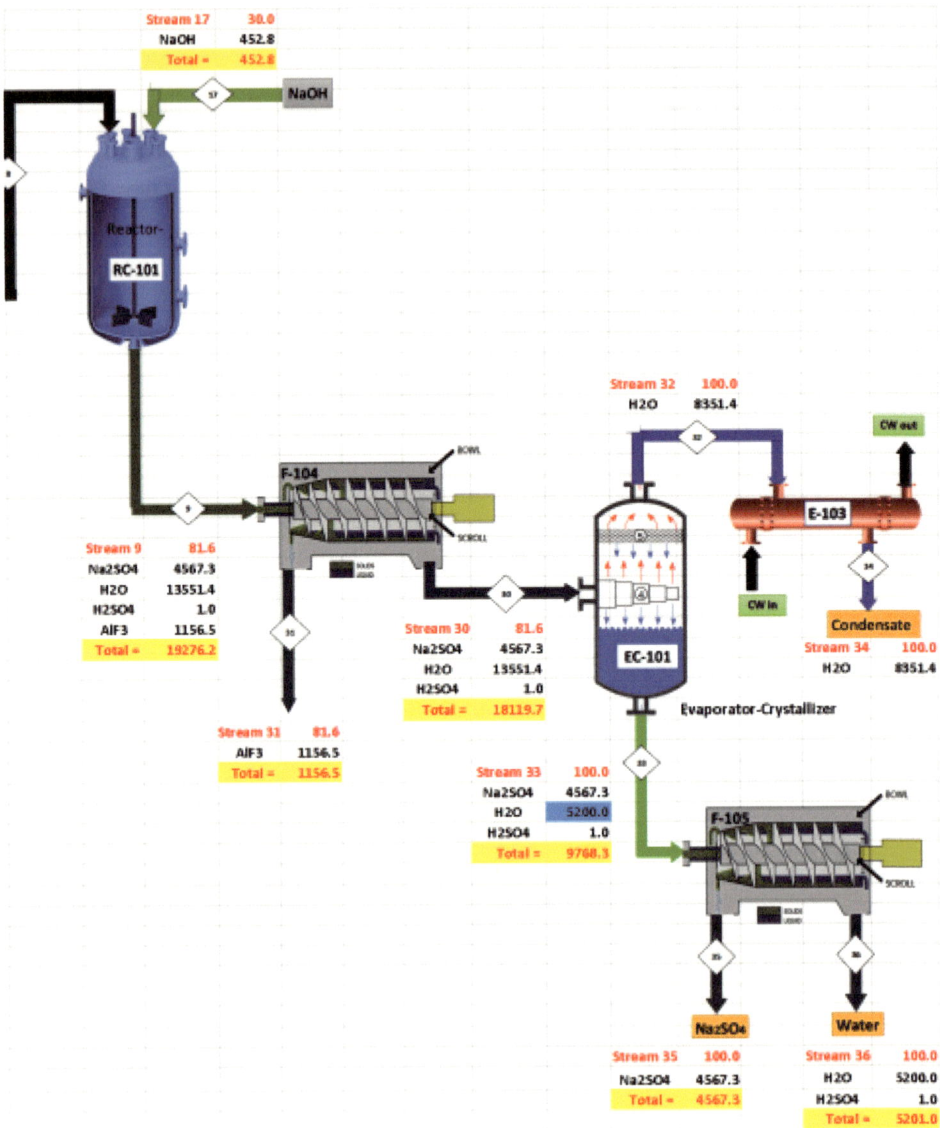

Figure 3.
Process flow diagram and material balance for the SPL treatment … concluded.

recovery unit. The recovered HF liquid is pumped (P-101) where part of it is used within the process to ensure that all the remaining aluminum sulfate is converted to AlF_3 (as explained below). The remaining part of the HF liquid can be sold as is or converted to potassium fluoride (KF); an important source of fluoride in many industries.

On the other hand, the normal boiling points of HF and HCN are 25.6°C and 19.5°C, respectively. Thus, one much better option (and much cheaper than scrubbing by $AgNO_3$ solution) is the condensation of the HF gas followed by the condensation of HCN gas at their partial pressures in the gas phase stream leaving reactor R-101. This option avoids using the very expensive $AgNO_3$ salt, but in this case, the condensed HCN must be destroyed by direct oxidation or it can be converted to a stable NaCN (soluble) salt by reacting HCN liquid with $NaNO_3$ (very cheap). But still a reactor and a separator are

Compound	MW, kg/kmol	Concentration, wt%	Mass Flow rate, kg/h	Molar Flow rate, kmol/h
NaF	41.99	15.61	925.70	22.05
Na_2CO_3	105.99	6.34	375.97	3.55
NaCN	49.01	1.95	115.64	2.36
$NaAlO_2$	81.97	9.76	578.78	7.06
Na_3AlF_6	209.94	13.66	810.06	3.86
$NaAlSiO_4$	202.14	6.83	405.03	2.00
CaF_2	78.07	6.83	405.03	5.19
C	12.00	39.02	2313.94	192.83
	Total =	100.00	5930.14	238.89

Table 11.
Normalized composition of the SPL main constituents used in this work.

needed. In either case, the resulting gas stream needs to be sent to the CO_2 recovery unit.

2. After completion of the leaching step, the slurry mixture is sent to filter F-101 where the insoluble solids (SiO_2, graphite and gypsum) are separated from the aqueous solution containing soluble intermediate and final products (Na_2SO_4, AlF_3 (and/or AlF_2OH), remaining $Al_2(SO_4)_3$, unreacted H_2SO_4, and water).

   The insoluble solids stream is sent to reactor R-103 where the SiO_2 is reacted with aqueous NaOH to produce soluble sodium silicate (Na_2SiO_3) according to reaction (27). See reaction R9 in **Table A.1**.

$$SiO_2(s) + 2NaOH(aq) \rightarrow Na_2SiO_3(aq) + H_2O \tag{27}$$

   which is then separated from the graphite-gypsum solid mixture via filter F-102. See **Figure 2**. The Na_2SiO_3 in the aqueous solution can then be saturated by evaporation and precipitated as Na_2SiO_3 crystals (not shown on the PFD).

   The graphite and gypsum can be then separated from each other in a froth flotation unit (FF-101) where an oil (e.g. 1–10 wt% kerosene) in water is used, along with air bubbling and slow agitation. See **Figure 3**. The recommended particle size for froth flotation lies between +25 and 75 μm [42]. The hydrophobic graphite along with kerosene floats up as a froth while the hydrophilic gypsum along with water settles to the bottom of the unit. The graphite-kerosene stream is sent to filter F-106 to recover the graphite and recycle the kerosene back to the froth flotation unit. Similarly, the gypsum-water stream is sent to filter F-107 to recover the gypsum and recycle the water back to the froth flotation unit.

   It should be mentioned that we have experimentally separated the graphite carbon from gypsum (using a kerosene/water volumetric ratio = 0.1 along with air bubbling at room temperature).

3. The aqueous phase from filter F-101 is cooled in E-102 and then sent to reactor R-102, where the remaining $Al_2(SO_4)_3$ is converted to AlF_3 (and/or AlF_2OH) by its reaction with part of the recovered HF liquid, according to the relatively high spontaneous Eq. (28) (ΔG_R = -196.65 kJ/mol). at 30°C. See reaction R10 in **Table A.1**.

$$Al_2(SO_4)_3(aq) + 6HF(l) \rightarrow 2AlF_3(aq) + 3H_2SO_4 \quad (28)$$

Due to the presence of fluoride ions in R-102, the dominant crystal species will be AlF_3. However, the reaction between Na_2SO_4 and HF(l) is much less competent than Eq. (28) since it is much less spontaneous (ΔG_R = -32.7 kJ/mol). See reaction R1 in **Table A.1**.

In order to recover the AlF_3 crystals, the contents of reactor R-102 are pumped through P-102 to the reactor-crystallizer RC-101, where the conditions required for AlF_3 crystallization have to be established. A controlled amount of NaOH has to be added to neutralize most of the remaining H_2SO_4 according to Eq. (29). See reaction R11 in **Table A.1**.

$$H_2SO_4 + 2NaOH \rightarrow Na_2SO_4(aq) + 2H_2O \quad (29)$$

and at the same time to maintain the solution in RC-101 at a pH of 4.5–5.5; required to saturate and precipitate AlF_3 [19], noting that the solubility of AlF_2OH (and AlF_3) decreases with the increase of the pH.

Any AlF_2OH produced can be easily converted to AlF_3 by its reaction with some of the HF liquid recovered earlier, according to the spontaneous presented by Eq. (30). See reaction R12 in **Table A.1**.

$$AlF_x(OH)_{3-x} + x\,HF(l) \rightarrow AlF_3 + x\,H_2O\ (x = 1 \text{ or } 2) \quad (30)$$

or,

$$AlF(OH)_2 + 2HF(l) \rightarrow AlF_3 + 2H_2O\ (\text{for } x = 1) \quad (31)$$

and

$$AlF_2OH + HF(l) \rightarrow AlF_3 + H_2O\ (\text{for } x = 2) \quad (32)$$

Thus, the reaction presented by Eq. (30) can be carried out before the addition of the NaOH solution.

The crystals produced in the reactor-crystallizer RC-101 are separated via filter F-104 as AlF_3 cake. To remove the impurities from the AlF_3, the stream needs to be washed with fresh water. The AlF_3 is then dried, cooled and stored.

The filtrate leaving filter F-104 is sent to the evaporator-crystallizer EC-101, where the Na_2SO_4 solution is saturated by flash evaporation under vacuum and Na_2SO_4 is crystallized and separated via filter F-105. See **Figure 3**. The Na_2SO_4 crystals can be further dehydrated and dried before being stored.

Lastly, the water vapor leaving EC-101 is condensed in E-103 and collected for reuse within the process, along with other recovered water from the various streams of the above described process.

5. Preliminary economic analysis

A preliminary economic analysis has been made on the above proposed process (assuming a theoretical 100% conversion and/or recovery) following the guidelines of ref. [43]. The amounts and costs of raw materials used as well as the amounts and

DOI: http://dx.doi.org/10.5772/intechopen.99055

market prices of the materials produced are listed in **Table 12**. The annual cost or price of a given material = amount (kg/h) x unit cost or price ($/kg) x 6475.2 (h/year). The 6475.2 factor comes from 0.77 x 24 x 365. We made a preliminary design for the process equipment and estimated the fixed capital cost of the plant excluding land, FCI_L, to be 27.32 M$.

The number of operators per job was estimated based on Eq. (33):

$$N_{OL} = \left(6.29 + 31.7\ P^2 + 0.23 * N_{np}\right)^{0.5} \qquad (33)$$

where P stand for particulate (solid) and N_{np} for non-particulate (fluid) handling equipment (P = 1 for FF-101, N_{np} = 15). The total number of operators required over the year = 4.47 N_{OL}. The salary per operator was assumed to be $49000.

The FCI_L along with the estimated annual costs of labor C_{OL}, raw materials C_{RM}, utilities C_{UT}, and waste treatment C_{WT} (given in **Table 13**) were used to calculate the cost of manufacturing excluding depreciation, COM_d, according to Eq. (34):

$$COM_d = 0.18\ FCI_L + 2.73\ C_{OL} + 1.23\ (C_{RM} + C_{UT} + C_{WT}) \qquad (34)$$

The calculated COM_d = 21.73 M$/year.

Now, assuming priceless produced HNO_3, Na_2SiO_3, CO_2 and output water, the income from main sales (revenue, R) was found to be 38.09 M$/year. Also, since $AgNO_3$ and AgCN are very expensive and sharply affect the profitably of the process, this option has been excluded in the economic analysis.

Raw Materials	Amount, kg/h	Value, $/kg	Products	Amount, kg/h	Value, $/kg
SPL	5930.1	0.12[a]	AlF_3	1156.5	1.6
H_2SO_4	3663.8	0.086	HF	284.7	0.9
NaOH	686.9	0.692	Graphite C	2213.9	0.9
			Na_2SO_4	4567.3	0.27
			$CaSO_4$	706.3	0.4
			HNO_3	134.0	0.243
			Na_2SiO_3	348.0	0.811
			CO_2	156.1	0.0
Input water	15041	$6.7x10^{-5}$	Output water	15716.7	$6.7x10^{-5}$

[a]*Estimated cost for crushing, grinding and handling of the SPL.*

Table 12.
Amounts of raw materials and products and their average prices [44].

Cost Item	M$/year
Operating labor cost, C_{OL}	1.421
Raw materials cost, C_{RM}	10.14
Utilities cost, C_{UT}	0.38
Waste treatment cost, C_{WT}	0.0
Cost of manufacturing excluding depreciation, COM_d	21.73

Table 13.
Estimated individual operating costs and COM_d.

Item	Value	Units
FCI_L	27.32	M$
Land cost, L = 0.1 FCI_L	2.732	M$
Working capital, WC = 0.2 FCI_L	5.464	M$
Salvage value, S = 0.1 FCI_L	2.732	M$
Construction period	2	years
Project life, n	10	years
Depreciation period, n_d	5	years
Depreciation, d = FCI_L/n_d	5.464	M$/year
Tax rate, t	20	%
Interest rate, i	10	%

Table 14.
Input data for discounted cumulative cash flow analysis.

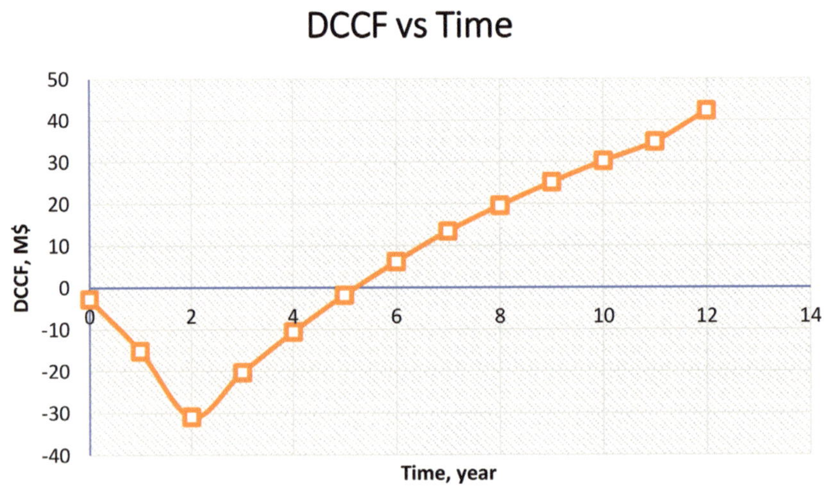

Figure 4.
Discounted cumulative cash flow diagram. (DCCF) for the above studied process.

The input data used for generating the cumulative cash flow analysis are presented in **Table 14**. The discounted cumulative cash flow diagram for the above process analysis is shown in **Figure 4**. Following [43] economic analyses and using the data presented above, and assuming an interest rate of 10%, a tax rate of 20%, the calculated net present value, NPV = 42.24 M$, the discounted payback period, DPBP = 2.38 years, and the discounted cash flow rate of return, DCFROR = 31.73%.

6. Conclusions

In this work an environmentally friendly process to recover the valuable elements contained in the SPL is presented and deeply analyzed. The decision to use H_2SO_4 as a leachant was justified through deep analysis. The proposed process along with the process flow diagram and complete material balance results have been explained and included.

The recovered materials include graphite carbon, aluminum fluoride (AlF_3), sodium sulfate (Na_2SO_4), and others when H_2SO_4 is used as the leaching agent. The level of emission of the hazardous gases such as HCN and HF are minimized. The recovered HF liquid is partially used within the process. The remaining HF can be used in production of potassium fluoride (KF). Also, CO_2 gas can also be recovered from the process gas streams.

The economic analyses indicate that the process will be profitable under the conditions stated in this work. The process net present value, NPV = 42.24 M$, the discounted payback period, DPBP = 2.38 years, and the discounted cash flow rate of return, DCFROR = 31.73%.

Acknowledgements

The authors are grateful for the United Arab Emirates University (Office of the Deputy Vice Chancellor for Research & Graduate Studies) who funded this work under Fund number G00003084, 2019/2020.

Appendix A

Note that reactions R8, R8*, R9, and R12 presented in **Table A.1** (for H_2SO_4) are common in all acid-leaching processes using HNO_3 (**Table A.2**), HCl (**Table A.3**), and $HClO_4$ (**Table A.4**)

#	Reaction	ΔH_R, kJ/mol	ΔG_R, kJ/mol
R1	$2NaF + H_2SO_4 \rightarrow Na_2SO_4 + 2HF(\mathbf{l})$	−20.3	−32.7
R1[a]	$2NaF + H_2SO_4 \rightarrow Na_2SO_4 + 2HF(\mathbf{g})$	**32.5**	−39.4
R2	$Na_2CO_3 + H_2SO_4 \rightarrow Na_2SO_4 + H_2O + CO_2(g)$	−122.6	−164.3
R3	$2NaCN + H_2SO_4 \rightarrow Na_2SO_4 + 2HCN(\mathbf{l})$	−181.8	−176.1
R3[a]	$2NaCN + H_2SO_4 \rightarrow Na_2SO_4 + 2HCN(\mathbf{g})$	−122.6	−170.5
R4	$2NaAlO_2 + 4H_2SO_4 \rightarrow Al_2(SO_4)_3 + Na_2SO_4 + 4H_2O$	−450.0	−419.7
R5	$2NaAlSiO_4 + 4H_2SO_4 \rightarrow Al_2(SO_4)_3 + Na_2SO_4 + 2SiO_2 + 4H_2O$	−348.7	−310.8
R6	$2Na_3AlF_6 + Al_2(SO_4)_3 \rightarrow 3Na_2SO_4 + 4AlF_3$	−119.9	−119.5
R7	$3CaF_2 + Al_2(SO_4)_3 \rightarrow 3CaSO_4 + 2AlF_3$	−209.1	−211.6
R8	$HCN(\mathbf{l}) + AgNO_3 \rightarrow AgCN + HNO_3$	−13.0	−15.8
R8[a]	$HCN(\mathbf{g}) + AgNO_3 \rightarrow AgCN + HNO_3$	−10.2	−4.4
R9	$SiO_2 + 2NaOH \rightarrow Na_2SiO_3 + H_2O$	−84.7	−88.9
R10	$Al_2(SO_4)_3 + 6HF(l) \rightarrow 2AlF_3 + 3H_2SO_4$	−223.3	−196.7
R11	$H_2SO_4 + 2NaOH \rightarrow Na_2SO_4 + 2H_2O$	−294.1	295.3
R12-a	$Al(OH)_2F(g) + 2HF(l) \rightarrow AlF_3 + 2H_2O$	−155.6	−140.7
R12-b	$Al(OH)F_2(g) + HF(l) \rightarrow AlF_3 + H_2O$	−140.7	−106.1

[a]*Stands for reactions involving HF(g) or HCN(g).*

Table A.1.
Calculated ΔH_R and ΔG_R at 30°C for the reactions of the main SPL constituents when leached with H_2SO_4 solution.

#	Reaction	ΔH_R, kJ/mol	ΔG_R, kJ/mol
R1	$NaF + HNO_3(l) \rightarrow NaNO_3 + HF(\mathbf{l})$	−16.8	−12.2
R1[a]	$NaF + HNO_3(l) \rightarrow NaNO_3 + HF(\mathbf{g})$	**9.6**	−15.5
R2	$Na_2CO_3 + 2HNO_3(l) \rightarrow 2NaNO_3 + H_2O + CO_2(g)$	−135.9	−155.9
R3	$NaCN + HNO_3(l) \rightarrow NaNO_3 + HCN(\mathbf{l})$	−97.6	−83.9
R3[a]	$NaCN + HNO_3(l) \rightarrow NaNO_3 + HCN(\mathbf{g})$	−68.0	−81.1
R4	$NaAlO_2 + 4HNO_3(l) \rightarrow NaNO_3 + Al(NO_3)_3(\mathbf{ia}) + 2H_2O$	−367.4	−268.2
R5	$NaAlSiO_4 + 4HNO_3(l) \rightarrow SiO_2 + Al(NO_3)_3(\mathbf{ia}) + NaNO_3 + 2H_2O$	−316.8	−213.8
R6-a	$Na_3AlF_6 + Al(NO_3)_3(\mathbf{ia}) \rightarrow 3NaNO_3 + 2AlF_3$	**55.8**	**15.3**
R6-b	$Na_3AlF_6 + Al(NO_3)_3.6H_2O \rightarrow 3NaNO_3 + 2AlF_3 + 6H_2O$	**32.5**	−28.05[c]
R6[b]	$Na_3AlF_6 + 3HNO_3(l) + Al(OH)_3 \rightarrow 3NaNO_3 + 2AlF_3 + 3H_2O$	−160.5	−135.0
R7	$1.5CaF_2 + Al(NO_3)_3(\mathbf{ia}) \rightarrow 1.5Ca(NO_3)_2 + AlF_3$	**83.6**	**40.3**
R7[b]	$1.5CaF_2 + Al(OH)_3 + 3HNO_3(l) \rightarrow 1.5Ca(NO_3)_2 + AlF_3 + 3H_2O$	−132.6	−110.03[d]
R10	$Al(NO_3)_3(ia) + 3HF(l) \rightarrow AlF_3 + 3HNO_3$	**24.0**	−35.9
R11	$HNO_3 + NaOH \rightarrow NaNO_3 + H_2O$	−153.6	−143.4

[a]*Stands for reactions involving HF(g) or HCN(g).*
[b]*Stands for alternative spontaneous reaction.*
[c]*ΔG_R at $T > 100°C$.*
[d]*ΔG_R at $T > 180°C$.*

Table A.2.
SPL reactions with HNO_3 and their ΔH_R and ΔG_R at 30°C.

#	Reaction	ΔH_R, kJ/mol	ΔG_R, kJ/mol
R1	$NaF + HCl(g) \rightarrow NaCl + HF(\mathbf{l})$	−41.9	−14.5
R1[a]	$NaF + HCl(g) \rightarrow NaCl + HF(\mathbf{g})$	−15.5	−17.9
R2	$Na_2CO_3 + 2HCl(g) \rightarrow 2NaCl + H_2O + CO_2(g)$	−186.1	−160.7
R3	$NaCN + HCl(g) \rightarrow NaCl + HCN(\mathbf{l})$	−122.6	−86.2
R3[a]	$NaCN + HCl(g) \rightarrow NaCl + HCN(\mathbf{g})$	−93.1	−83.5
R4	$NaAlO_2 + 4HCl(g) \rightarrow NaCl + AlCl_3 + 2H_2O$	−185.6	−35.4
R5	$NaAlSiO_4 + 4HCl(g) \rightarrow SiO_2 + AlCl_3 + NaCl + 2H_2O$	−134.9	19.0
R6	$Na_3AlF_6 + AlCl_3 \rightarrow 3NaCl + 2AlF_3$	−226.2	−226.7
R7	$3CaF_2 + 2AlCl_3 \rightarrow 3CaCl_2 + 2AlF_3$	−311.8	−322.1
R10	$AlCl_3 + 3HF(l) \rightarrow AlF_3 + 3HCl$	−432.3	−363.9
R11	$HCl(a) + NaOH \rightarrow NaCl + H_2O$	−95.5	−114.7

[a]*Stands for reactions involving HF(g) or HCN(g).*

Table A.3.
SPL reactions with HCl(g) and their ΔH_R and ΔG_R at 30°C.

#	Reaction	ΔH_R, kJ/mol	ΔG_R, kJ/mol
R1	$NaF + HClO_4(g) \rightarrow NaClO_4 + HF(\mathbf{l})$	−111.0	−71.9
R1[a]	$NaF + HClO_4(g) \rightarrow NaClO_4 + HF(\mathbf{g})$	−84.6	−75.2
R1[c]	$NaF + HClO_4(\mathbf{ia}) \rightarrow NaClO_4 + HF(\mathbf{g})$	**50.4**	24.8[d]

#	Reaction	ΔH_R, kJ/mol	ΔG_R, kJ/mol
R2	$Na_2CO_3 + 2HClO_4(g) \rightarrow 2NaClO_4 + H_2O(l) + CO_2(g)$	−324.4	−275.3
R3	$NaCN + HClO_4(g) \rightarrow NaClO_4 + HCN\mathbf{(l)}$	−191.8	−143.6
R3[a]	$NaCN + HClO_4(g) \rightarrow NaClO_4 + HCN\mathbf{(g)}$	−162.1	−140.7
R4	$NaAlO_2 + 4HClO_4(g) \rightarrow NaClO_4 + Al(ClO_4)_3\mathbf{(ia)} + 2H_2O$	−743.6	−523.7
R5	$NaAlSiO_4 + 4HClO_4(g) \rightarrow SiO_2 + Al(ClO_4)_3\mathbf{(ia)} + NaClO_4 + 2H_2O$	−693.0	−469.3
R6	$Na_3AlF_6 + Al(ClO_4)_3\mathbf{(ia)} \rightarrow 3NaClO_4 + 2AlF_3$	**55.2**	32.1[e]
R6[b]	$Na_3AlF_6 + 3HClO_4(g) + Al(OH)_3 \rightarrow 3NaClO_4 + 2AlF_3 + 3H_2O$	−443.1	−314.1
R7	$1.5CaF_2 + Al(ClO_4)_3\mathbf{(ia)} \rightarrow 1.5Ca(ClO_4)_2 + AlF_3$	**141.7**	32.1[f]
R7[b]	$1.5CaF_2 + Al(OH)_3 + 3HClO_4(g) \rightarrow 1.5Ca(ClO_4)_2 + AlF_3 + 3H_2O$	−356.5	−224.9
R10	$Al(ClO_4)_3\mathbf{(ia)} + 3HF(l) \rightarrow AlF_3 + 3HClO_4(g)$	**305.8**	159.8[g]
R10[a]	$Al(OH)_3 + 3HClO_4(g) \rightarrow Al(ClO_4)_3\mathbf{(ia)} + 3H_2O$	−498.3	−346.1
R11	$HClO_4(g) + NaOH \rightarrow NaClO_4 + H_2O$	−247.7	−203.0

[a]*Stands for reactions involving HF(g) or HCN(g).*
[b]*Stands for alternative spontaneous reaction.*
[c]*Stands for $HClO_4(ia)$.*
[d]*ΔG_R at $T > 240°C$.*
[e]*ΔG_R at $T > 225°C$.*
[f]*ΔG_R at $T > 550°C$.*
[g]*ΔG_R at $T > 350°C$.*

Table A.4.
SPL reactions with $HClO_4(g)$ and their ΔH_R and ΔG_R at 30°C.

#	Reaction	ΔH_R, kJ/mol	ΔG_R, kJ/mol
27	$NaCN + 4H_2O \rightarrow [Na(H_2O)_4]^+ + CN^-$	—	—
28	$NaCN + H_2O \rightarrow NaOH + HCN$	**56.1**	**59.6**
29	$HCN + NaOH \rightarrow NaCN + H_2O$	−56.1	−59.6
30	$NaF + 4H_2O \rightarrow [Na(H_2O)_4]^+ + F^-$	—[a]	—[a]
31	$F^-(l) + H_2O \rightarrow HF(l) + OH^-$	—	—
32	$NaAlO_2 + 2H_2O \rightarrow Na(Al(OH)_4)$	−26.1	−23.8
33	$NaAlO_2 + 2H_2O \rightarrow NaOH + Al(OH)_3$	**2.6**	**25.6**
34	$Na_2CO_3 + H_2O \rightarrow 2Na^+ + (CO_3)^{2-} + H_3O^+ + (OH)^-$	—	—
35	$\mathbf{Al_2(SO_4)_3 + 6H_2O \leftrightharpoons 2Al(OH)_3 + 6H^+ + 3SO_4^{2-}}$	—	—
36	$3NaAlSiO_4 + 3H_2SO_4 \rightarrow 3SiO_2 + Na_2SO_4 + NaAl_3(SO_4)_2(OH)_6$	−527.5	−451.4
37	$2NaAl_3(SO_4)_2(OH)_6 + 6H_2SO_4 \rightarrow 3Al_2(SO_4)_3 + Na_2SO_4 + 12H_2O$	**8.9**	−29.7
38	$Na_3AlF_6 + 4NaOH \rightarrow NaAl(OH)_4 + 6NaF$	−164.8	−169.3
39	$2NaAl(OH)_4 + 4H_2SO_4 \rightarrow Al_2(SO4)_3 + Na_2SO_4 + 8H_2O$	−397.8	−372.1
40	$2Al(OH)_3 + 3H_2SO_4 \rightarrow Al_2(SO_4)_3 + 6H_2O$	−161.1	−175.6
41	$2Na_3AlF_6 + 2Al(OH)_3 + 3H_2SO_4 \rightarrow 3Na_2SO_4 + 4AlF_3 + 6H_2O$	−140.5	−147.6
42	$CaF_2 + H_2SO_4 \rightarrow CaSO_4 + 2HF(l)$	+57.6	−11.6
43	$3CaF_2 + 2Al(OH)_3 + 3H_2SO_4 \rightarrow 3CaSO_4 + 2AlF_3 + 6H_2O$	−185.1	−193.6

[a]*Ionic reactions have no specific ΔH_R or ΔG_R.*

Table A.5.
Calculated ΔH_R and ΔG_R at 30°C for other potential reactions taking place during the SPL leaching process.

#	Reaction	ΔH$_R$, kJ/mol	ΔG$_R$, kJ/mol
44	$Li_3AlF_6 + Al(OH)_3 + 1.5H_2SO_4 \rightarrow 1.5Li_2SO_4 + 2AlF_3 + 3H_2O$	−152.1	−158.3
45	$2LiF + H_2SO_4 \rightarrow Li_2SO_4 + 2HF(g)$	**64.6**	−5.9
46	$2Na_3Fe(CN)_6(ia) + 6H_2SO_4 \rightarrow 3Na_2SO_4 + Fe_2(SO_4)_3 + 12HCN(l)$	−279.4	−365.3
47	$Na_4Fe(CN)_6(ia) + 3H_2SO_4 \rightarrow 2Na_2SO_4 + FeSO_4 + 6HCN(l)$	−123.3	−214.7
48	$1.5MgF_2 + Al(OH)_3 + 1.5H_2SO_4 \rightarrow 1.5MgSO_4 + AlF_3 + 3H_2O$	−76.9	−83.7
49	$Al_2O_3(s) + 3H_2SO_4 \rightarrow Al_2(SO_4)_3 + 3H_2O$	−180.8	−158.7
50	$Al_4C_3 + 6H_2SO_4 \rightarrow 2Al_2(SO4)_3 + 3C + 6H_2(g)$	−1784.2	−1858.8
51	$Fe_2O_3 + 3H_2SO_4 \rightarrow Fe_2(SO_4)_3 + 3H_2O$	−178.0	−164.9
52	$SiO_2 + 4HF \rightarrow SiF_4(g) + 2H_2O$	−77.6	−101.7
53	$SiO_2 + 6HF \rightarrow H_2SiF_6(ia) + 2H_2O$	−260.7	−190.8

Table A.6.
Calculated ΔH_R and ΔG_R at 30°C for the reactions of the SPL trace constituents when subjected to H_2SO_4 leaching.

Author details

Samir I. Abu-Eishah*, Manal D.M. Raheem, Fatma A.S. Aljasmi,
Fatima M.O. Alameri, Amna G.R. Alblooshi and Intesar F.R. Alnahdi
Department of Chemical and Petroleum Engineering, United Arab Emirates University, Al-Ain, UAE

*Address all correspondence to: s.abueishah@uaeu.ac

References

[1] Courbariaux Y, Chaouki J, Guy C. Update on spent potliners treatments: Kinetics of cyanides destruction at high temperature. Ind. Eng. Chem. Res. 2004;43:5828-5837

[2] Lossius L, Øye H. Melt penetration and chemical reactions in 16 industrial aluminum carbon cathodes. Metallurgical and Materials Transactions B, 2000;31:1213-1224.

[3] Mikša D, Homšak M, Samec N. Spent potlining utilisation possibilities. Waste Management & Research 2003;21: 467-473.

[4] Tabereaux A. The SPL waste management challenge in primary aluminum, Light Metals Age, March 16 2021. Available at: https://www.lightmetalage.com/news/industry-news/smelting/the-spl-waste-management-challenge-in-primary-aluminum/

[5] World Aluminum, 2017. http://www.world-Aluminum.org/statistics/

[6] Broek S, Øye HA. Fundamentals of managing spent potlining (SPL), Travaux 47, Proceedings of the 36th International Committee for Study of Bauxite, Alumina & Aluminum (ICSOBA) Conf., Belem, Brazil. Oct. 29-Nov. 1, 2018.

[7] Sørlie M, Øye HA. Cathodes in aluminum electrolysis, 3rd ed., Aluminum-Verlag, Dusseldorf, 2010; 662 pages, ISBN: 3942486075, 9783942486071

[8] Cooper BJ. Considerations for dealing with spent potlining, 11th Australasian aluminum smelting technology conference, Dubai, UAE, Dec 6-11, 2014.

[9] Lossius LP, Øye HA. Melt penetration and chemical reactions in carbon cathodes during aluminum electrolysis. II. Industrial Cathodes, Light Metals. 1993; 331-340.

[10] Brisson PY, Soucy G, Fafard M, Darmstadt H, Servant G. Revisiting sodium and bath penetration in the carbon lining of aluminum electrolysis cell. Light Metals. 2005;727-732.

[11] Brilloit P, Lossius L, Øye H. Penetration and chemical reactions in carbon cathodes during aluminum electrolysis: Part I. laboratory experiments. Metallurgical & Materials Transactions B. 1993;24:75-89.

[12] Lisbona DF. Development of a process for the treatment of spent pot-lining with recovery of an Aluminum hydroxyfluoride product, PhD Thesis. The University of Nottingham, Australia, 2009.

[13] US5164174A. Detoxification of aluminum spent potliner by thermal treatment, lime slurry quench and post-kiln treatment, 1992.

[14] Holywell G, Breault R. An overview of useful methods to treat, recover, or recycle spent pot lining. Journal of the Minerals, Metals and Materials Society. 2013;65:1441–1451.

[15] Chanania F, Eby E. Best demonstrated available technology (BDAT) background document for spent aluminum potliners–K088, US Environmental Protection Agency, office solid waste, Washington, DC; May 31, 2000.

[16] Augood DR. Some handling considerations for spent potlining, Light Metals. 1986;979-992.

[17] Aldaco R, Garea A, Fernández I, Irabien Á. Fluoride reuse in aluminum trifluoride manufacture: Sustainability criteria, AIChE annual meeting, Cincinnati, OH, USA; Nov 1, 2005.

[18] Zhao X, Ma L. Hazardous waste treatment for spent pot liner, IOP Conf. Series: Earth Environ. Sci. 2018;108: 042023. doi: 10.1088/1755-1315/108/4/042023

[19] Ntuk UU. Crystallization of $AlF_2OH.H_2O$ for application in the recycle of fluoride from spent pot lining (SPL), PhD Thesis. The University of Queensland, Australia; 2015.

[20] Thonstad J, Fellner P, Haarberg GM, Hives J, Kvarde H, Sterten A. Aluminum Electrolysis: Fundamentals of the Hall-Héroult Process, Aluminium-Verlag, 2001. 359 pages

[21] Mansfield K, Harple, J, Swayn G. SPL treatment and fluoride recycling project. 2002;315-327.

[22] Grobelny M. High temperature crystallization of aluminum fluoride. Course, hydrolysis, solid phases. Journal of Fluorine Chemistry. 1977;9: 187-207.

[23] Pong TK, Adrien RJ, Besida J, O'Donnell TA, Wood DG. Spent Potlining – A hazardous waste made safe. Process Safety and Environmental Protection. 2000;78:204-208.

[24] Sanjuan B, Michard G. Aluminum hydroxide solubility in aqueous solutions containing fluoride ions at 50° C. Geochimica et Cosmochimica Acta. 1987;51:1823-1831.

[25] Grolman RJ, Holywell GC, Kimmerle FM. Recycling of spent pot linings. US Patent 5470559A , 1995.

[26] Lisbona DF, Steel KM. Treatment of spent pot-lining for recovery of fluoride values. Light Metals-Warrendale-Proceedings-the Minerals, Metals & Materials Society. 2007;843–848. ISSN: 0147-0809

[27] Lisbona DF, Somerfield C, Steel KM. Leaching of spent pot-lining with aluminum nitrate and nitric acid: Effect of reaction conditions and thermodynamic modelling of solution speciation, Hydrometallurgy. 2013; 134-135:132-143

[28] Bishoyi N. Treatment of Spent Pot Lining by Chemical Leaching Using Nitric Acid for Enrichment of its Fuel Value and Optimization of the Process Parameters, B.Sc Thesis; National Institute of Technology, Rourkela, India, 2015.

[29] Shi Z-N, Li W, Hu X-W, Ren B-J, Gao B-L, Wang Z-W. Recovery of carbon and cryolite from spent pot lining of aluminum reduction cells by chemical leaching. Trans. Nonferrous Met. Soc. China. 2012;22:222-227.

[30] Parhi SS, Rath P. Gainful Utilization of Spent Pot Lining - a Hazardous Waste from Aluminum Industry, Master Thesis, National Institute of Technology, Rourkela, Odisha, India; August 2014.

[31] Zhao LC. Process for recovering waste liner of Aluminum electrolyzer. CN1320491A, 2001-11-07.

[32] Cao X-Z, Shi Y-Y, Zhao S, Xue X-X. Recovery of valuable components from spent pot-lining of aluminum electrolytic reduction cells, Journal of Northeastern University. 2014;35:1746–1749. DOI: 10.3969/j.issn.1005-3026.2014.12.017

[33] Li X, Yin W, Fang Z, Liu Q, Cui Y, Zhao J, Jia H. Recovery of carbon and valuable components from spent pot lining by leaching with acidic aluminum anodizing wastewaters, Metallurgical and Materials Transactions B, 2019;50: 914–923.

[34] Ntuk U, Tait S, White ET, Steel KM. The precipitation and solubility of aluminum hydroxyfluoride hydrate between 30-70°C, Hydrometallurgy. 2015;155:79–87.

[35] Lisbona D, Steel K. Recovery of fluoride values from spent pot-lining: Precipitation of an aluminum hydroxyfluoride hydrate product. Separation and Purification Technology. 2008;61:182–192.

[36] Ketcheson K, Fingas M. Sodium cyanide: Properties, toxicity, uses and environmental impacts. In: The Handbook of Hazardous Materials Spills Technology, New York: McGraw Hill; 2002. p. 39.1-39.22. ISBN: 007135171X 9780071351713

[37] NCABrasil Expert Auditors Ltd. Cyanide Production Operation, Summary Audit Report for Proguigel Chuimica S/A, CandeiasUnit/2016. Available at: https://www.cyanidecode.org/sites/default/files/pdf/ProquigelCandeiasSAR2017.pdf

[38] Available at Inorganic acids and bases - pKa values (engineeringtoolbox.com

[39] Stoffregen RE, Alpers CN, Jambor JL. Alunite-Jarosite crystallography, Thermodynamics, and Geochronology, Reviews in Mineralogy and Geochemistry. 2000;40:453-479.

[40] Annoni R, Souza PS, Petrániková M, Miskufova A, Havlik T, Mansur MB, Submerged-arc welding slags: Characterization and leaching strategies for the removal of aluminum and titanium, Journal of Hazardous Materials. 2013;244-245:335-341. DOI: 10.1016/j.jhazmat.2012.11.053

[41] Kaaber H, Mollgaard M. Process for recovering aluminum and fluorine from fluorine containing waste materials. US Patent 5,558,847, 1996.

[42] Chelgani SC, Rudolph M, Kratzsch R, Sandmann D, Gutzmer J. A review of graphite beneficiation techniques, Mineral Processing and Extractive Metallurgy Review. 2016;37: 58-68.

[43] Turton R, Bailie RC , Whiting WB, Shaeiwitz JA, Bhattacharyya D. Analysis, Synthesis and Design of Chemical Engineering Processes, 4th ed., NY: Prentice Hall; 2012. 1104 p. ISBN-13: 978-0-13-261812-0

[44] Indicative Chemical Prices A-Z. Available at: https://www.icis.com/explore/chemicals/channel-info-chemicals-a-z/

Section 3

Role of the Society in Recycling and Circular Economy

Chapter 13

Urban Mining of e-Waste and the Role of Consumers

Dimitris Georgantzis Garcia and Sven Kevin van Langen

Abstract

This chapter adds to the body of literature on the Circular Economy (CE), urban mining, and their intersection with consumer behaviour, by first providing a review of existing and emergent EU regulations aimed towards enhancing the collection rate of household WEEE. The fast growth of the EEE waste stream and its potential for Urban Mining as well as the inability of WEEE collection to keep up with the growth of the EEE industry is showcased with statistical data. The final section critically analyses the literature the intersection between consumer behaviour and closed-loop supply chains for EEE, identified through a systematic keyword search to ensure replicability. The findings point at a lack of theoretical, methodological and product-case heterogeneity among the identified sources, with most of them employing the Theory of Planned Behaviour and survey methods and focusing on mobile phones or general WEEE. While the literature suggests important behavioural differences across EEE categories, this was not representatively explored. The final section contributes to filling this gap by developing a taxonomy of EEE categories based on characteristics that may predispose consumer behaviour. The identified dimensions are: size, involvement, long-term reliability expectations, value type, internet access, multifunctionality, the quality of being outdated and social meaning.

Keywords: circular economy, urban mining, consumer behaviour, waste electrical and electronic equipment (WEEE), e-waste, electrical and electronic equipment (EEE), EU regulation

1. Introduction

As an increasing number of countries is focusing on CE [1–4], so too does the treatment of WEEE get more attention in academics [4–8]. A topic that has seen little attention is the role of consumers, and specifically consumer behaviour, in the retrieval of household WEEE from anthropogenic stocks, a process called urban mining. This chapter will first provide a short background as to why the proper retrieval of WEEE is important, followed by an overview of the state-of-the-art in WEEE regulations and collection. The main focus of this chapter will be on consumer behaviour in Closed Loop Supply Chains for Electrical and Electronic Equipment (*e*CLSC) to better understand issues with household WEEE collection.

WEEE that is not properly collected and treated often gets dumped or finds its way into the informal sector where it is processed with methods bad for the environment and human health [9]. Informal WEEE dumping grounds and processing sites are known to contaminate soils, air, and water, including major rivers [10]. This is the result from using unsafe processing methods such as mechanical treatment (leading to a lot of hazardous dust), open burning (releasing toxic fumes), and leaching processes (with waste acid then dumped and toxic fumes being released), having significant effects to the environment and on human health. Health effects especially affect children that are often forced to labour on informal WEEE processing sites, developing changes in typhoid function, a lower forced vital capacity, changes in cellular expression, changes in behaviour and temperament [11], furthermore, exposure to WEEE can later lead to reduced virility, stillbirths, spontaneous abortions, premature births, reduced birth lengths, reduced birthweights and other birth defects showing that improper WEEE handling can damage more than one generation of people at exposure. WEEE is also known to damage DNA and lower educational outcomes.

The urban mining of EEE, which can be defined as a process to retrieve (raw) materials and energy from urban areas, specifically from anthropogenic stocks of WEEE, has only recently started to garner attention in academics [12]. Urban mining is seen as a must to achieve ambitious circular economy targets and an effective method to reduce supply risks of critical raw materials, such as rare earth metals in the EEE industry, as well as for base metals, such as copper (of which an estimate stock in the EU of 82 Megatons exists) [13]. The urban mining of WEEE has the potential to be an industry worth 57 billion USD a year [4]. Besides the economic potential, urban mining is typically better for the environment as compared to the traditional mining of natural resources [14]. The procurement of sufficient volumes of WEEE is currently a prohibiting factor for urban mining. Improvements in the collection rates of household WEEE from consumers are deemed essential to develop the WEEE recycling industry and curb the problems with the rapidly growing EEE waste stream [15]. The growth of the WEEE waste stream is further evidenced in **Table 1**. In 2019, at the global scale, only 17.4% of WEEE got properly collected, with an estimated 44.3 megaton of WEEE likely to be dumped and/or illegally traded to be recycled in a non-environmentally and health damaging way on the informal market, a growth of 7.4 megaton when compared to the 36.9 megaton that saw its way into the informal market or was illegally dumped in 2014 [9].

In this chapter we aim to contribute to the growing body of literature on household WEEE collection and urban mining in the EU and the role of consumers in *e*CLSCs. **Section 2** provides an overview of regulations and directives relevant to the collection of WEEE from consumers in the EU, as well as recent trends in the amount of household EEE put to market in the EU and the amount of household WEEE collected in the EU. **Section 3** provides an in-depth study of the role of consumers in *e*CLSCs. **Section 4** provides a discussion of our findings. **Section 5** presents concluding remarks pertaining to this chapter.

2. WEEE regulation developments in the EU

The EU was one of the first major economic blocks to implement far reaching regulations to curb the growth of WEEE. The first big steps were taken by the EU with the WEEE Directive that set collection, recycling, and recovery targets for WEEE from 2003 onwards, and the RoHS directive that set limitations on the use of hazardous substances from 2004 and onwards. Since then, the body of regulations concerning WEEE has grown considerably and is still being expanded upon, most recently because of aims set out in the European Green Deal and Circular Economy Action Plans. The EU is seen as having the most advanced and progressive

Household WEEE collection compared to Household EEE put to market			
	2011	**2017**	**Delta**
Household EEE put to market per capita in the EU.	17.36 kg	20.27 kg	2.91 kg
Household WEEE collected per capita in the EU.	5.88 kg	7.37 kg	1.49 kg
Percentage of household WEEE collected relative to the amount of household EEE put to market in the same year in the EU.	29%	36%	7%
Household WEEE per capita that remains uncollected in the EU.	11.48 kg	12.9 kg	1.42 kg
	2014	**2019**	**Delta**
Household EEE put to market per capita globally.	6.4 kg	7.3 kg	0.9 kg
Household WEEE collected per capita globally.	1.1 kg	1.3 kg	0.2 kg
Percentage of household WEEE collected relative to the amount of household EEE put to market in the same year globally.	17%	17.4%	0.4%
Household WEEE per capita that remains uncollected globally.	5.3 kg	6 kg	0.7 kg

Table 1.
Both the amount of household EEE put to market and the amount of household WEEE collected is rising in the EU27, as well as the relative amount of household WEEE when put to the amount of household EEE put to market in the same year. However, the uptake in household WEEE collected is lower than the uptake in household EEE put to market, causing a growing amount of uncollected WEEE based on data retrieved from the Eurostat env_waselee database on WEEE (http://ec.europa.eu/eurostat/product?code = env_waselee&language = en). A similar trend is seen on the global level where WEEE collection cannot keep up with the growth of the EEE industry based on Data from the Global e-Waste Monitor 2020 [9].

regulations in the world and its regulations are often followed by other countries in the world [4]. In this section, a background is provided on existing and proposed EU regulations that directly or indirectly affect the role of consumers in *e*CLSCs, so as to provide the state-of-the-art and future scenarios for WEEE regulation. Furthermore, recent trends in household WEEE collection in the EU are presented and compared to the global trend.

2.1 WEEE Directive

The EU's WEEE Directive, which first came into force in 2003, caused a paradigm shift by placing the responsibility for WEEE on the producers and distributors, instead of solely on municipalities, the so-called Extended Producer Responsibility (EPR) [16, 17]. At its first implementation, besides shifting responsibility to producers, the EU also set WEEE collection targets for its member states [18]. At first the collection target for member states was set at 4 kg of household WEEE per capita or the same weight as the average of collected WEEE in the previous three years by a member state, whichever is higher. In later revisions of the WEEE directive [19], the target collection rates were increased to 45% of EEE put to market (taking the average in the three preceding years) for 2016 till 2018. Since 2019, either 65% of EEE put on the market must be collected or 85% of the WEEE generated within a member state. Furthermore, for each category of household WEEE, separate recycle and reuse target rates are given by the WEEE directive, putting extra pressure to facilitate collection on for example large household appliances or gas discharge lamps.

As previously mentioned, under the directive the producers or distributors of EEE are responsible for the collection of resulting WEEE. This includes a right of consumers to return WEEE, free of charge, to where they originally bought a product [19]. EEE producers and distributors often opt to join a collaborative take-back system to which they pay a fee, based on the average processing costs of a product group, which will handle the management of WEEE for companies [3, 20]. EPR policy is often badly implemented, and municipalities still play an important and costly role in the collection of WEEE, in part because consumers are often not knowledgeable in the options available to them [21].

2.2 RoHS & REACH

The EU's RoHS Directive was the first directive to restrict the use of specific hazardous substances in EEE for many types of products, including most consumer products, and is often seen as a supplement to the WEEE Directive, first instuted in 2004 [22]. The directive has later been updated to provide clarity [23], and to add new substances to the list of restrictions [24]. More additions are currently being considered, such as brominated flame retardants, chlorinated flame retardants, and PVC. Other countries have since followed with similar regulations. Sometimes, as is the case with the China RoHS, the regulation does not prohibit the listed hazardous substances but require products to be marked for having none or minimal amounts of certain hazardous substances. Furthermore, because of the size of the EU's market, products offered globally are often compliant with the EU's RoHS regulation, even if offered outside of the EU. Products exported from the EU also have to comply with RoHS regulation.

The EU's REACH regulation also limits to use of hazardous substances but extends beyond the EEE sector, covering 209 different substances [25]. While similar to the RoHS directive, its implementation differs a lot. Besides covering all products imported or produced in the EU and not just EEE products, being a regulation instead of a directive means the law is the same across every member state. Furthermore, while the RoHS directive specifically lists the substances to which it pertains, the REACH regulation refers to an external list that is easier to update, the substance of very high concern list maintained by the European Chemicals Agency. When listed, authorization is required to include said substance in a product on the EU market, customers and consumers also have the right to request a safety data sheet and to be provided information regarding the safe use and disposal of the product.

2.3 Battery Directive

The EU's Battery Directive affects both batteries and accumulators, which are seen as separate from EEE in regulations but are often included in EEE. It limits the use of specific chemicals, requires proper waste management pertaining to the recycling and collection of batteries and accumulators, sets collection targets for batteries and assigns financial responsibility. As of this moment the annual collection target for batteries is 45% of annual battery sales.

In December 2020 the European Commission proposed a new batteries regulation to replace the old Battery Directive and would come into effect in 2025 if accepted [26]. The proposed regulation aims to make batteries and accumulators used more sustainable. The repurposing of used batteries is heavily encouraged, a new collection target of 65-70% is proposed for portable batteries, a new reporting system for electric vehicle and industrial with a new collection target (yet to be set) would come into effect, lead-acid and li-ion batteries would get a new recycling efficiency target as well as new material recovery rates for cobalt, nickel, lithium, copper, and lead, a carbon footprint declaration is to be provided, minimal performance and durability requirements are set, non-rechargeable batteries are to be phased out, clear EPR specifications are to be developed, more obligations on removability are given for the product design phase, an electronic information exchange system and product passport scheme must be implemented, and a mandatory supply chain due diligence must come into effect. As of now, EU member states are still debating the approval of the proposed regulation[1].

2.4 Eco-design Directive, energy labelling regulation, and the EU ecolabel

The EU's Eco-design Directive is different from most directives in that it is a framework directive, it does net set any standards in itself but provides a framework for regulations that now cover over 30 product groups, mostly within in the EEE sector[2,3] [27]. While most legislations under the Eco-design Directive are mandatory, some are voluntary agreements. Most of the requirements stemming from these legislations regard the sustainability of products and can cover a wide range of rules, such as material requirements, ease of disassembly, the availability of service manuals and spare parts, cleaner production processes, and (further) rules on EPR.

The EU's Energy Labelling Regulation, which came into force this year and replaced Directive 2010/30/EU, requires products from certain product groups to carry energy labels on their packaging [28]. The label assigns an energy class to products so consumers can easily compare products on their energy consumption as well as some other product-group specific information. Labels also carry a QR code so consumers can easily find more information regarding a product in the European Product Registry for Energy Labelling database.

The EU's Ecolabel Regulation, most recently updated in 2009, provides a voluntary scheme so producers can market their products as best of class [29]. Ecolabel requirements are set up for specific product groups and aim to identify the top 10% most sustainable products within a product group. The Ecolabel certification of a product often involves performing a life cycle assessment and life cycle costing analysis. The development of requirements is an ongoing process performed by the European Commission's Product Bureau[4].

The eco-design, energy labelling, and Ecolabel regulations often set high voluntary standards. These standards provide a way for producers to differentiate their products. Furthermore, they are often used as requirements in green public procurement within the EU in addition to other product group specific criteria[5]. As public procurement forms 14% of the EU's market purchases, green public procurement has become a key policy instrument for the European Green Deal and the EU's Circular Economy Action Plans [2, 28, 29].

2.5 WEEE collection trends in the EU

As of 2017, only 2 EU countries managed to reach the 65% household WEEE collection target, of which one country (Bulgaria) only achieved it by using a different definition in their reporting according to the Eurostat env_waselee

---

1 https://www.euractiv.com/section/batteries/news/brussels-in-balancing-act-to-gain-eu-support-for-battery-regulation/

2 https://ec.europa.eu/energy/topics/energy-efficiency/energy-efficient-products/list-regulations-product-groups-energy-efficient-products_en

3 https://ec.europa.eu/info/energy-climate-change-environment/standards-tools-and-labels/products-labelling-rules-and-requirements/energy-label-and-ecodesign/energy-efficient-products/voluntary-agreements-under-eco-design-legislation_en

4 https://susproc.jrc.ec.europa.eu/product-bureau/product-groups

5 https://ec.europa.eu/environment/gpp/eu_gpp_criteria_en.htm

database on WEEE[6]. 18 out of the EU27 countries, as well as the UK, did reach the 45% collection rate target that was at that time still the requirement. Over time, the amount of household WEEE collected per capita, within the EU27, has gone up from 5.88 kg in 2011 to 7.37 kg per capita in 2017 according to Eurostat estimates, a total increase of 1.49 kg, as is shown in **Table 1**. While this means a higher amount of household EEE put to market has been collected as WEEE, the total amount of household WEEE that is not collected has also increased. In 2011, 17.36 kg of household EEE was put to market, while in 2017 20.27 kg of household EEE was put to market, an increase of 2.91 kg, nearly double amount of the WEEE that was additionally collected in 2017 over 2011, as is shown in **Table 1**. Thus, the amount of uncollected WEEE in the EU is still rising year on year. Take note that for the collection targets under the WEEE directive, the rate of household WEEE collected in a year is set against the average amount of EEE in the preceding three years, as the amount of EEE put to market is rising this results in a lower percentage in our calculations compared to those used by the EU. Nonetheless, globally Europe performs as the best region in WEEE collection, both in absolute and relative metrics [9]. Globally, the amount of WEEE generated has grown from 44.4 Megaton in 2014 to 53.6 Megaton in 2019 and is expected to grow to 74.7 Megaton in 2030, meanwhile the amount WEEE collected has only grown from 7.5 Megaton (17%) in 2014 to 9.3 Megaton (17.4%) in 2019, meaning that currently 44.3 Megaton (82.6%) of household WEEE globally ends up dumped or in the informal recycling sector [9].

3. Consumers in a closed loop supply chain for EEE (*e*CLSC)

Considering the timeliness of understanding consumer behaviour in *e*CLSC, in this section, we aim to contribute to the said literature by (1) providing an overview of the topic, (2) building a holistic conceptualisation of the role of the consumer in *e*CLSC and (3) developing a consumer-behaviour-focused taxonomy of EEE categories.

Consumers are not only the engine that can fuel the success of *e*CLSC through their behaviour and decision making, they are also one of the core receptors of the health, environmental and even ethical benefits that *e*CLSC can help achieve. The case of WEEE magnifies this relationship given the widely observed tendency of consumers to stockpile (i.e. hoard) replaced equipment, even in cases when the products are broken, due to a combination of overvaluing discarded EEE and a lack of WEEE management infrastructure. Given the typical hazardous and toxic elements contained in most EEE, this tendency constitutes an important health risk, directly for them, and an environmental risk to ecological systems and society. As such, it is not surprising that EEE stockpiling has been reported to score among the top urgent and important issues within the context of WEEE management [4].

Addressing this issue can be viewed as an opportunity to progress in other global issues, such the Sustainable Development Goals (SDG), by enhancing proper WEEE recovery and reuse at end-of-use (EoU)[7], particularly on the consumers' side. The latter has also been identified as a key challenge for WEEE management [4] and its connection to SDG highlights the potential environmental and social benefits of improving recovery of discarded EEE from consumers. Moreover, since there is space for improvement in the retrieval of discarded EEE from consumers in Europe,

---

[6] http://ec.europa.eu/eurostat/product?code = env_waselee&language = en.

[7] By end-of-use we mean the time when some EEE goes from being in-use, to being out-of-use. Conversely, end-of-life is taken to identify the moment when the EEE in question ceases to function properly, leading to its end-of-use.

the stock of WEEE in urban environments is currently underexploited. Along these lines, urban mining, which relates to the recovery of materials from discarded EEE, could improve significantly if consumer behaviour shifts towards the extension of products', parts' and materials' useful lifetimes.

Stockpiling of EEE by consumers is envisaged to worsen as more appliances become smart and are given internet-access capabilities. These raise data protection considerations which, if not studied and communicated correctly, can pose a further barrier for consumer returns of WEEE due to fear of personal data breaches. Additional ethical considerations are attached to the issue of increasing technical obsolescence, whereby equipment becomes functionally impaired, due to the fast technological advancements and marketing which continuously contributes to the creation of new consumer needs. This leads to the premature replacement of EEE which is also more likely to be overvalued and stockpiled. These examples highlight the fast pace at which consumer behaviour evolves with technological advancement.

Finally, empirical findings suggest that the same theoretical models applied to different EEE categories have substantially different results, e.g. [30, 31]. Despite the important implications of these findings, there are no studies synthesising the characteristics of EEE categories that account for substantial differences in consumer behaviour. Specifically, there is no framework for the systematic differentiation between EEE categories with respect to their physical, functional and symbolic characteristics' influences on consumer behaviour.

Therefore, in this section, we aim to answer the following research question: What EEE characteristics are most significant in pre-conditioning *e*CLSC-relevant consumer behaviour and the processes behind it?

3.1 Materials and methods

3.1.1 Literature review methods

We conducted a keyword search following the PRISMA framework [32] for reporting systematic literature reviews (up to article selection) which is detailed in the *Supplementary Materials* document[8] (see **Appendix**). The identified sources were critically analysed to identify significant research gaps. A first worthy observation is the low number of sources we were able to identify, with only 40 articles being identified after removal of the duplicates from both utilised databases, Scopus and Web of Science. After exclusion of articles based, first, on screening of the abstracts, and second, on the content of the full articles, the final dataset was comprised of 24 studies, based on whether the consumer behavioural role was central or not, to their analyses. Additionally, a second dataset, built under less strict conditions, i.e. including articles whose main focus is not understanding consumer behaviour in *e*CLSC. The latter dataset was comprised of 31 articles. Moreover, the oldest study our search was able to capture was from 2013 [33]. These results alone illustrate well the fact that consumer behaviour in the context of *e*CLSC is still a nascent field.

3.1.2 Taxonomy building methods

Research has found some confusion around the definitions of typology and taxonomy. While the main differences emerge from the approach used, i.e.

[8] The *Supplementary Materials* document is available upon request from the authors.

inductive (empirical-to-conceptual; objects to dimensions/characteristics) vs. deductive (conceptual-to-empirical; theory to dimensions/characteristics), they have often been used interchangeably [34]. We employ the term taxonomy more generally to refer to the classification system, without requiring that it is constructed only inductively or deductively, as it has been suggested in the literature [34]. Specifically, we use the following definition: "A taxonomy T is a set of n dimensions D_i ($i = 1, \ldots, n$) each consisting of k_i ($k_i \geq 2$) mutually exclusive and collectively exhaustive characteristics C_{ij} ($j = 1, \ldots, k_i$) such that each object under consideration has one and only one C_{ij} for each D_i" [34] (p. 340). That is:

$$T_i = \{D_i, i = 1, \ldots, n | D_i = \{C_{ij}, j = 1, \ldots, k_i; k_i \geq 2\}\}. \tag{1}$$

In **Section 3.3**, we develop a taxonomy of EEE categories which aims to facilitate the identification of potential EEE-characteristic-related changes in behaviours that are relevant to *e*CLSC. We draw from [34], to employ robust and transparent methods in the construction of the taxonomy. The resulting taxonomy is presented in **Table 2** and discussed in more depth in Section 3.3. We offer a more detailed account of the procedure we applied, which went through five iterations before reaching its final version, in the *Supplementary Materials* file (see **Appendix**).

3.2 Conceptualisation and literature review

3.2.1 The consumer's role in CLSC

While previous research tends to consider consumers' involvement in CLSC in terms of consumption phases (i.e. purchase, use, lifetime extension, dispose etc.) [35], in addition to this, we take a more direct approach to exploring the extant interactions. We begin by considering how consumers' decision making may influence CLSC implementation. This leads to the distinction between direct influences, where consumers decide on whether to engage or not as suppliers of discarded equipment, and indirect influences, where consumers make decisions based on addressing their functional and emotional needs by acquiring some EEE or repairing/upgrading already owned equipment that no longer fulfils the consumers' expectations. In other words, direct-influence behaviours (DIB) have to do with the supply or lack thereof of discarded EEE in initiatives which aim to re-introduce out-of-use equipment back into the economy, a.k.a. *reversing* behaviours, while indirect-influence behaviours (IIB) pertain to choices that affect the sales volumes of outputs from CLSC that aim to create value. We provide a more in-depth description of our conceptualisation in the *Supplementary Materials* (see **Appendix**). For our purposes, we present the resulting classification of behaviours in **Tables 3** and **4**.

The tables show three levels (layers) of behaviours with each level breaking down its preceding levels into several other, more concrete, behaviours (as numbering increases). Behaviours within the same level are mutually exclusive, such that adoption of one implies non-adoption of the others (see *Supplementary Materials* and **Appendix**). Therefore, the benefits associated with behaviours within a given level, that are not adopted, form the opportunity cost of the adopted behaviour. The point we aim to highlight is that all these behavioural layers, based on the situations that lead consumers to engage or not with an *e*CLSC, are of interest to consumer research in this area. Some studies may explore more general behaviours, such as discarding [33, 36], or specific behaviours such as the purchase of remanufactured/recycled products, see [35] for an overview. In doing so, results offer more, or less, abstract perspectives on behaviours which may lead to different results. Here, our classification provides a basis for understanding findings on

EEE category	Size				Involvement		Long-term reliability expectations			Value type			Internet access		Multifunctionality			Outdated		Social meaning (status, etc.) identity etc.)		
	S	M	L	XL	L	H	L	M	H	H	M	F	N	Y	L	M	H	N	Y	N	M	HS
Washing machine				X		X			X			X	X		X			X		X		
Coffee machine		X			X			X				X	X		X			X		X		
Mobile phone	X				X			X			X		X			X			X		X	
Smartphones	X					X		X			X			X			X	X				X
LCD-TV			X			X		X			X		X			X		X				X
Hair dryer	X				X		X					X	X		X			X		X		
DVD player		X			X			X		X			X		X				X	X		
Electric shaver	X				X			X				X	X		X			X		X		
Radio receiver		X			X			X		X			X		X				X	X		
Refrigerator				X		X			X			X	X			X		X		X		
Smart TV			X			X		X			X			X			X	X			X	
Electric iron	X				X		X					X	X		X			X		X		
VCR		X			X			X		X			X		X				X	X		
CRT-TV			X		X			X		X			X		X				X	X		
Desktop comp.			X			X			X		X			X			X	X			X	
Laptop		X				X		X			X			X			X	X				X
Flat irons	X				X		X					X	X		X			X		X		
Blender		X			X		X					X	X		X			X		X		
Stereo system			X			X			X	X			X		X			X				X
Printer			X		X			X				X	X		X			X		X		

EEE category	Size				Involvement		Long-term reliability expectations			Value type			Internet access		Multifunctionality			Outdated		Social meaning (status, etc.) identity etc.)		
	S	M	L	XL	L	H	L	M	H	H	M	F	N	Y	L	M	H	N	Y	N	M	HS
MP3/MP4 player	X				X			X		X			X			X			X		X	
Electric fan		X			X		X					X	X		X			X		X		
Smart MP3/MP4	X				X			X			X			X			X	X			X	
Smart printer		X				X		X			X			X		X		X			X	
Lamp		X			X		X					X	X		X			X		X		
Air conditioner			X			X			X			X	X		X			X		X		
Batteries	X				X		X					X	X		X			X		X		
Vacuum cleaner		X				X		X				X	X		X			X		X		
Microwave		X			X		X					X	X		X			X		X		

Table 2.
Taxonomy iteration 5. Size(Small, Medium, Large), Involvement(Low, High), Long-term reliability expectations(Low, Moderate, High), Value type (Hedonic, Mixed, Functional), Internet access(No, Yes), Multifunctionality (Low, Moderate, High), Social meaning (Negligible, Moderate, Highly significant).

Direct-influence behaviours (DIB)		
Layer 1	**Layer 2**	**Layer 3**
Discard	Reuse	Second-hand selling. Donating/Leasing.
	Recycle	Selling as scrap. Return for incentives. Disassembly and reuse of parts. ...
	Disposal in ordinary waste	
Stockpiling (out-of-use storage)		

Table 3.
By direct-influence behaviours (DIB) we refer to those that stem from decisions about whether to dispose of, and how to dispose of, equipment that is no longer in-use. The behaviours in the table are presented in three levels, and behaviours from the same level are mutually exclusive since they represent the options between which consumers may choose. In layer 1, consumers decide to either discard or stockpile. In layer 2, if they choose to discard, they may choose between reversing behaviours (reuse, recycle etc.) or discard in the ordinary waste routes. Finally, layer 3 provides some examples of reuse- and recycle-based reversing behaviours. See Supplementary Materials document, available upon request from the authors.

Indirect-influence behaviours (IIB)		
Layer 1	**Layer 2**	**Layer 3**
Acquisition, previously unowned EEE category	Alternative to new EEE purchase	Second-hand purchasing. Borrowing/leasing/renting. Remanufactured/recycled product purchasing. ...
	New product purchase	
Replacement of already owned	Alternative to new EEE purchase	Second-hand purchasing. Borrowing/leasing/renting. Remanufactured/recycled product purchasing. ...
	New product purchase	
Repair/upgrade		

Table 4.
By indirect-influence behaviours (IIB) we refer to those that stem from decisions about how an underfilled need should be addressed. The behaviours are presented in three levels, and behaviours from the same level are mutually exclusive since they represent the options between which consumers may choose. In layer 1, consumers decide to either acquire some product from a previously unowned EEE category, to replace some EEE they already own or to repair/upgrade some already owned EEE. In layer 2, provided that the consumer chooses one of the two former options, consumers may acquire a new product or engage in some behavioural alternative. Finally, layer 3 provides some examples of behaviours alternative to the purchase of new products. See Supplementary Materials document, available upon request from the authors.

concrete behaviours in relation to their higher-level (abstract) analogues, eventually allowing for between-level comparisons with respect to changes in EEE categories and institutional and cultural contexts.

Additionally, an important gap exists in the literature with respect to explorations into similar mechanisms that may occur in different consumption phases. For example, there are no empirical considerations of whether the emotional reasons for which consumers replace their EEE prematurely, i.e. before its end-of-life (EoL), to which marketing strategies tend to appeal, are partly the same as the reasons behind consumers' attachment to the replaced products, which eventually leads to overestimation of their value and subsequent stockpiling. Some studies suggest that this may be the case, while some others seem to assume that stockpiling tends to occur for overestimation of future functional needs. Our conceptualisation is better suited to mixed context considerations such as this one, since it groups all behaviours into two unambiguous groups, highlighting the potential interactions between behavioural contexts that may otherwise seem unrelated.

3.2.2 The consumer behaviour perspective: drivers and barriers

Two distinguishable research foci can be identified in the literature: Consumer behaviour regarding the sales volume of secondary and primary EEE, as an engine for the success of *e*CLSC [37–39], and consumer *reversing* and out-of-use storage (stockpiling) behaviours (see **Tables 3** and **4**). Recognising this, a recent systematic literature review [35], aimed to synthesise extant findings from the consumer behaviour literature on determinants of purchase, extension of life and EoU/EoL management of EEE. In contrast, our keyword search was restricted to the CLSC and reverse logistics (RL) literature. Consequently, the studies identified hereby focused primarily on the consumer's role regarding WEEE management, such as recycling and reversing behaviours, as shown in **Tables 5** and **6**. Surprisingly, while [35] identify some additional studies focusing on purchasing and lifetime extension behaviours, our search was able to identify further sources regarding disposal and reversing behaviours. This highlights the need for consumer behaviour research on all the phases of consumption of EEE to align etymologically with research on CLSC, RL and CE, to explicitly account for the importance of all the roles of consumers in relation to *e*CLSC. This result indicates that consumer behaviour research on purchase and lifetime extension of EEE is not fully aware of its importance to the context of *e*CLSC.

Despite these differences regarding the sample of articles identified, our findings in terms of theoretical frameworks and behavioural predictors are in coherence with [35]. In particular, as we depict in **Tables 5** and **6**, the identified literature is dominated by theoretical frameworks that are built around the Theory of Planned Behaviour (TPB). While the results support the TPB's utility in understanding the influences of some cognitive factors behind behaviour, they do come with some

Study	Dependent construct	Main predictors
[36][a]	Discharge behaviour	Subjective norms, perceived behavioural control, size of EEE
[40][a]	Willingness to recycle online	Perceived behavioural control, subjective norms, attitudes, economic motivation
[41]	Disposal for reuse	Barriers: attachment and frugality. Drivers: infrequent use and emotional reward.
[42][a]	Intention to recycle	Attitude, perceived behavioural control, subjective norm and individual responsibility. In turn, sense of duty found to predict attitudes.
[43]	Recycling behaviour	Attitudes, moral norms, awareness of consequences, perceived convenience.
[44]	Acceptability of incentive schemes for take-back	Information conditions, education, gender, age, income. Type of product found to influence usage habits

[a]*The study uses (extended) Theory of Planned Behaviour as a theoretical framework.*

Table 5.
Target construct and identified predictors for studies exploring no specific WEEE categories. The "Dependent construct" column contains the construct of interest for each study, be it behavioural, attitudinal or otherwise. The "Main predictors" column contains the main predictors of the dependent construct, identified by each study.

Study	EEE category	Dependent construct	Main predictors
[31][a]	ICT	Reversing behaviour	Attitudes, moral norms, perceived behavioural control, subjective norms.
[30][a]	Mobile phone	Reversing behaviour	Attitudes, moral norms, perceived behavioural control, subjective norms.
[45]	Mobile phone, washing machine, coffee machine	Preferences for collection services	Price, reluctance for incineration, nature of product type
[39]	Smartphone	Recycling behaviour	Incentives, convenience, information security.
[46][a]	Mobile phone	Intention to recycle	Attitude, perceived behavioural control, subjective norm. In turn, perceived benefits and sense of duty found to predict attitudes.
[47][a]	Mobile phone	Returning to companies	Attitude, perceived behavioural control, subjective norm.
[48][a]	Mobile phone	Recycling behaviour	Barriers: lack of proximity and convenience of waste management systems.
[33]	Mobile phone	Post-consumption disposal	Age, gender, income, place of residence.

[a]*The study uses (extended) Theory of Planned Behaviour as a theoretical framework.*

Table 6.
Target construct and identified predictors for studies exploring specific WEEE categories explicitly. The "EEE category" column displays the product category addressed by the study. The "Dependent construct" column contains the construct of interest for each study, be it behavioural, attitudinal or otherwise. The "Main predictors" column contains the main predictors of the dependent construct identified by each study.

limitations. These include the tendency to focus on the formation of intentions under the assumption that these will be strong predictors of actual behaviour, e.g. [42, 46], and the lack of space for the identification of new processes, mechanisms and factors which can be relevant to behavioural outcomes, such as institutional context, personality traits, habitual behaviours or cultural dispositions, among others. The assumption that reported intentions reflect actual behaviour has been identified as an important limitation in the wider sustainable and ethical consumer behaviour literature due to a phenomenon known as the intention-behaviour gap, whereby consumers over-report on their intentions and attitudes when compared to observations of how much they really adopt sustainable-ethical behaviours [49, 50]. Finally, the lack of heterogeneity on theoretical (and methodological) frameworks results in a very good understanding of a very small portion of the plurality of mechanisms, processes and factors that determine consumer behaviour. Therefore, there is a need for theoretical and conceptual innovation regarding consumer behaviour for *e*CLSC, particularly across EEE categories and characteristics as we illustrate in the following sections.

In **Table 5**, we include the articles that do not focus on behaviours regarding no specific EEE category. Conversely, **Table 6** contains the identified studies that explore behaviours focusing on specific EEE categories. In both tables each study is presented with the main construct of interest and the main identified predictors, while in **Table 6**, the EEE categories studied are also specified. A quick inspection of these two tables provides some interesting insights. Specifically, the literature at the intersection between consumer behaviour and *e*CLSC, is dominated by studies on mobile phones specifically (6 studies) and on no specific WEEE category (6 studies). One article was also found to consider ICT equipment more widely [31], another study looked at smartphones more specifically [39], and only one article considered more than one EEE category explicitly [45].

Moreover, in the context of our discussion of different levels of behaviours (see **Tables 3** and **4**), we found most studies to fall within levels 2 and 3 as shown in **Tables 5** and **6**, and only one study considered post-consumption disposal behaviours more generally, but did so by breaking them down into some lower level, less abstract, behaviours [33]. As such, the research literature offers much space for improvement in this sense.

In conclusion, there is a need for innovation regarding the theoretical frameworks employed to enrich academic knowledge on consumer behaviour relevant to *e*CLSC, beyond the constructs hypothesised by TPB and similar frameworks. Moreover, results from studies focusing on the formation of intentions should be interpreted with care due to their susceptibility to biases. There is a need for empirical research to explore consumer behaviours pertaining to a wider variety of behavioural layers and specific EEE categories to establish the fundamental behavioural differences across categories of varying characteristics, the different perspectives that can be achieved through framings of varying abstraction, and the reasons behind them. This is further highlighted below.

3.2.3 EEE characteristics and pre-conditioning of consumer behaviour

Studies have used consumer surveys to profile the purchasing, usage, stockpiling, replacing and discarding of EEE in different national and sub-national settings (e.g. [51, 52]). For instance, [51] conducted a survey, distributed among a representative sample (n = 395) of households in Sao Paolo, Brazil. The study considers all the stages of consumption and, given the large variety of product categories that fall within the scope of EEE, the authors account and collect data for 26 separate EEE categories. Not surprisingly, their results vary significantly from one EEE category to another, as well as in comparison to similar results from different geographical contexts.

When considering in-use EEE, they found that each household had, on average, 17 items, where mobile phones and cathode-ray-tube TVs (CRT-TV) lead the ranking with about 2 per household. However, a closer look at how long the equipment had been owned, revealed that 60% of the mobile phones had been owned for less than 2 years at the time of reporting, while almost 85% of CRT-TVs had been owned for longer than 2 years, and about 25% had been owned for over 10 years. The latter, however, was true only for a negligible percentage of in-use mobile phones. Similar results are obtained by [44] from a sample of Portuguese consumers, where LCD-TVs are found to be typically in use for more than 10 years, while mobile phones were in use typically for 3-5 years. Given that CRT-TVs were found to have been replaced by LCD-TVs, since only 16% of the former were less than 2 years old while this applied to 74% of LCD-TVs [51], the differences and parallels of these studies likely indicate that CRT-TVs had already been replaced a lot earlier in Portugal. As illustrated by these examples, there are significant differences in consumer purchasing and use behaviour between different EEE categories, which persist across social contexts.

When considering out-of-use EEE that was being stored, i.e. stockpiling, the study finds these to represent about 12% of the total amount of EEE that is present in the surveyed households. The authors attribute this finding to the "treasure effect", a phenomenon whereby consumers tend to overvalue out-of-use products, and consequently do not discard them but instead keep them in a drawer or storage room over the belief that at some point in the future it will be needed and used again. This idea was supported by their findings which reveal that more than 50% of the out-of-use EEE was fully functional for the majority of EEE. However, for washing machines, microwave ovens, electric drills and DVD players, over 50% of the out-of-use EEE is functionally damaged. Therefore, their findings, when it comes to stockpiling behaviour, are significantly different for different EEE categories.

The authors find similar results concerning the acquisition of EEE from different routes (second hand, new ...), reasons for acquiring the EEE and disposal routes (reuse, recycling ...). Furthermore, [36] conduct a survey in South Korea (n = 2000) where they focus on identifying the current state of adoption of WEEE disposal behaviours and the cognitive factors behind it. The study too finds that products of different characteristics lead to contrasting results and dedicates one section to exploring how their results vary between EEE categories of varying sizes (small, medium and large). Namely, their findings suggest, among other differences, that while take-back initiatives only represent 10.24% of the disposition routes for small and medium sized appliances, it is the leading route for large EEE as it represents 34.5% of the total. However, the authors do not go beyond these characteristics as it is not the focus of their study.

As illustrated by the examples provided above, when considering consumer behaviour, the EEE category under study plays an important role in determining the needs and wants of the consumer, and hence their behaviour. Additionally, very significant differences emerge in similar studies in different national contexts due to important institutional and cultural differences. However, this does not change the fact that there are product characteristics which fundamentally influence consumer perception and behaviour in their decision-making process [45, 53].

Despite that, [36]'s attempt to explicitly assess the behavioural differences among different EEE clusters, namely by size, is one of the only available accounts. Other interesting findings in support of the importance of considering the types of EEE categories include [45]'s article which finds product type to be one of the most critical factors in determining behavioural differences. However, the study fails to unveil what the factors are that differentiate the EEE categories they explore, providing little information on how these EEE characteristics may be operating in leading to different behavioural outcomes. Finally, [30, 31] conduct two studies with the same sample (and dataset) but targeting ICT products in general and mobile phones, respectively (see **Table 4**). Both studies use the TPB to explore the adoption of recycling behaviours for ICT equipment and mobile phones separately. The results change substantially for the behavioural adoption, as intentions are only able to explain about 9% of the variation in the case of mobile phones, in contrast to 15% for ICT in general, while the opposite happens in the case of intentions to recycle, of which about 36% of the variation can be explained for mobile phones, but only 30% is captured in the general case. Additionally, when comparing structural equation modelling results with artificial neural network outcomes, the results agreed in the case of mobile phones, but in the case of general ICT equipment, the two analyses result in differences regarding the predictors' significances relative to one another. This further illustrates the idea that different EEE categories may better fit certain theoretical and methodological frameworks than others.

While all this evidence highlights the importance of developing an understanding of the EEE category characteristics that influence consumer behaviour and the reasons behind these influences, the literature is currently missing a comprehensive framework through which to do so. As [45] conclude, there is a need "[...] for the refinement of EEE classifications used for collection operations to encompass consumers' preferences, and not recycling requirements only". Moreover, we further extend their claim to all consumer-behavioural aspects of *e*CLSC, such as acquiring, using and storing, besides the already mentioned disposal.

3.3 A consumer behaviour focused taxonomy of EEE categories

We begin by stating the intended use of our taxonomy which is the first step in taxonomy building [34]. The intended users of our taxonomy include researchers in the area of WEEE more generally, as well as more specifically consumer researchers in this area. In particular, while there are increasing accounts of behavioural differences among EEE categories, e.g. [51, 54], there are no frameworks that synthesise the main characteristics that may lead to such differences. Our taxonomy aims to provide a foundation on which to build subsequent knowledge regarding the behavioural differences that arise between EEE categories of differing characteristics and that are critical for *e*CLSC success.

Our taxonomy's dimensions and characteristics (see **Section 3.1.2**) are drawn from empirical and conceptual observations in the literature. However, one of the basic goals of a taxonomy is to be easily extendible [34]. In this sense, we provide hereby a starting point on which to further extend our knowledge regarding the main EEE characteristics that lead to behavioural differences among consumers. To construct the taxonomy, we used an initial pool of 29 EEE categories to test the relevance of the taxonomy. The method and the classification of the initial sample of EEE categories is presented in full detail in the *Supplementary Materials* to this chapter, see **Appendix**. The final taxonomy, which we present in **Table 2**, reads as follows:

T_4 = {*Size* (Small; Medium; Large; Extra-large); *Involvement* (Low; High); *Long-term reliability expectations* (Low; Moderate; High); *Value type* (Hedonic; Mixed; Functional); *Internet access* (No; Yes); *Multifunctionality* (Low; Moderate; High); *Outdated* (No; Yes); *Social meaning* (Negligible; Moderate; Highly significant)}.

Next, we provide an explanation of each dimension and their characteristics together with some of the potential behavioural differences that may be expected.

3.3.1 Size

The *Size* dimension contains three characteristics: small (e.g. mobile phones, smartphones, hair drier, electric shaver ...), medium (e.g. coffee machine, DVD player, radio receiver, VCR, laptop ...), large (e.g. LCD-TV, desktop computer, stereo system ...) and extra-large (e.g. washing machine, refrigerator ...). We conceptualise it as having four characteristics since we found the typical *small-medium-large* system to be ambiguous when trying to classify our pool of items at the taxonomy building stage. This has interesting implications for consumer research aiming to elicit attitudinal or other differences in consumer perceptions through self-reports. In particular, through disambiguation, grouping EEE categories in four clusters of size, rather than three, could improve discriminant validity of the studies. In other words, while the relative differences between EEE sizes decrease, less of the EEE categories fall within the "boundaries" of the size characteristics, making clustering more natural for respondents of these studies. Size of the EEE is likely to lead to differences in out-of-use storage, use and discarding behaviours [36, 51, 55] of consumers.

3.3.2 Involvement

The *Involvement* dimension contains two characteristics: low (e.g. batteries, hair drier, electric shaver ...) and high (e.g. washing machine, smart TV, air conditioner ...). This dimension can be broken down into two intertwined aspects: **price** and **risk**. Hence, low-involvement EEE categories fall within a lower price range and have less risk associated with their purchase, while high-involvement ones fulfil the opposite. As the name suggests, these differences in what we call involvement, invoke different levels of interest and importance, quantity and type of information required to reach decisions for the consumption process at hand, in other words this captures how involved the decision making is expected to be. Less involved decision making tends to be dominated by price and routine considerations, while high-involvement decision making involve premeditation and information seeking leading to a more conscious sequence of decisions.

3.3.3 Long-term reliability expectations: essentiality of EEE

The *Long-term reliability expectations* dimension contains three characteristics: low (e.g. hair drier, electric iron, blender ...), moderate (e.g. coffee machine, mobile phone, DVD player) and high (e.g. washing machine, refrigerator, air conditioner ...). We identify differences in the assessed EEE categories regarding how essential or necessary they are to consumers' day-to-day lives. Moreover, some products are more susceptible to considerations regarding the expected longevity of the equipment. Quite clearly, these two considerations are intertwined in the sense that consumers are likely to strongly consider longevity when purchasing items that they expect to need to use very often. However, they are not equal, since some, e.g. high involvement or large EEE, could include strong considerations of longevity for other reasons, i.e. without it being an essential piece of equipment to one's everyday life. As such we conceptualise this dimension which compositely considers the importance of **longevity considerations** when acquiring the EEE and how **essential** the equipment is to consumers' day-to-day life.

There is a trivial effect of this dimension on purchasing and use behaviours, since we aim to differentiate between EEE that invokes more considerations for longevity and is more frequently needed/used, by definition. It is worth mentioning that this is similar to the influence discussed for involvement of EEE. Despite that, the behavioural outcomes of interest here differ substantially since they are restricted to involvement regarding considerations of longevity and necessity, while the previous dimension is more general. Moreover, these differences are particularly relevant to considerations about repair, restoring and second-hand markets. This is because having longevity in mind develops some expectations that the consumer will tend to meet. In turn, this results in better kept equipment that may be fit for further reuse in remanufacturing or even second-hand markets.

3.3.4 Value type

The *Value type* dimension contains three characteristics: hedonic (e.g. stereo system, DVD player, MP3/MP4 player ...), mixed (e.g. laptop, smartphone, smart TV ...) and functional (e.g. washing machine, refrigerator, air conditioner ...). In other words, our taxonomy distinguishes between EEE in which hedonic value dominates, those in which functional value dominates and those in which they are both comparable. We refer to fun and entertainment value which consumers associate with some EEE category as hedonic, while we refer to utilitarian, service or use value that consumers extract from using some EEE as functional value. Equipment with hedonic value type is more likely to give rise to emotional responses [56] making them more susceptible to premature replacement (i.e. before EoL) and stockpiling, particularly if they are also large. On the other hand, EEE categories with functional value are likely to be replaced at EoL or repaired [54]. Finally, mixed value EEE categories are likely to give rise to more consumer-dependent responses depending on whether they perceive the hedonic or the functional value of the equipment more.

3.3.5 Internet access

The *Internet access* dimension contains three characteristics: no, i.e. no internet access capabilities, (e.g. washing machine, batteries, lamp ...) and yes, i.e. with internet access, (e.g. laptop, smart TV ...). It is becoming increasingly common for EEE to be upgraded into its smart version, i.e. a version with internet access capabilities. This raises concerns for consumers regarding their personal data. Therefore, consumers are more likely to stockpile EEE with internet access, and hence less likely to return for reuse or recycle, in order to avoid facing the additional cost incurred due to concerns and further involvement that may be viewed as unnecessary. Additionally, the smart versions of EEE categories invoke changes in other dimensions considered in our taxonomy when compared to their conventional counterparts, like size (due to optimization of component sizes), involvement (due to increases in price) and multifunctionality – which emerges naturally from internet access.

3.3.6 Multifunctionality

The *Multifunctionality* dimension contains three characteristics: low (e.g. washing machine, coffee machine, hair drier ...), moderate (e.g. mobile phone, LCD-TV ...) and high (e.g. laptop, smartphone ...). We distinguish between three levels of multifunctionality. In particular, multifunctionality refers to the attribute of offering more than one function. While every product can be considered multifunctional provided some creativity is in place, e.g. a washing machine or a refrigerator doubling up as a table and a notice boar, respectively, this is not what our multifunctionality aims to capture. Conversely, it aims to distinguish between EEE whose intended function is very specific, often singular; such as washing machines, lamps or electric shavers; somewhat multifunctional, most often offering a couple of functions; such as classic mobile phones which can serve for communication but also for reproduction of sounds or taking pictures; and highly multifunctional EEE, which is usually designed to offer a large variety of capabilities resulting in both hedonic and functional values; such as laptops and smartphones.

More multifunctional EEE is less likely to be kept unused, since it fulfils many needs, making it likely to be used frequently. Therefore, equipment that is more single-functional offers a greater opportunity for sharing, leasing, and borrowing initiatives. In addition, there may be interesting implications on stockpiling behaviour since more multifunctional equipment is more susceptible to the thought that it may be needed in the future, even more so for moderately multifunctional EEE which is not necessarily being used frequently in the first place. Similarly, since highly multifunctional EEE tends to offer mixed value (see **Section 3.3.4**) it may also turn out to be more susceptible to stockpiling behaviours and premature replacement.

3.3.7 Outdated EEE

The *Outdated* dimension contains three characteristics: no, i.e. not outdated, (e.g. stereo system, DVD player, MP3/MP4 player ...) and yes, i.e. outdated, (e.g. washing machine, refrigerator, air conditioner ...). We regard as outdated EEE categories, those which have been widely replaced by newer, functionally more advanced EEE. A tangible example is that of classic mobile phones vs. smartphones. In other words, outdated EEE displays great losses in sales in the last years and are often discontinued (but not always). Outdated equipment may be stored out-of-use or in-use due to emotional attachment, as a collectable or vintage item. This type of equipment may also be more susceptible to being discarded in conventional waste routes, with all its health and environmental implications. Finally, since often discontinued, outdated EEE is unlikely to be acquired through the purchase of new EEE but is more likely to enter the second-hand and recycled product markets.

3.3.8 Social meaning (identity and status)

The *Social meaning* dimension contains three characteristics: negligible (e.g. stereo system, DVD player, MP3/MP4 player ...), moderate (e.g. laptop, smartphone, smart TV ...) and highly significant (e.g. washing machine, refrigerator, air conditioner ...). This dimension aims to capture some indication of the level of social meaning typically attached to a given EEE category. Material possessions are often attached to social meaning, in other words, they communicate belonging to a certain social group [57]. For example, cars and houses have been symbols of wealth and success, i.e. social status, for decades. A similar situation can be observed increasingly prominently with EEE, such as mobile phones [54, 58]. It is straightforward to see that this social meaning has little to do with the functionality of the products (see high heels and fast cars, for example), making products that are highly stereotyped and to which a great deal of social meaning is attached, much more susceptible to emotional responses. Such EEE categories are more likely to awake the desire to replace products before their EoL, but also potentially to lead to an overestimation of the replaced equipment [54]. This establishes the level of social meaning attached to a given EEE category as an important indicator of susceptibility to being prematurely replaced and kept out-of-use, hence affecting all consumer behaviour phases.

Not surprisingly, given that the EU's classification system [4] does not consider consumer behaviour but rather operational and material differences, we found their classification to poorly differentiate among EEE categories with respect to consumer behaviour. As shown in **Table 2** in bold, in the last iteration of the taxonomy, we classified items that the EU classification system regards as pertaining to different classes. The classification of these under our taxonomy's dimensions and characteristics revealed that they only differed on three out of the eight proposed dimensions, indicating that with respect to our taxonomy's dimensions and under considerations of consumer behavioural outcomes, the EEE categories were actually very similar. In other words, any behavioural differences could easily be attributed to one of the three dimensions on which they differ, but behavioural differences would be less likely to emerge. This suggests that EU regulation could benefit from the perspective offered by our taxonomy in accounting for the associated differences. Additionally, this supports the utility of our taxonomy as an alternative compass for understanding the differences across EEE categories for consumer behaviour and *e*CLSC research.

4. Discussion

In this chapter, we have provided an overview of the current state of WEEE regulation and collection trends in the EU and analysed and contributed to the consumer behaviour literature in the context of *e*CLSC by building a taxonomy of EEE categories. In this section we offer a discussion of implications, limitations, and avenues for future research.

In Section 2, discussing the EU's household WEEE related regulations, which is considered the state-of-the-art [4], the main implication from our research is that it is not enough. The increase in household WEEE collection falls behind the increase in new EEE put to the market, resulting in an ever-growing stream of WEEE that ends up being dumped or processed in the informal sector. This is not just an issue in the EU, but a global issue as is evidenced in **Table 1** [9]. Considering that Europe is the region most effective in household WEEE collection, it is a suitable object of study when researching regulation and identifying novel ways of growing the urban mining sector. Bad WEEE collection is not just a missed financial opportunity [13], dumped or ill processed WEEE forms serious environmental and health hazards, hitting developing countries especially hard [10]. Based on these findings, we consider it critical that improvements are made in engaging consumers with proper and responsible WEEE collection and the role of consumers in *e*CLSCs should be well understood.

Some limitations of our work are related to the use of systematic keyword search for the consumer behaviour literature on *e*CLSC. In particular, by restricting our search to the context of CLSC, it identified mostly studies about the post-consumption or disposal phase. However, the existence of a recent literature review that identifies studies relating to all phases of consumption [35], reduces the negative impacts of the limitation discussed hereby. This is also supported by the similarity of our findings. Moreover, our search is able to identify a significant number of articles that are not identified by [35], hence offering a complementary view, rather than a biased one. Finally, to reduce potential negative impacts, we employed snowball mapping to identify further studies from the reference lists of the original set of articles.

We conducted a classification of consumer behaviours for *e*CLSC based on the consideration of direct, i.e. pertaining to considerations about equipment that has reached its EoU phase, and indirect, i.e. emerging from considerations about fulfilling some need(s), influences of consumer behaviour on *e*CLSC (see **Tables 3** and **4**). These have highlighted different levels of abstraction with which behaviours may be conceptualised. Future research could consider the concreteness of their conceptualisations to balance between reliability and generalisability of their results. While less concrete conceptualisations may offer less confident real-life predictions on a specific-case basis, results may be applicable to more scenarios. Moreover, abstract conceptualisations of behaviours (e.g. disposal) can be broken down into more concrete ones (e.g. return for incentives, sell second-hand, sell as scrap etc.) based on the objectives of the specific studies. While there are studies employing such techniques, e.g. [33], they do not do so explicitly. We have not encountered studies that also measure the overarching abstract behaviour in addition to its concrete constituents. Therefore, our findings suggest an interesting avenue for future research could be to account for these conceptualisations, of varying levels of abstraction, in order to allow for comparison of the results. In other words, *how do results change between the concrete and the more abstract conceptualisations of behaviours?*

The role of consumer behaviour in *e*CLSC was found to be critical to their success, but poorly understood by the surveyed literature. Specifically, there is a lack of variety concerning theoretical and methodological frameworks resulting in a significant amount of knowledge generation but pertaining to a very specific portion of the many simultaneous factors and processes leading to consumer behaviour. As such, institutional and cross-cultural perspectives, among others, have been overlooked in exchange for repeated applications of frameworks built around the TPB. Moreover, some studies employing this framework tend to assume that findings about behavioural intentions are realistically representative of actual behaviour, which has been widely rejected in the context of sustainable and ethical consumer behaviour. Therefore, we call for careful interpretations of such results.

In relation to methodological and theoretical innovation, we propose that, given the already existing literature on CLSC from a game theoretic perspective, experimental economics [59] may offer an interesting framework for the empirical testing of extant and future models. Additionally, in this proposed direction, future research could focus on modelling and optimising of consumer decisions, on top of all other nodes in the modelled CLSC network. Moreover, the development of further structural models based on institutional and cultural considerations offers a natural continuation to the set of TPB-based studies.

Our review identified a sector of studies aiming to characterise the waste streams of certain national contexts through household surveys. While these studies offer little empirical explanation for the reasons behind observed behaviours, they do identify significant differences between the concrete behaviours, EEE categories and geographical contexts. This highlights the need for future research to work towards the elucidation of the most significant reasons for behavioural differences to emerge across all the aforementioned domains. To this end, we offer a conceptual starting point as discussed below.

In **Section 3.3** we developed a taxonomy of EEE categories by considering how their characteristics may predispose consumer behaviour towards different directions. While there had been some attempts to consider how EEE categories' characteristics tend to predispose consumers in specific behavioural areas [54], to the best of our knowledge, ours is the first attempt at capturing a large pool of dimensions through which to differentiate among EEE relating to all phases of consumption. The main benefits associated with our methods are those of replicability, extendibility and systematicity. Specifically, by following a fully transparent method and reporting on every iteration and decision made leading up to the final taxonomy [34], see Supplementary Materials document (see **Appendix**), we allow for plasticity and advancement of the taxonomy based, for example, on case-specific applications, theoretical and empirical criticism and extension or modification due to increase in data availability.

The classification of the sample of EEE categories used in the construction of the taxonomy was necessarily carried out *ad hoc* in some instances, due to unavailability of product-specific studies and data leading to potential ambiguity or disagreement among users. Therefore, it should be highlighted that the goal is not to set said classification in stone, but rather to provide a conceptually sound starting point for research to build upon as knowledge is generated. Despite its limitations, however, our taxonomy highlights what we have determined to be the most relevant characteristics that predispose consumers to adopt certain behaviours. We expect this to aid the inclusion of dimensions other than size in future comparisons of behavioural outcomes between EEE categories.

The main implications of our taxonomy can be summarised into the following points: (1) There are many physical, functional and symbolic dimensions and characteristics on which EEE categories may differ, leading to differences in behaviour. (2) There is a lack of research explicitly trying to consider and understand these differences, other than superficial explorations of those related to size and price heterogeneity. (3) Our result calls for further research to use, explore, expand and modify our taxonomy such that its value to researchers can increase. (4) There are interesting interconnections between some of the dimensions, which can be understood as moderation effects, which offer interesting avenues for future research. For example, value type may be able to explain some of the tendency to stockpile, but this may apply more strongly to small EEE categories than the other sizes, i.e. a moderation effect from size on the relationship between value type and stockpiling behaviour. (5) The EU would benefit from complementing extant classifications of EEE categories with our taxonomy's perspective in order to explicitly consider how consumers relate to EEE categories of different characteristics.

5. Conclusion

In this chapter, we first developed an understanding of the current state of household WEEE regulation in the EU. The EU currently has the state of the art in household WEEE regulation and achieves the world's highest collection rates. Nonetheless, the EU, and the rest of the world, cannot cope with the growth of the EEE sector, and WEEE collection falls behind, leading to a vast amount of valuable resources that is removed from the formal economy and instead causes further harm to the environment and human health. Based on these findings, which establish the return of consumer products as one of the most underexploited resources when it comes to the sustainable recirculation of products and materials into the economy, we build an understanding of the role of consumers in *e*CLSC. Our analysis establishes that the intersection between the consumer behaviour and the *e*CLSC literatures is still nascent and offers many interesting routes for research. Finally, we develop a taxonomy of EEE categories based on characteristics, physical, functional and symbolic, that predispose consumers to some behaviours. We identify 8 key dimensions: size, involvement (price and risk), long-term reliability expectations (essentiality and requirement for longevity), value type, internet access, multifunctionality, the quality of being outdated and social meaning. The chapter concludes by discussing limitations, which mostly emerge due to the systematic literature search, and how we addressed them. Additionally, implications and avenues for future research, including the necessary methodological and theoretical innovation beyond the TPB framework and survey methodologies are discussed.

Acknowledgements

This research has received funding from the European Union's Horizon 2020 research and innovation programme under the Marie Skłodowska-Curie Innovative Training Networks (H2020-MSCA-ITN-2018) scheme, grant agreement number 814247 (ReTraCE project).

Conflict of interest

The authors declare no conflict of interest.

Nomenclature

CE	Circular Economy
CLSC	Closed Loop Supply Chain(s)
*e*CLSC	Closed Loop Supply Chain(s) for Electrical and Electronic Equipment
EEE	Electrical and Electronic Equipment
EoL	End of Life
EoU	End of Use
EPR	Extended Producer Responsibility
REACH	Registration, Evaluation, Authorisation and Restriction of Chemicals
RL	Reverse Logistics
RoHS	Restriction of Hazardous Substances
WEEE	Waste Electrical and Electronic Equipment

Appendix

Due to space limitations, we wrote a *Supplementary Materials* document that is available upon request from the authors. Its contents can be summarised as follows: First, it provides further insight into our conceptualisation of *e*CLSC-relevant consumer behaviours. Second, it provides complementary details about the literature search and literature identification phase conducted as part of **Section 3** in the chapter. Finally, it offers an in-depth account of our taxonomy development strategy, including explanations for all the decisions and iterations leading to the final taxonomy presented in **Table 2**.

Author details

Dimitris Georgantzis Garcia[1,2*†] and Sven Kevin van Langen[3,4‡]

1 Sheffield University Management School, University of Sheffield, Sheffield, FL, UK

2 South East European Research Centre (SEERC), Thessaloniki, Greece

3 International PhD Programme/UNESCO Chair "Environment, Resources and Sustainable Development", Department of Science and Technology, Parthenope University of Naples, Naples, Italy

4 Supply Chain Department, Olympia Electronics, Eginio, Greece

*Address all correspondence to: dgarcia@seerc.org; dgeorgantzisgarcia1@sheffield.ac.uk

†Dimitris Georgantzis Garcia is the main author of sections 3 & 4, with minor contributions to sections 1 & 2. Both authors contributed equally to Section 5.

‡Sven Kevin van Langen is the main author of sections 1 & 2, with minor contributions to sections 3 & 4. Both authors contributed equally to Section 5.

References

[1] European Commission. A new Circular Economy Action Plan For a cleaner and more competitive Europe. Luxembourg; 2020.

[2] European Commission. The European Green Deal. Brussels; 2019.

[3] Andersen T. A comparative study of national variations of the European WEEE directive: manufacturer's view. Environmental Science and Pollution Research. 2021;1–20.

[4] Shittu OS, Williams ID, Shaw PJ. Global E-waste management: Can WEEE make a difference? A review of e-waste trends, legislation, contemporary issues and future challenges. Waste Management. 2021; 120:549–563.

[5] Buekens A, Yang J. Recycling of WEEE plastics: a review. Journal of Material Cycles and Waste Management. 2014;16(3):415–434.

[6] Anandh G, PrasannaVenkatesan S, Goh M, Mathiyazhagan K. Reuse assessment of WEEE: Systematic review of emerging themes and research directions. Journal of Environmental Management. 2021;287:112335.

[7] Bressanelli G, Saccani N, Pigosso DCA, Perona M. Circular Economy in the WEEE industry: a systematic literature review and a research agenda. Sustainable Production and Consumption. 2020;23:174–188.

[8] Ongondo FO, Williams ID, Cherrett TJ. How are WEEE doing? A global review of the management of electrical and electronic wastes. Waste Management. 2011;31(4):714–730.

[9] Forti V, Baldé CP, Kuehr R, Bel G. The Global E-waste Monitor 2020. United Nations University (UNU), International Telecommunication Union (ITU) & International Solid Waste Association (ISWA), Bonn/Geneva/ Rotterdam. 2020;

[10] Vaccari M, Vinti G, Cesaro A, Belgiorno V, Salhofer S, Dias MI, et al. WEEE treatment in developing countries: Environmental pollution and health consequences—An overview. International journal of environmental research and public health. 2019;16(9): 1595.

[11] Grant K, Goldizen FC, Sly PD, Brune M-N, Neira M, van den Berg M, et al. Health consequences of exposure to e-waste: a systematic review. The Lancet Global Health. 2013;1(6):e350–e361.

[12] Zhang L, Zhong Y, Geng Y. A bibliometric and visual study on urban mining. Journal of Cleaner Production. 2019;239.

[13] Tercero L, Rostek L, Loibl A, Stijepic D. The Promise and Limits of Urban Mining. Karlsruhe; 2020.

[14] Xavier LH, Giese EC, Ribeiro-Duthie AC, Lins FAF. Sustainability and the circular economy: A theoretical approach focused on e-waste urban mining. Resources Policy. 2019;

[15] Tesfaye F, Lindberg D, Hamuyuni J, Taskinen P, Hupa L. Improving urban mining practices for optimal recovery of resources from e-waste. Minerals Engineering. 2017;111:209–221.

[16] Widmer R, Oswald-Krapf H, Sinha-Khetriwal D, Schnellmann M, Böni H. Global perspectives on e-waste. Environmental Impact Assessment Review. 2005;25(5 SPEC. ISS.):436–58.

[17] OECD. Extended Producer Responsibility, A guidance Manual for Governments. Paris: OECD Publishing; 2001.

[18] Européenne U. Directive 2002/96/EC on waste electrical and electronic equipment (WEEE). JO EU. 2003;37: 24–38.

[19] Directive EC. Directive 2012/19/EU of the European Parliament and of the Council of 4 July 2012 on waste electrical and electronic equipment, WEEE. Official Journal of the European Union L. 2012;197:38–71.

[20] Paleari S. The EU waste electrical and electronic equipment directive: The implementation of producer responsibility across the EU-27. Journal of Solid Waste Technology and Management. 2015;41(2):173–188.

[21] Corsini F, Rizzi F, Frey M. Extended producer responsibility: The impact of organisational dimensions on WEEE collection from households. Waste Management. 2017;59:23–29. Available from: https://www.sciencedirect.com/science/article/pii/S0956053X16306158

[22] Label PID, oHS Compliance Engineer R. Directive 2002/95/EC of the European Parliament and of the Council of 27 January 2003 on the restriction of the use of certain hazardous substances in electrical and electronic equipment. 2005;

[23] Directive EC. Directive 2011/65/EU of the European Parliament and of the Council of 8 June 2011, on the restriction of the use of certain hazardous substances in electrical and electronic equipment (recast). Official Journal of the European Communities. 2011;

[24] Directive EU. Restriction of the use of certain hazardous substances in electrical and electronic equipment (RoHS). Off J Eur Communities. 2013; 46:19–23.

[25] Commission E. Regulation (EC) No 1907/2006 of the European Parliament and of the Council of 18 December 2006 concerning the registration, evaluation, authorisation and restriction of chemicals (REACH), establishing a European Chemicals Agency, amending Directive 1999/4. Official Journal of the European Union. 2006;396:1–849.

[26] European Commission. COM(2020) 798/3 2020/353 (COD). Brussels; 2020.

[27] Directive EC. Directive 2009/125/EC of the European Parliament and of the Council of 21 October 2009, establishing a framework for the setting of ecodesign requirements for energy related products (recast). Official Journal of the European Communities. 2009;

[28] European Parliament, Council of the European Union. REGULATION (EU) 2017/1369 OF THE EUROPEAN PARLIAMENT AND OF THE COUNCIL of 4 July 2017 setting a framework for energy labelling and repealing Directive 2010/30/EU. Official Journal of the European Union L 198/1. 2017;L 198(1).

[29] European Parliament, Council of the European Union. Regulation (EC) No 66/2010 of the European Parliament and of the Council. Offical Journal of the European Union. 2009;L–27.

[30] Najmi A, Kanapathy K, Aziz AA. Exploring consumer participation in environment management: Findings from two-staged structural equationmodelling-artificialneural network approach. Vol. 28, Corporate Social Responsibility and Environmental Management. 2021. p. 184–95.

[31] Najmi A, Kanapathy K, Aziz AA. Understanding consumer participation in managing ICT waste: Findings from two-staged Structural Equation Modelling-Artificial Neural Network approach. Vol. 28, Environmental Science and Pollution Research. 2021. p. 14782–96.

[32] Moher D, Liberati A, Tetzlaff J, Altman DG. Reprint-Preferred Reporting items for systematic reviews

and meta-analyses: The PRISMA statement. Physical Therapy. 2009;89 (9):873–880.

[33] Dixit S, Vaish A. Sustaining environment and organisation through e-waste management: A study of post consumption behaviour for mobile industry in India. Vol. 16, International Journal of Logistics Systems and Management. 2013. p. 1–15.

[34] Nickerson RC, Varshney U, Muntermann J. A method for taxonomy development and its application in information systems. Vol. 22, European Journal of Information Systems. 2013. p. 336–59.

[35] Corsini F, Gusmerotti NM, Frey M. Consumer's Circular Behaviours in Relation to the Purchase, Extension of Life, and End of Life Management of Electrical and Electronic Products: A Review. Vol. 12, Sustainability. 2020.

[36] Park J, Kim SD, Choi SO. Demonstrating the effects of behavioural control beliefs on the actual WEEE discharge routes: A case study in South Korea. Vol. 163, Resources Conservation and Recycling. 2020. p. 11.

[37] Shekarian E. A review of factors affecting closed-loop supply chain models. Vol. 253, Journal of Cleaner Production. 2020. p. 119823. Available from: internal-pdf://1.38.191.244/ Shekarian-A review of factors affecting closed.pdf

[38] Wang Z, Huo J, Duan Y. Impact of government subsidies on pricing strategies in reverse supply chains of waste electrical and electronic equipment. 2019/07/29. Vol. 95, Waste Manag. 2019. p. 440–9.

[39] Bai H, Wang J, Zeng AZ. Exploring Chinese consumers' attitude and behaviour towards smartphone recycling. Vol. 188, Journal of Cleaner Production. 2018. p. 227–36.

[40] Wang B, Ren C, Dong X, Zhang B, Wang Z. Determinants shaping willingness towards on-line recycling behaviour: An empirical study of household e-waste recycling in China. Vol. 143, Resources, Conservation and Recycling. 2019. p. 218–225.

[41] Simpson D, Power D, Riach K, Tsarenko Y. Consumer motivation for product disposal and its role in acquiring products for reuse. Vol. 65, Journal of Operations Management. 2019. p. 612–35.

[42] Kumar A. Exploring young adults' e-waste recycling behaviour using an extended theory of planned behaviour model: A cross-cultural study. Vol. 141, Resources, Conservation and Recycling. 2019. p. 378–389.

[43] Kochan CG, Pourreza S, Tran H, Prybutok VR. Determinants and logistics of e-waste recycling. Vol. 27, International Journal of Logistics Management. 2016. p. 52–70.

[44] Botelho A, Ferreira Dias M, Ferreira C, Pinto LMC. The market of electrical and electronic equipment waste in Portugal: Analysis of take-back consumers' decisions. Vol. 34, Waste Management and Research. 2016. p. 1074–80.

[45] Mansuy J, Verlinde S, Macharis C. Understanding preferences for EEE collection services: A choice-based conjoint analysis. Vol. 161, Resources Conservation and Recycling. 2020. p. 12.

[46] Kumar A. Extended TPB model to understand consumer "selling" behaviour. Vol. 29, Asia Pacific Journal of Marketing and Logistics. 2017. p. 721–42.

[47] Dixit S, Badgaiyan AJ. Towards improved understanding of reverse logistics - Examining mediating role of return intention. Vol. 107, Resources Conservation and Recycling. 2016. p. 115–28.

[48] Yla-Mella J, Keiski RL, Pongracz E. Electronic waste recovery in Finland: Consumers' perceptions towards recycling and re-use of mobile phones. Vol. 45, Waste Management. 2015. p. 374–84.

[49] Georgantzis Garcia D, Kipnis E, Vasileiou E, Solomon A. Consumption in the Circular Economy: Learning from Our Mistakes. Vol. 13, Sustainability. 2021. p. 601.

[50] Carrington MJ, Neville BA, Whitwell GJ. Why Ethical Consumers Do not Walk Their Talk: Towards a Framework for Understanding the Gap Between the Ethical Purchase Intentions and Actual Buying Behaviour of Ethically Minded Consumers. Vol. 97, Journal of Business Ethics. 2010. p. 139–58.

[51] Rodrigues AC, Boscov MEG, Günther WMR. Domestic flow of e-waste in São Paulo, Brazil: Characterisation to support public policies. Waste Management. 2020;102: 474–485.

[52] Chi X, Wang MYL, Reuter MA. E-waste collection channels and household recycling behaviours in Taizhou of China. Vol. 80, Journal of Cleaner Production. 2014. p. 87–95.

[53] Sabbaghi M, Esmaeilian B, Raihanian Mashhadi A, Behdad S, Cade W. An investigation of used electronics return flows: a data-driven approach to capture and predict consumers storage and utilisation behaviour. 2014/12/24. Vol. 36, Waste Manag. MA 0; 2015. p. 305–15.

[54] Cox J, Griffith S, Giorgi S, King G. Consumer understanding of product lifetimes. Vol. 79, Resources, Conservation and Recycling. 2013. p. 21–29.

[55] Mishima K, Nishimura H. Requirement analysis to promote small-sized E-waste collection from consumers. 2015/11/27. Vol. 34, Waste Manag Res; 2016. p. 122–8.

[56] Claeys C, Swinnen A, vanden Abeele P. Consumer's means-end chains for "think" and "feel" products. Vol. 12, International Journal of Research in Marketing. 1995. p. 193–208.

[57] Dittmar H. Material possessions as stereotypes: Material images of different socio-economic groups. Vol. 15, Journal of Economic Psychology. 1994. p. 561–85.

[58] Srivastava L. Mobile phones and the evolution of social behaviour. Vol. 24, Behaviour & Information Technology. 2005. p. 111–29.

[59] Croson R, Gächter S. The science of experimental economics. Vol. 73, Journal of Economic Behaviour & Organisation. 2010. p. 122–31.

Chapter 14

Compost, Social Sustainability, and Circular Economy in Guatemala

Peter A. Kumble

Abstract

The research presented in this chapter explores a variety of objectives: first, what are the dynamics and associated requirements for initiating a new start-up composting business that would embrace the principles of Circular Economy? Secondly, is there a market for compost both in an urban environment and for farmers regionally in a development world economy such as Guatemala? With this, how can employment opportunities for at-risk youth from the most impoverished neighborhood in Guatemala City be created while adhering to the tenants of social sustainability? And finally, what were the requirements involved in making compost in the challenging high altitude climatic conditions of Guatemala City?

Keywords: compost, social sustainability, circular economy, Guatemala

1. Introduction

This chapter reports on the success and failure of making compost for soil amendment using green waste from a large fruit and vegetable market situated in Guatemala City, Central America. Perhaps the term "failure" is inappropriate and should be referred to as, "factors that contributed to limiting the intended success of a start-up business." For anyone who has initiated a new business, you are continuously faced with innumerable problems complicating its success—such as the availability of skilled labor, licensure, market dependability for your product, managing the costs for materials, adapting to unpredictable weather and climate, marketing or branding of your product, and 100 s of other variabilities beyond your control. Perhaps the more successful ventures are those who can remain flexible and adapt quickly to externalities. Starting a new business and managing it on a daily basis is not for the faint of heart, however if the passion for what you have intended to accomplish is great and if it can bring significant improvement for people, the environment, and local economies, then this alone is a measurement of success. The challenges and "failings" of this work are discussed in the Results and Discussion section of this chapter.

Yet, how does one measure success in a broader context and how does one know that their efforts were productive and effective? For this research project, creating change within the lives and future prospects of perhaps the most disadvantaged youth from one of the poorest neighborhoods in the capital city of a developing world nation—Guatemala City—became the key measurement of success.

IntechOpen

Certainly, developing a successful business that embraced the principles of circular economy while not losing money and remaining solvent were equally important. However as stated above, achieving measurable social sustainability was the driver in defining success, perhaps even more so than simply being viable and achieving a measure of environmental and economic sustainability.

Guatemala is situated in the northern portion of Central America with the Pacific Ocean to the west and the Caribbean Sea to the east, with international borders of Mexico to the north, to the southeast Honduras and El Salvador, and Belize on its eastern frontier (**Figure 1**). The capital, Guatemala City, is centrally positioned in the southern portion of the country at 1500 meters in elevation (nearly 5000 feet). Thus, nighttime temperatures are moderately cool in the summer. This is in stark contrast to the very warm and humid climate of the eastern portion of

Figure 1.
Guatemala situated in the northern portion of Central American shares borders with Mexico, Belize, Honduras and El Salvador (Guia Geografico).

the country, known as the Peten, home to Tikal and the Mayan Biosphere reserve. Guatemala City has a population of nearly 2.9 million residents with more than 4 million or greater migrating to the capital city during each business week from rural areas [1]. Many low-income people live within the city and are concentrated into 22 neighborhood and employment zones based upon social and economic class separation. Zone 3 is the most impoverished where squat housing has been constructed upon early remnants of the nearby landfill, referred to locally as the *basurero* (**Figure 2**). People who work in the *basurero*—not for a living wage but rather to pick through the unsorted trash for subsistence—have erected a neighborhood of shanty houses using concrete blocks and corrugated metal as construction materials [2]. Most of the inhabitants here are fixed or rather trapped in conditions best characterized as abject poverty. They mostly subsist upon trash from the city as it is dumped from garbage trucks (**Figure 3**). They pick through the trash to collect plastic and glass which they then turn in for cash at recycling facilities, perhaps discarded food to eat, or cast-off household items to use; no doubt this is a grim existence for their families [3]. Guatemala City experiences a rainy season that can last for up to five or six months (May through October). The rainy season can and does cause the unstable terrain at the landfill to collapse [4], literally swallowing the people who are sorting through the trash. It is a tragedy that their bodies are rarely recovered due to the high levels of methane gas generated from the organic waste. Further hazards come from the greenhouse gas released from the decomposing waste which can cause cancerous tumors. Many of these people who are picking through the trash have marginable options for receiving basic health care and their injuries go untreated.

The research reported in this chapter set out to establish if compost could be produced to amend the marginal soil in the City's parks and landscaped planting beds using green waste typically discarded from a large City-managed fruit and

Figure 2.
Dwellings have been constructed upon portions of a former landfill using concrete blocks and corrugated metal.

Figure 3.
Families sort the recently dumped trash from garbage trucks to gather recyclables, discarded household items and edible food.

Figure 4.
Trucks that deliver farm produce to the central market could be used to deliver compost back to the farm for soil enrichment.

vegetable market known as CENMA. Diverting the additional organic waste from disposal in the landfill made logical sense; why throw away something that could be used to create something else of value? However, what was unknown were the challenges in using the organic waste to produce compost in this high altitude and temperately cool environment during the spring and summer months. In addition to selling the finished product locally, the research attempted to determine if the finished compost might have a regional market, meaning might there be a broader client for the newly prepared compost? Could the compost also be delivered to the farms who had originally grown the fruit and vegetables to be used for enrichment of their soil, employing the same trucks (**Figure 4**) that brought commodity to the market on their return trip?

2. Circular economy, closed loop systems, and agricultural byproducts

Whereas the author of this chapter is not an expert scholar within the burgeoning field of Circular Economy (CE), it is important to point out similarities of CE to what was once referred to as a closed loop system; the creation of a closed-loop system for producing compost, coffee or clothing, for example, should be based upon the principles of (CE). It is important to mention that the term circular economy first appeared in the literature as part of a study by Pearce and Turner (1990) [5]. This earlier research, referenced by Anderson (2007) and later Kumble (2019), worked to establish a link between production activities in industry [6, 7]. Thus, CE is recognized as a good strategy that can minimize any unneeded waste by increasing manufacturing efficient [8–10]. In general terms, this was called a closed-loop system, and was initially introduced by Boulding (1966) and later refined by Leontief (1991). The concept is based upon using raw materials and the superfluous waste contained within a closed loop [11, 12], meaning that you recycle the end product and its associated components in the manufacturing of a new item. Obviously, this is perhaps a standard used within the agricultural industry, but it can also be applied at a much smaller scale. An excellent example of this is where organic waste is converted into fertilizer and recycled into the soil. A closed loop system is a bit more complicated to achieve within the manufacturing industries given the huge variety of the product being produced. Nevertheless, if one were to try to explain CE as it might relate to ecological efficiency as demonstrated in agricultural production, things become perhaps less clear due to the diversity of what we make and how it is both distributed and consumed, as mentioned above. The literature has established that sophisticated cultures demand more resources to meet both their social and economic needs [13–15] at the cost of resource depletion, many of which are finite and ultimately not sustainable.

As presented in an extensive literature review of more than 500 articles, Merli et al. (2018), the authors established that CE can overcome what they referred to as the "take–make–disposal linear pattern of production and consumption" [16]. Whereas this may sound like a mouthful, pun intended, the principle behind this phrase aimed to preserve raw materials resources, within the production system as long as possible. Although one might think that all manufacturing aims to do this as a way to cut costs and improve efficiency, the sheer volume of industrial waste that ends up being disposed of in a landfill or elsewhere suggests otherwise. Merli et al. established a connection between the temporal scale with product production and reuse—the longer and greater diversity of how raw products are used play to the benefit of sustainable efficiency. Unfortunately, scholars seldom consider both the social and institutional inferences of CE at the environmental and economic level. Professor of landscape architecture John T. Lyle demonstrated this dysfunction as part of his applied research nearly 40 years ago, launching a movement which later became more broadly referred to as *regenerative design*. This work was published in the book, *Design for Human Ecosystems* [17], which eloquently demonstrated how closed-loop systems are used for waste water recycling, integrated pest management, renewable energy production, and efficient use of finite resources. Professor Lyle's applied work can be found today at the Center for Regenerative Studies on the campus of Cal Poly Pomona, California, USA.

It is interesting yet not surprising that Circular Economy is a contested title and description [18, 19]. I believe that this is attributed to the fact that it is very much an interdisciplinary topic, with feet in different scholarly and professional fields. Such definitive controversy or tension is not unique to CE. For example, ecotourism experiences a similar level of confusion and uncertainty; is it part of the tourism industry or simply a movement that promotes natural or cultural resource conservation with a focus upon nature-based experiences while also accounting for the intended educational and learning experience of the place visited? Perhaps even the often and over used terminology—sustainability—shares the dubious distinction of uncertainly, different meanings, and "gray" clarity. Regardless, ascribing CE within the context of compost production in a developing-world economy is not flawed or inappropriate, nor does it create confusion. In fact, it might even help to provide a clearer category for how recycling through compost production fits within the broader business of manufacturing. In an effort to lend clarity to this topic, Kumble (2019), reported on how Kirchherr et al. (2017) evaluated 114 definitions of the term CE, which led to the creation of the following definition: "*...CE describes an economic system that is based on business models which replace the end of life concept with reducing, and alternatively reusing and recycling materials in production, distribution, and consumption processes...*" [20]. Perhaps the confusion does not exist within trying to actually define what truly is CE, but rather which through what business or policy models to use [7]. As with ecotourism or even the popularized term sustainability, it is perhaps better expressed as a verb and not a noun; it is about action and should not be boxed-in with a one-fits-all place-based definition. What is exciting is that now CE can be perceived as a possible means by which achieve the principles of sustainable development [21], and more specific for the work in Guatemala contained in this chapter, achieving social sustainability.

More recent literature describes the need for quantifiable factors associated with the lifespan of a product to best determine the efficiency of CE. It seems that this current trend is attempting to alleviate the uncertainty mentioned above. This then introduces a new set of questions, such as what those indicators should be and of course how to account for variability? Again, this raises the question of boxing-in, a trend with rigid definitions. A plausible definition of lifespan of a product might be related to the number of times or repetition that a something is used and reused while also considering the longevity or duration of that use. Research by Figge et al. (2018) contend that the duration (temporal) and circularity (complexity) are necessary for sustainable resource use, but how should one clearly ascribe measurements that combine both approaches such as temporal complexity [22]? Again, Figge's research team argued for a complex matrix to measure both, which is not surprising given their background in economic studies. If then one were to use this model and apply it toward the production of compost from green waste and brown carbon such as cardboard or wood chips, this production technique would achieve temporal longevity. What now becomes significant is that compost production suggests quantitative and qualitative factors that are key concepts of CE.

Sama, et al. (2018) described how the food industry has successfully made the more toward the production of fair-trade products and socially responsible consumption, both which are a critical measurement of sustainability within CE [23]. With that, the world has been moving, albeit slowly, toward circular economy with the demand to become more sustainable its daily life with the production of coffee, clothing, or perhaps compost. With the popularizing of fair-trade products, consumer demand for these goods produced in developing countries such as Guatemala can be found at a worldwide scale. This trend is evident in the move toward "green projects" that supports environmentally sustainable investments, as reported by Falcone el al. (2018) [24] on ethical socially responsible projects. They reported on the trend for funding radical green innovation. It could be argued that this is simply *green washing*, or it might suggest a new paradigm shift in how business is being conducted due to reduced costs, reduced energy consumption, and the added benefit of producing a positive and sustainable result. George, et al. (2015) discussed the connection between green finance within circular economy [25] specific to the biomass production sector; this is akin to the production of compost.

The United Nations Environmental Program (UNEP) published a report in 2011 reporting on the trend toward an economy based upon low carbon outputs, signaling the move toward green businesses [26]. Has this been a true shift in response to climate change or as mentioned earlier, simply green washing? With regard toward the aforementioned biomass production, the movement toward renewable energy and associated industries—such as making compost from green waste in Guatemala—is no doubt significant. Again, the UNEP report found that the money required worldwide could be 2% of global GDP between 2010 and 2050 [27]. Although a significant sum, are their other viable alternatives?

When weighing the costs and long-term benefits of the global movement toward being more green, not because it is marketable but because it is necessary, what are then the implications of making compost from market waste, and how can this small action by the municipal government in Guatemala City be a model for other communities to follow? When trying to apply a change in how business is conducted, some world economies have adopted a top-down centralized approach [28], while others believe that a community-based bottom-up movement is more appropriate [29]. It is difficult to generalize which is more appropriate and perhaps local conditions and the size or scale of the problem is the main determinant.

No doubt, CE contains numerous complexities in both how it might be defined and quantified, likely due to the various disciplines associated, as argued above. With this understanding, or perhaps the uncertainty of how to best demonstrate the circular nature of making compost, in addition with how does it in fact represent a closed loop system applicable in Guatemala or any other culture, the research presented in this chapter attempted to achieve a variety of applied and theoretical objectives:

- What is necessary for starting a business that demonstrates the principles of CE using the production of compost as an outcome?

- Could this action trigger a break from what clearly appears to be dim prospects for teenagers from the Zone 3 neighborhood in Guatemala City thus providing an alternative for their future livelihood?

- When considering the precepts of social sustainability [30] and basic human rights [31], what role could or should green investment in compost production play?

- Are there unknown obstacles toward making compost in the high in altitude climatic conditions of Guatemala City with cool nighttime temperatures during the summer, periods of low rainfall during portions of the year, and inundation from rain during other times?

- Could a simplified technique of making compost produce enough end product equal to the more industrial windrow commercial production approach?

The work reported in this chapter did in fact have multiple objectives as described above. Initiating a startup business and its associated challenges of balancing economic, environmental, and social objectives was not to be taken lightly, however altruistic as they may have seemed. These three pillars or three Es are the foundations of sustainability [32], but how should one bridge gap between pedagogical theory and real-world working conditions while factoring in a myriad of political and social challenges? Perhaps a brief revisit of how sustainability became part of the world dialog is in order.

3. Discussion of circular economy and sustainability in practice

Recognition of the critical importance for sustainable development within our lives began nearly 30 years ago in 1992 at the Earth Summit in Rio de Janeiro and again 20 years ago at the World Summit for Sustainable Development in Johannesburg [33]. Ironically, for some readers, these two important events predate their birth, yet they are no less significant as both signaled a paradigm shift in global awareness. Of the three Es as pillars of sustainability, one could argue that achieving social sustainability is particularly complex due to the constant changes or variability within localized society. As described by Kumble (2019) [7], Boyer et al. (2016) [30] enumerated and analyzed the particular difficulties in understanding social sustainability, however they citied the variable definitions and gaps due to the interdisciplinary nature of the topic. This is not necessarily a failing in the general research but more likely attributed to the complexity of interdisciplinary topics; each field is interpreted and understood differently by its associated scholars and proponents. In retrospect, social sustainability along with environmental factors and drivers of any economy are very much place based [7]. Boyer's research team appropriately used the comparison of a three-legged stool and the 3Ps—prosperity, planet and people—for understanding complex problems withing the world and to not inventing new paradigms which would only busy an already crowded field of understanding.

Working on the principle of simplicity in action, the research presented in this chapter would appear to be a positive and logical situation whereby the compost that was created from organic waste diverted from a landfill could then be used to amend marginal soil, train workers, create jobs, mitigate an ecological and environmental crisis, and provide a future of skilled employment. Making compost would embrace the theories of CE, would mitigate the terrible environmental impacts from the batsuro while triggering new opportunities in the business market, consistent with the bottom-up model proposed by Ghisellini el al., 2016 [29] described above. As mentioned earlier in this chapter, the idea of the closed loop system of manufacturing has been in use for many generations in the industrial sectors of the world associated with manufacturing [34, 35]. The startup business for making compost wished to explore if this could be done in Guatemala City and not be hindered by the numerous complex cultural, environmental, and economic obstacles.

4. Challenges of Guatemala City

Although making compost from green market waste was one of key goals of the project reported in this chapter, it was really based on the foundational intention of creating future employment opportunities for the disenfranchised and poorly educated youth from the squat neighborhood in Zone 3. Perhaps if one of these goals could be accomplished—compost production—it would trigger the success of the other—future opportunities for the youth who had little future prospects. With this clear objective, the work aimed to explore how to achieve social sustainability. Minica and France (2008) postulated that social sustainability is in fact composed of the following four key objectives: 1. education and training; 2. promoting human health; 3. winning the fight against poverty; 4. creating a equitable and just working environment [36]. Yes, self-empowerment can be achieved through education, however with only 69.1% of the population in Guatemala who can read and write, it is perhaps the most illiterate nation in Central America. Similarly, 8 out of 10 people will never graduate from high school, not because they are lazy or lack ambition but because they must leave school while they are still kids to find work in support of

Figure 5.
Approximately 80 percent of Guatemalans will never complete high school. This is particularly acute in Zone 3.

their family (**Figure 5**). Finding jobs for this age group from Zone 3 is very difficult due to their insufficient education [37]. Thus, as mentioned above, the key research initiative of this work determined that the social pillar of sustainable development should be the most important because of the need to create opportunities for the future of the youth from Zone 3.

The CENMA fruit and vegetable market generates a huge amount of green waste each day as part of the trimming and packaging of the commodity for local and international markets. Not surprisingly, some 115 cubic meters (150 cubic yards) of organic waste is trucked daily to the landfill; yet it need not be. The head for Public Works of Guatemala City agreed to collaborate with the author and his graduate students from the University of Massachusetts, Department of Landscape Architecture and Regional Planning, for a variety of reasons: technical knowhow and enthusiastic students with a strong adherence to environmental justice. The City agreed to a student-initiated start-up business to produce useable compost on a small tract of land situated immediately adjoining CENMA. The site was unfortunately very small, 0.48 hectares (1.2 acres), but it would allow the team to attempt to test the principles of CE and pillars of sustainability.

The Municipal Parks Department in Guatemala City were investing nearly $300,000 USD each year to amend the soil on their land at the parks and landscaped planting beds. The director of Public Works agreed to the purchase of compost from the new startup business for their public-sector landscaping projects.

5. Summary review of techniques for producing compost

Whereas the intention of this chapter is not to explain new technologies for the production of compost, it is useful to review of how one makes compost, such as the careful mixing of brown and green raw material. It is important to point out that no animal manure or carcasses are used because animal waste can spread diseases. The composting process presented in this chapter describes the use of organic plant material often referred to as brown and green material (**Table 1**) [38]. Brown material is comprised of shredded wood chips, dry grass stalks, or cardboard and

Brown material	Green material
Straw or hay	Fruit
Woodchips	Vegetables
Cardboard	Egg shells
Dry leaves or grass	Coffee grounds
Tree bark	Freshly cut grass
Sawdust	

Table 1.
The differences between brown and green organic material.

thus does not decompose as rapidly as green material. So why use brown material? It provides the finished compost product with a light texture. In comparison, green material refers to more recently cut or harvested wet waste such as vegetables or plant biomass and will decompose quickly. In the mixing of brown and green, the brown material is more stable, meaning that the amount of time required for it to break down or decompose is more predictable [39], likely due to the fact that it contains much less moisture.

Organic waste is comprised of the leaves, stems, and bark of plants and insects. Interestingly enough, manure or animal feces is also a fertilizer that is referred to as organic. Brown materials such as wood chips, sawdust and cardboard, although processed by man, comes from trees and is organic unless it contains dyes or is coated with plastic. Man-made pesticides are natural and are not considered as organic material. However, it is important to understand that sometimes pesticides are organic because some plants will create chemicals naturally in their leaves to protect against insects [7].

The natural process of decomposition of organic material can be described as the breaking down of organic material. We can observe this process in the forest understory, within the leaf litter or dead wood from trees and shrubs. Nutrient-enriched humus returns organic material to the soil providing essential minerals supporting and accelerating plant growth; it should be thought of as enriched food stock for root system of plants [40]. Much of the decomposed humus is often in the top layers of soil—typically the O-layer—and is the darker color that can be seen when inserting a soil probe and extracting a sample. Non-organic waste is very slow to decompose and can take hundreds of years to break down into useful material [41].

A commercial or production compost operation manages the decomposition of organic material in a more controlled environment, allowing the process to occur more rapidly to produce a consistent and useable quality product. It does this by regulating the amount of oxygen, water, and brown to green material intentionally. When we refer to a 'compost pile' it suggests a mound or pile of organic measurable waste that is undergoing decomposing [42]. This finished end-product we call 'compost' (**Figure 6**), and can be used to amend existing soil, making that soil healthier or more alive and better suited for retaining soil moisture content and thus the growing of plants. When one tills the soil and harvests fruit and vegetables, the soil can become less vital or degraded. By adding compost or barnyard waste into soil, it becomes replenished with fresh nutrients, contributing to increased soil fertility [43].

In a healthy forest, decomposition of organic matter occurs as part of the digestive processes enabled by a variety of microorganisms [41] that feed on dead or dying plant material and animals. The organisms reproduce, die, and recycle themselves as new organic material through the process of decomposition. These

Figure 6.
This large pile of newly produced compost is mixed with existing soil to increase fertility and plant growth.

tiny creatures are contained in decomposing organic material and do not need to be added to a commercial composting pile. As with any alive material, these microorganisms require food, water, and air to live, consuming some of the organic material found in a compost pile. Whereas insects, worms, and even snails are valuable for making compost, they actually perform less work to in the decomposition process than do the microorganisms. Thus, microorganisms are an essential and necessary component for the production of compost.

As mentioned above, air, water, and the appropriate mix of organic material will allow the decomposition of organic material to reach its finished state in a predictable amount of time yielding healthy and useable compost. Typically, the composting process for organic material (in a compost bin, windrow, or pile) requires 90 to 120 days to occur [44], provided that the organic waste is receiving the needed combination of oxygen and moisture, and most importantly and that it is turned or churned regularly to allow air and moisture to effectively enter the pile [45]. Of course, this entire process can be accelerated significantly by increasing the amount of oxygen that enters the composting material; some operations can produce useable compost in very short time of 30 to 45 days, although the energy and financial costs of doing this may not be realistic.

How should one then choose to produce useable compost from organic brown and green waste? There are two commonly used technologies employed today; commercial operations often use a approach commonly referred to as windrows [46], which are basically very long and narrow piles of compost (**Figure 7**). A windrow is at a minimum 1.5 m (5 feet) tall with equal width and are difficult to manage with only manual labor [7]. As mentioned, a commercial operation often will use the windrow method due to their efficiency in accommodating a larger mass or volume of organic waste material. Due to their size and particularly their length, the windrow technology typically require many hectares of useable surface area and expensive commercial machinery such as a tractor that can effectively pull the mechanized windrow turner (**Figure 8**) which creates a uniform shape of the windrow pile while also churning or mixing the compost allowing necessary oxygen to enter and accelerate the decomposing process. Trial and error have determined that the tractor must use a 'creeper' gear whereby it moves very slowly yet allows the PTO (power take off drive) to spin the turner at a fast speed.

Figure 7.
This is an example of a municipal compost facility using wind-rows for large-scale production.

Figure 8.
This mechanized device is pulled by a tractor and is used to shape, turn, or churn the compost in the windrows.

As an alternative to windrows, smaller operations will use what is typically referred to as the in-cell technique. As its name suggests, this approach uses modular structures that hold the compost in place and look somewhat like a large cube [7]. The biggest advantage of the in-cell or 3-cell technology are that they can be maintained using manual labor and do not require expensive machinery such as a tractor and windrow turner for turning or churning the compost mix while it is in the process of decomposition. There were some obvious reasons why the in-cell composting technique was employed at the compost operation in Guatemala. These include the following:

- By using manual labor to move the compost from one cell to another and thus accelerating the decomposition process, more youths could be employed;

- Cells do not involve the need for expensive equipment typical of the windrow system; and

- Cells do not require training of employees to operate mechanized machinery or do they require maintenance and upkeep.

The in-cell compost technique looks much more like the compost "bin" that a homeowner might use for decomposing kitchen and yard organic waste (**Figure 9**). Home-composting typically has one compost bin/cell. Yes, some households can and do have multiple compost bins, however the contents are seldom mixed or shoveled into an adjoining bin. For this reason, three cells should be employed to be more effective (**Figure 10**). But why three cells? One should begin with a new compost mix that is started in cell A: once the contents begin to shrink in size as the green waste decomposes, all of the contents from cell A should then be shoveled into cell B and a new batch started in the now available cell A. The act of moving the mix from cell A to cell B adds oxygen and mixes the contents, similar to what a windrow turning machine might accomplish. Later, the contents from cell B are moved into cell C for completion, and a new batch is begun in cell A, which has seen its material moved into cell B.

There are perhaps many ways to construct cells for producing compost, all based on the availability of cost-effective materials and creativity of the builder. Wooden shipping pallets were selected for use at the CENMA site because they are inexpensive and were easily obtained. Each of the shipping pallets were simply fastened together using long screws or nails to join one to another (**Figures 10** and **11**). Small sections of chain link fencing were used to enclose the front opening for each cell. Initially, the intention was to use steel fencing post which would be hammered into the ground forming each of the corners, coupled with welded wire fencing fastened to the posts to form the enclosed sides. However, the hard and rocky surface area found at the test site rendered this approach unfeasible as it was impossible to drive the metal posts into the ground. The wooden shipping pallets were readily available, easy to fasten together, and inexpensive.

For a newly established compost pile (in cell A), one must regularly monitor the internal temperature during the initial weeks to determine the rate of

Figure 9.
A compost bin or cell is used to make compost from household green waste, fallen leaves, and grass clippings.

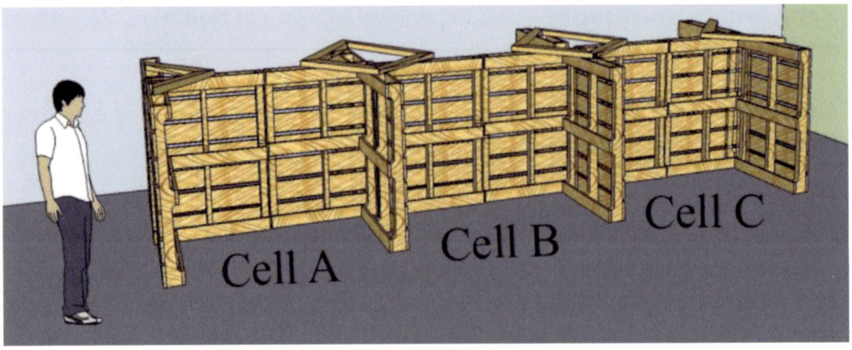

Figure 10.
This schematic illustrates how wooden pallets can easily fastened together as a three-cell compost system (illustration by Seth Morrow).

Figure 11.
Wooden shipping pallets are readily available and can be simply fastened together to replicate the three-cell compost system.

decomposition (or cooking as it is referred to). At the same time, the size of the pile decreases or shrinks in size as the green material breaks down. Similar to that of a windrow, a thermometer with a long one-meter probe is used to measure the internal temperature (**Figure 12**) to determine if and how fast the compost process is successfully occurring. **Table 2** illustrates cell-monitoring data for one of the two test cells constructed. The data was monitored in order to chart the time and temperature on a weekly basis, in addition to observed odor (smelliness). Approximately one month after the pile is made, the temperature cease increasing while the shrinkage of the pile should also decrease in rate. When this occurs, the pile should be shoveled into cell B. This process of turning the pile simply involves taking the material out of the cell and mixing it up, meaning that it is important to move the decomposing material from the middle of the pile to the outside layers of the relocated pile, now in the second cell (cell B).

Figure 12.
Decomposition temperature is monitored regularly using this thermometer with a 1 meter long probe.

The microorganisms should be actively feeding on the organic material, meaning that they are now consuming the composting mass of material. Obviously, the microorganisms do not have the ability to move very far, so it is very important to adequately mix the pile, putting the microorganisms into direct contact with organic material to continue the decomposition process. Whereas the turning of a pile will introduce oxygen, it is quickly consumed, and it is not the primary function of turning or mixing the contents. Oxygen should enter a compost pile through proper ventilation and pore space (voids in the mix). Some people will also include perforated plastic pipe, similar to that used for stormwater under-drainage, and place the pipe across the bottom and then extending perpendicularly and vertically up through the pile to better allow the movement of oxygen.

Date	Days	Height	Change	Temp	Change	Smell	Change	Humidity	Change
Wendell									
22-Mar		30		75		3		4	
27-Mar	5	26	-4	120	45	3	0	4	0
3-Apr	7	23	−3	130	10	2	−1	3	−1
13-Apr	10	23	0	130	0	1	−1	1	−2
Wes									
22-Mar		30		75		3		4	
27-Mar	5	27	−3	120	45	3	0	3	−1
3-Apr	7	26	−1	75	−45	2	−1	1	−2
13-Apr	10	26	0	75	0	1	−1	1	0

The units for Height are in inches, Temperature is in F, and Humidity is based upon relative %. Smell was subjective with 1 being low and 3 being high.

Table 2.
The rate of decomposition, changes in temperature, humidity and odor were monitored weekly in each of the test compost cells.

6. Compost cells at the test site in Guatemala

The useable compost production space provided at CENMA could physically accommodate approximately 400 cells and produced approximately 1480 cubic meters of compost annually. The three-cell system, constructed from wooden shipping palettes, were 1.8 meters (6 feet) deep, 1.5 meters (5 feet) wide, and 1.2 meters (4 feet) tall. An estimated 140 cubic meters of newly produced compost was stored on-site each month [7]. The new compost was stored on-site to facilitate loading it onto transport truck for distribution to its final destination.

Much of the green waste deposited into collection barrels at the CENMA market actually contained a large amount of non-compostable garbage, such as cans, bottles, dirty diapers, etc. (**Figure 13**). It was an awful mess to clean and sort and necessitated a different approach; clearly those using the market mistook the collection

Figure 13.
Non organic waste is unfortunately disposed of into the compost collection barrels in the market.

DOI: http://dx.doi.org/10.5772/intechopen.100280

Figure 14.
Clearly marked barrels were relocated close to fruit and vegetable processing.

barrels as suitable for general waste disposal. Clearly marked barrels were later placed in strategic locations close to where fruit and vegetables were being processed for sale (**Figure 14**). The prototype compost facility was able to receive 20 cubic meters of raw materials daily: (1/3 organic waste, 1/3 wood chips, 1/3 cardboard).

7. Technical and organizational challenges

Whereas then proposed composting business was to be situated next to the market on a large flat tract of land used by transport trucks, the market manager

Figure 15.
This schematic illustration shows the alternating layers of brown and green material in addition to the perforated plastic pipes to allow for oxygen to move more freely (illustration by Seth Morrow).

was concerned that the compost would smell and become a hinderance to market vendors and the general public. He was fearful that it would smell as bad as a landfill and would only allow the use of a narrow tract of land adjacent to CENMA. Unfortunately, this site was not ideal for two reasons: 1. it was not large enough to produce the volume of required useable compost to meet municipal demand by the City's landscaping division, and 2. it was much too small to accommodate the available organic waste generated daily by the market.

The teenage employees from the Zone 3 neighborhood were hired, trained, and were responsible for many of the chores such as sorting waste from the collection barrels and mixing the compost from cell A to B and then to C (**Figure 15**). These young workers were also trained in how to monitor the moisture and temperature of the new compost piles in production. Ten to twelve workers were initially employed, selected from youth from Zone 3 who needed jobs. Only youth who were enrolled and remained in a secondary school were eligible and they had to remain in school to be employed in the compost business. Each of the youth were trained in how to maintain a bank account to receive their weekly pay. Sponsors from the USA participated in a cost-share program, matching the money earned by the Guatemalan youth with a matching donation, doubling the money earned.

8. Results and discussion

The startup composting business discussed in this chapter aimed to provide useable compost for the Guatemala City's municipal government (MUNI) who were using on an annual basis nearly 15,000 cubic meters of soil for landscaping the along the roads and parks. Because the base soil was of such poor quality, the mix of compost to soil would need to be 1:1, creating a demand for up to 7500 cubic meters each year. As stated earlier in this chapter, the 0.48-hectare site adjacent to the CENMA market was inadequately small and could not meet the needs of the City for soil amendment; a larger production workspace had to be provided. This study found that in order to produce 7500 cubic meters of useable compost each year, nearly 625 cubic meters would need to be generated each month, or approximately 30 cubic meters on a daily basis. To meet the demand for just the compost needed by the City government, the necessary site had to be nearly six times larger than that of the CENMA site, or approximately 2.6 hectares (6.4 acres). Interestingly enough, if one were to adopt the wind-row method for compost production, discussed earlier in this chapter, the area needed to meet the municipal demand is estimated to be approximately 1.6 hectares (4.0 acres) because windrows are more efficient in their use of space and production. Discussions with the head for public works in Guatemala City promised space for a larger scale production facility situated below (to the south) of CENMA; unfortunately, this expansion never occurred. Ironically, the City requested an even greater volume of compost from the startup company if it could be produced. The positive element here was that a broader market demand existed for the compost, and at that point in time, no one else was able to or willing to step up and make it.

Also ironic was the volume of organic waste trucked to the landfill each day, equaling roughly 138 cubic meters. When one considers that the initial small production space, or for that matter the 2.6 hectares site discussed above, both were incapable to accommodate all of the organic waste generated by the market assuming that it could be converted into useable compost; to do so would require a site of approximately 5.8 hectares. Unfortunately, production space limitations resulted in unacceptable shortfalls of the volume of compost that was produced, and as such, the project could not live up to its potential. Yet from a more positive perspective, the raw unprepared product was available—free of charge—with a willing client and inexpensive labor, suggesting that the failure of the business idea was not due to a flawed business plan but rather necessary space. With some abandoned brownfield sites (former industrial manufacturing facilities) nearby, this could be readily overcome.

Referring back to the success and failures of this research projected mentioned in the introduction of this chapter, the startup business was never able to meet the real demand for compost by the Municipal Government of Guatemala City; however, with a larger working production site, this could be achieved in the future. And with that, the ability to achieve the intended goal for a business that would demonstrate the principles of circular economy and social sustainability could be achieved.

Monitoring data conducted weekly at the two test cells that were constructed as a control experiment (**Table 2**) reveled that the high altitude and dry climate of Guatemala City caused much of the moisture in the newly mixed compost (cell A) to dry out prematurely, resulting in a very slow or even stalled rate of decomposition. This was unexpected and required altering the brown to green mix to increase the green organic volume during the initial mixing of the new compost piles in cell A.

9. Conclusions

Initiating a startup venture Poverty in Guatemala City will never be completely overcome and improved living conditions for those from Zone 3 achieved through a commercial composting business. However, each step toward this goal can and will make a tangible, and most importantly a sustainable difference in the future of children who live in Zone 3 and who have minimal future opportunities for a prosperous life.

Such as the one presented in this chapter demonstrated that one must find their own unique skills to contribute, whether it is the desire for developing programs to address social justice, expanding the knowledge-base of composting methods in different environments, or finding creative fund-raising opportunities. Each part or component of the program for recycling green waste from CENMA, putting to

Figure 16.
This diagram depicts the three-pillars of sustainability in action.

work at-risk youths of Zone 3 neighborhoods, and creating a useable product that was economically viable adhered to the three pillars of sustainably (**Figure 16**).

Sustainability in practice applies here, regardless of whether one is a proponent of social justice, an entrepreneur in search of starting a sustainable company to help the poor, or a public official determining the level of feasibility of a project, or even a potential financial donor. One need not look for complex operations that utilize high technology. Simplicity in action and techniques that achieve multiple objectives simultaneously are often the most effective and resilient.

Acknowledgements

The author wishes to acknowledge the support and encouragement of Susana Asensio, Director of the Department of Urban Construction; Brady Greene, regional director for Vida Joven Guatemala; Rosario Burgos, Environmental Coordinator; Antonio Peña, Director of Nurseries; Lazaro Zamora, Director of the Central Wholesale Market (CENMA); and Kevin Gervais. In particular, this research effort could not have been possible without the hard work and commitment by eight outstanding graduate students from the University of Massachusetts; they are: Travis Shultz, Dan Shaw, Tamzeena Hutchinson, Adam Monroy, Brian Giggey, Seth Morrow, Megan Regan, and Jason Dell'Orfano.

Funding

This research received no external funding.

Conflicts of interest

The author declares no conflict of interest.

Photographic credits

All of the photographs depicted herein were made by the author using a Nikon D800 digital camera, unless otherwise noted.

Author details

Peter A. Kumble
Faculty of Environmental Sciences, Department of Landscape and Urban Planning, Czech University of Life Sciences, Prague, Czech Republic

*Address all correspondence to: kumblep@fzp.czu.cz

References

[1] Ishiyama, L. Recycled Life. AllGlad: Salida, CA, USA, 2006.

[2] Vice. The Humanitarian Crisis in Guatemala City's Immense Garbage Dump. 2014. Available online: https://www.vice.com/en_us/article/vdpbxm/the-basurero-is-burning-life-at-the-gates-ofhellinguatemala-city (accessed on 12 May 2019).

[3] Endeavors. Guatemala Cities Dirtiest Secret. 2008. Available online: http://endeavors.unc.edu/win2008/guatemala.php (accessed on 10 April 2009).

[4] Reuters News. Garbage dump landslide kills 4 in Guatemala. Available from: 2008. Available online https://www.reuters.com/article/idUSN20477164 (accessed on 13 April 2009).

[5] Pearce, D.; Turner, R.K.; Economics of Natural Resources and Environment; Johns Hopkins University Press. Baltimore, MD, USA, 1990.

[6] Anderson, M.S. An introductory note on the environmental economics of the circular economy. Sustain, Sci. 2007 2, 1, 133-140.

[7] Kumble, P.A. Reflections on service learning for a circular economy project in a Guatemalan neighborhood. Sustainability. 2019, 11, 4776.

[8] Geissdoerfer, M.; Savaget, P.; Bocken, N.M.P.; Hultink, E.J. The Circular Economy: A new sustainability paradigm? J. Clean. Prod. 2016, 143, 757-768.

[9] Petit-Boix, A.; Leipold, S. Circular Economy in cities: Reviewing how environmental research aligns with local practices. J. Clean. Prod. 2018, 195, 1270-1281.

[10] Genovese, A.; Acquaye, A.A.; Figueroa, A.; Koh, S.C. Lenny. Sustainable supply chain management and the transition towards a circular economy: Evidence and some applications. Omega 2017, 66, 344-357.

[11] Boulding, K.E. Econ. Coming Spaceship Earth; Daly, H.D., Ed.; towards a steady-state economy. Freeman California: San Francisco, CA, USA, 1966.

[12] Leontief, W. The economy as a circular flow. Struct. Chang. Econ. Dyn. 1991, 2, 1.

[13] Figge, F.; Givry, P.; Canning, L.; Franklin-Johnson, E.; Thorpe, A. Eco-efficiency of virgin resources: a measure at the interface between micro and macro levels. Ecol. Econ. 2017, 138, 12-21.

[14] Krausmann, F.; Gingrich, S.; Eisenmenger, N.; Erb, K.H.; Haberl, H.; Fischer-Kowalski, M. Growth in global materials use: GDP and population during the 20th Century. Ecol. Econ. 2009, 68, 2696-2705.

[15] York, R.; Rosa, E.A.; Dietz, T. Stirpat, Ipat and ImPact: Analytic tools for unpacking the driving forces of environmental impacts. Ecol. Econ. 2003, 46, 351-365.

[16] Merli, R.; Preziosi, M.; Acampora, A. How do scholars approach the circular economy? A systematic literature review. J. Clean. Prod. 2018, 178, 703-722.

[17] Lyle, J.T. Design for Human Ecosystems: Landscape, Land Use and Natural Resources; Van Nostrand Reinhold Company: New York, NY, USA, 1985.

[18] Kirchherr, J.; Piscielli, L.; Bour, R.; Kostense-Smit, E.; Muller, J.; Huibrechtse-Truijens, A.; Hekkert, M. Barriers to the circular economy:

Evidence from the European Union (EU). Ecol. Econ. 2018, 150, 264-272.

[19] Skene, K.E. Circles, spirals, pyramids and cubes: Why the circular economy cannot work. Sustain. Sci. 2017, 13, 1-14.

[20] Kirchherr, J.; Reike, D.; Hekkert, M. Conceptualizing the circular economy: An analysis of 114 definitions. Resour. Conserv. Recycl. 2017, 127, 221-232.

[21] Geissdoerfer, M.; Savaget, P.; Bocken, N.; Jultink, E.J. The circular economy—A new sustainable paradigm? J. Clean. Prod. 2017, 143, 757-768.

[22] Figge, F.; Stevenson-Thorpe, A.; Givry, P.; Canning L.; Franklin-Johnson, E. Longevity and circularity as indicators of eco-efficient resource use in the circular economy. Ecol. Econ. 2018, 150, 278-306.

[23] Sama, C.; Crespo-Cebada, E.; Diaz-Caro, C.; Escrbana, M.; Mesias, F. Consumer preferences for foodstuff produced in social-environmentally responsible manner: A threat to fair trade producers? Ecol. Econ. 2018, 150, 290-296.

[24] Falcone, P.M.; Imbert, E. Social life cycles approach as a tool for promoting the market uptake of bio-based products from a consumer perspective. Sustainability 2018, 10, 131.

[25] George, D.A.R.; Chi, B.; Lin, A.; Chen, Y. A circular economy model of economic growth. Environmental Modeling. Software 2015, 73, 60-63.

[26] UNEP. Towards a Green Economy: Pathways to Sustainable Development and Poverty Eradication; United Nations Environment Program: Nairobi, Kenya, 2011.

[27] Falcone, P.M.; Sica, E. Assessing the opportunities and challenges of green finance in Italy: An analysis of the biomass production sector. Sustainability 2019, 11, 517.

[28] George, D.A.R.; Chi, B.; Lin, A.; Chen, Y. A circular economy model of economic growth. Environmental Modeling. Software 2015, 73, 60-63.

[29] Geng, Y.; Fujita, T.; Park, H.; Chiu, A.S.F.; Huisingh, D. Recent progress on innovative eco-industrial development. J. Clean. Prod. 2016, 114, 1-10.

[30] Boyer, R.H.W.; Peterson, N.D.; Arora, P.; Caldwell, K. Five approaches to social sustainability and an Integrated way forward. Sustainability 2016, 8, 1-18.

[31] Purvis, B.; Mao, Y.; Robinson, D. Three Pillars of Sustainability: In Search of Conceptual Origins; Springer: 2018. ISSN 1862-4065.

[32] Duran, D.C.; Gogan, L.M.; Artene, A.; Duran, V. The components of sustainable development: A possible approach. Procedia Econ. Financ. 2015, 26, 806-811.

[33] Kidd, C. The evolution of sustainability. J. Agric. Environ. Ethics. 1992, 5, 1-26.

[34] Korhonen, J.; Honkasalo, A.; Seppala, J. Circular Economy: The concept and its limitations. Ecol. Econ. 2018, 143, 37-48.

[35] Desrochers, P. Industrial symbiosis: The case for market coordination. J. Clean. Prod. 2004, 12, 8-10.

[36] Minica, M.; Frant. F. The dimensions of durable development. Ann. Uni. Craiova Econ. Sci. 2008, XXXV17, 3432-3429.

[37] Knoth, Gretchen. Guatemalan Poor: Garbage Dump Education System. 2009. Available online: http://www.scoop.co.nz/stories/WO0908/SOO066.htm (accessed on 8 May 2019).

[38] Dougherty, M. Field Guide to on Farm Composting; Natural Resource, Agriculture and Engineering Service (NRAES): Ithaca, NY, USA, 1999; 128 pp.

[39] Dougherty, M. Composting for Municipalities: Planning and Design; Natural Resource, Agriculture and Engineering Service (NRAES): Ithaca, NY, USA, 1998; 136 pp.

[40] Dick, W.A.; McCoy, E.L. Enhancing soil fertility by addition of compost. In Science and Engineering of Composting: Design, Environmental, Microbiological and Utilization Aspects; Hoitink, H.A.J., Keener, H.M., Eds.; Renaissance Publications OH: Worthington, DC, USA, 1993; pp. 622-624.

[41] Epstein, E. The Science of Composting; Technomic Publishing Co. Inc.: Basel, Switzerland, 1997.

[42] Baldwin, K.; Jackie, G. Composting on Organic Farms; Center for Environmental Farming Systems. North Carolina State University Cooperative Extension Service: Raleigh, NC, USA, 2006; 22 pp.

[43] Parnes, R. Fertile Soil: A Grower's Guide to Organic and Inorganic Fertilizers; agAccess: Davis, CA, USA, 1990.

[44] Cornell University. 1995. Available online: http://cwmi.css.cornell.edu/Composting.htm (accessed on 12 March 2009).

[45] Noyes, N. Easy Composters You Can Build; Storey Publishing, LLC: North Adams, MA, USA, 1995.

[46] Houck, N.J.; Robert, E G. Composting Operations; Penn State University, College of Agricultural Sciences, Housing and Food Services, Office of the Physical Plant: State College, PA, USA, 2001.